LOGIC, LANGUAGE AND COMPUTATION

CSLI Lecture Notes
Number 96

LOGIC, LANGUAGE AND COMPUTATION

VOLUME 2

edited by
Lawrence S. Moss
Jonathan Ginzburg
Maarten de Rijke

CSLI
PUBLICATIONS
Center for the Study of
Language and Information
Stanford, California

Center for the Study of Language and Information
Leland Stanford Junior University
Printed in the United States
03 02 01 00 99 5 4 3 2 1

Library of Congress Cataloging-in-Publication Data

Logic, language and computation / edited by Jerry Seligman & Dag Westerståhl.
p. cm. — (CSLI lecture notes ; no. 55)
Includes bibliographical references and index.

ISBN 1-57586-181-x (v. 2)
ISBN 1-57586-180-1 (pbk.)

1. Language and logic. 2. Logic, Symbolic and mathematical. 3. Computational linguistics. I. Seligman, Jerry, 1964– . II. Westerståhl, Dag, 1946– . III. Series.
P39.L593 1995
410′.285—dc20 95-50832
CIP

∞ The acid-free paper used in this book meets the minimum requirements of the American National Standard for Information Sciences—Permanence of Paper for Printed Library Materials, ANSI Z39.48-1984.

CSLI was founded early in 1983 by researchers from Stanford University, SRI International, and Xerox PARC to further research and development of integrated theories of language, information, and computation. CSLI headquarters and CSLI Publications are located on the campus of Stanford University.

CSLI Publications reports new developments in the study of language, information, and computation. In addition to lecture notes, our publications include monographs, working papers, revised dissertations, and conference proceedings. Our aim is to make new results, ideas, and approaches available as quickly as possible. Please visit our web site at
http://csli-www.stanford.edu/publications/
for comments on this and other titles, as well as for changes and corrections by the author and publisher.

Contents

Contributors

Jon Barwise Departments of Mathematics, Philosophy, and Computer Science, Indiana University, Bloomington IN 47405 USA. barwise@cs.indiana.edu.

David Beaver Linguistics Department, Stanford University, Stanford, CA 94305 USA. dib@csli.stanford.edu.

Richard Cox School of Cognitive and Computing Sciences, University of Sussex, Falmer, Brighton BN1 9QH, UK richc@cogs.susx.ac.uk.

Tsutomu Fujinami School of Knowledge Science, Japan Advanced Institute of Science and Technology, 1-1 Asahidai Tatsunokuchi Nomi Ishikawa, 923-1292 Japan. fuji@jaist.ac.jp.

Jelle Gerbrandy ILLC/Department of Philosophy, Nieuwe Doelenstraat 15, 1012 CP, Amsterdam, the Netherlands. gerbrand@philo.uva.nl.

Sheila Glasbey Department of Applied Computing University of Dundee DD1 4HN, UK. srg@computing.dundee.ac.uk.

Edward Hermann Haeusler Departamento de Informática, PUC-Rio, Rua Marquês de São Vicente 225, 22453-900 - Rio de Janeiro, RJ Brazil. hermann@inf.puc-rio.br.

Wiebe van der Hoek Department of Computer Science, Utrecht University, P.O. Box 80089, 3508 TB Utrecht, the Netherlands. wiebe@cs.ruu.nl

David Israel Artificial Intelligence Center, SRI International, 333 Ravenswood Ave., Menlo Park, CA 94025-3493 USA. israel@ai.sri.com.

Yoshiki Kinoshita Electrotechnical Laboratory, Tsukuba, 305 Japan. yoshiki@etl.go.jp.

Oliver Lemon Computer Science Department, Trinity College, The University of Dublin, Ireland. ojlemon@cs.tcd.ie.

Elizabeth Macken Center for the Study of Language and Information, Stanford University, Stanford, CA 94305 USA. macken@csli.stanford.edu.

Edwin D. Mares Department of Philosophy, Victoria University of Wellington, P.O. Box 600, Wellington, New Zealand. Edwin.Mares@vuw.ac.nz.

Arthur Merin Lehrstuhl für Formale Logik und Sprachphilosophie, Institut für Maschinelle Sprachverarbeitung, Universität Stuttgart, Azenbergstr. 12, 70174 Stuttgart, Germany. merin@ims.uni-stuttgart.de.

Jon Oberlander Division of Informatics, University of Edinburgh, 2 Buccleuch Place, Edinburgh EH8 9LW, UK J.Oberlander@ed.ac.uk.

Denise Aboim Sande e Oliveira Departamento de Informática, PUC-Rio, Rua Marquês de São Vicente 225, 22453-900 - Rio de Janeiro, RJ Brazil. denise@inf.puc-rio.br.

Rohit Parikh Department of Computer Science, CUNY Graduate Center, 33 West 42nd Street, New York, NY 10036. ripbc@cunyvm.cuny.edu.

John Perry Center for the Study of Language and Information, Stanford University, Stanford, CA 94305 USA. john@csli.stanford.edu.

Ian Pratt Department of Computer Science, University of Manchester, Oxford Road, Manchester, M13 9PL. ipratt@cs.man.ac.uk.

Maarten de Rijke ILLC, University of Amsterdam, Plantage Muidergracht 24, 1018 TV Amsterdam, The Netherlands. mdr@wins.uva.nl.

Atsushi Shimojima Japan Advanced Institute of Science and Technology. ashimoji@mic.atr.co.jp.

Paul Skokowski Symbolic Systems, Bldg. 460, Rm. 127E, Stanford University, Stanford, CA, 94305-2150. paulsko@turing.stanford.edu.

Clarisse Sieckenius de Souza Departamento de Informática, PUC-Rio, Rua Marquês de São Vicente 225, 22453-900 - Rio de Janeiro, RJ Brazil. clarisse@inf.puc-rio.br.

Keith Stenning Division of Informatics, University of Edinburgh, 2 Buccleuch Place, Edinburgh EH8 9LW, UK. K.Stenning@ed.ac.uk.

Koichi Takahashi Electrotechnical Laboratory, Tsukuba, 305 Japan. takahasi@etl.go.jp.

Steven Vickers Department of Computing, Imperial College of Science, Technology and Medicine, 180 Queen's Gate, London SW7 2BZ. s.vickers@doc.ic.ac.uk.

Thomas Ede Zimmermann Institut für Maschinelle Sprachverarbeitung, Universität Stuttgart, Azenbergstr. 12, D-70174 Stuttgart, Germany. ede@ims.uni-stuttgart.de.

R. Zuber Universite Paris 7, Linguistique, Boite 7003, 2 Place Jussieu, 75005 Paris, France. rz@ccr.jussieu.fr.

Preface

This volume contains papers presented at the Second Conference on Information-Theoretic Approaches to Logic, Language, and Computation. The program committee for the meeting consisted of the editors of this volume, together with Robert Koons and Hideyuki Nakashima. As usual with such proceedings, the papers here are revisions and expansions of papers presented at the conference.

The ITALLC conferences originated as conferences on the theme of Situation Theory and its Applications. The name change reflects the broader orientation of the participants and the more inclusive nature of the material. But as with the earlier meetings, the papers contain contributions to areas of linguistics, philosophy, and computer science which contribute to, or depend on, various notions of information.

The ITALLC conference was held at Regent's College in London in July 1996. We are grateful to Nick Braisby and to the Department of Psychology of London Guildhall University for their help and generosity in supporting the conference.

Lawrence S. Moss
Jonathan Ginzburg
Maarten de Rijke

LOGIC, LANGUAGE AND COMPUTATION

State spaces, Local Logics, and Non-Monotonicity

Jon Barwise

Departments of Computer Science,
Mathematics, and Philosophy
Indiana University

Abstract This paper uses the information-theoretic machinery developed in *IF* [2] to relate the way error is handled in the sciences to the problems of non-monotonic reasoning.

In Parts 1 and 2 of *Information Flow in Distributed Systems* [2] (hereinafter *IF*) Jerry Seligman and I have developed what seems to us a promising mathematical theory of information. One problem tackled there is that of accounting for the sorts of exceptions to law-like regularities associated with information flow that are traditionally associated with non-monotonic reasoning. Another problem addressed is that of reconciling everyday common sense theories with scientific theories. Again, non-monotonicity becomes an issue one has to face since if human reasoning is essentially non-monotonic but scientific reasoning is classical (hence monotonic), there would seem to be a serious obstacle to reconciling the two. In this paper I want to use the tools from *IF* to further explore these issues.

Introduction

A bumper sticker recently in vogue advocated "Think globally, act locally." While the sentiment is admirable, if taken literally the injunction is impossible to follow. Thinking and reasoning are inherently local, located in space and time, always taking place against prevailing background conditions. We are reliant on these background conditions being certain ways, a reliance that seems to be almost hard-wired.

Sometimes we have to revise our reasoning by recognizing that the background conditions we have been relying on, typically unwittingly, have

---

Logic, Language, and Computation, Vol. II, edited by Lawrence S. Moss, Jonathan Ginzburg, and Maarten de Rijke. Copyright ©1999, CSLI Publications.

changed. For example, I have a certain common sense understanding of our home's heating system, the furnace, thermostat, vents, and the way they function to keep the house warm. My understanding gives rise to inferences like the following.

(p_1) The thermostat is set between 65 and 70.

(p_2) The room temperature is 58.

$\vdash$ (q) Hot air is coming out of the vents.

The inference $p_1, p_2 \vdash q$. takes place against a set of background conditions and, for most of the winter it does just fine. But sometimes the reasoning breaks down. If during a blizzard I add to the above inference the premise

(p_3) The power is off

then the inference fails:

$$p_1, p_2, p_3 \nvdash q.$$

In fact, we would expect the inference

$$p_1, p_2, p_3 \vdash \neg q.$$

In this paper I explore a formal, pragmatic account of non-monotonicity. The idea is that reasoning is classical but "local," based on certain background conditions that hold locally. Various things can happen, including the arrival of addition information, that cause us to alter—not the classical forms of reasoning—but rather our understanding of prevailing background conditions. In particular, I develop a rigorous model called "weakening the background" that blocks the undesired inference and instead yields the above desired inference and many other like it.

A review of local logics

This section reviews (with slightly different terminology) some material from Part 2 of *IF*.

A *Boolean classification* $\boldsymbol{A} = \langle S, \Sigma, \models, \wedge, \neg, \rangle$ consists of a non-empty set S of objects called *situations*, a set Σ of objects called *propositions*, a binary relation $\models$ on $S \times \Sigma$, a binary operation $\wedge : \Sigma \times \Sigma \to \Sigma$, and a unary operation $\neg : \Sigma \to \Sigma$ satisfying the following conditions:

1. $s \models p_1 \wedge p_2$ iff $s \models p_1$ and $s \models p_2$.
2. $s \models \neg p$ iff $s \not\models p$.

In *IF*, propositions and situations are called "types" and "tokens," respectively. *IF* adopts the basic Gentzen sequent calculus approach to logic, so that $\Gamma \vdash \Delta$ means that the conjunction of the propositions in Γ entails the disjunction of the propositions in Δ in whatever logic is at hand. By a *sequent*, we mean simply a pair $\langle \Gamma, \Delta \rangle$ where Γ and Δ are sets of propositions.

A *(Boolean) local logic* $\mathfrak{L} = \langle \boldsymbol{A}, \vdash, N \rangle$ consists of a Boolean classification $\boldsymbol{A} = \langle S, \Sigma, \models, \wedge, \neg, \rangle$ together with a binary relation $\vdash$ relating sets of propositions, and a set $N \subseteq S$ of situations called the *normal situations* satisfying the following additional conditions:

Entailment The relation $\vdash$ satisfies all the usual structural and propositional rules associated with classical propositional logic, including the rules:

Identity: $\alpha \vdash \alpha$,

Weakening: If $\Gamma \vdash \Delta$ then $\Gamma, \Gamma' \vdash \Delta, \Delta'$,

Global Cut: If $\Gamma, \Sigma_0 \vdash \Delta, \Sigma_1$ for each partition $\langle \Sigma_0, \Sigma_1 \rangle$ of some set Σ', then $\Gamma \vdash \Delta$.

Normal situations If s is a normal situation, $\Gamma \vdash \Delta$, and $s \models p$ for all $p \in \Gamma$ then $s \models q$ for some $q \in \Delta$.

A local logic $\mathfrak{L}$ is *sound* if every $s \in S$ is normal. $\mathfrak{L}$ is *complete* if for all sets Γ, Δ of propositions, if $\Gamma \not\vdash \Delta$ then there is a normal situation s such that $s \models p$ for every $p \in \Gamma$ and $s \models \neg q$ for every $q \in \Delta$.

The use of normal situations in a local logic $\mathfrak{L}$ allows us to model the implicit background assumption that the situation we are reasoning about is normal, relative to $\mathfrak{L}$.

Definition 1 A partial ordering $\sqsubseteq$ on local logics $\mathfrak{L}_1$ and $\mathfrak{L}_2$ on a fixed classification $\boldsymbol{A}$ is defined by: $\mathfrak{L}_1 \sqsubseteq \mathfrak{L}_2$ iff

1. For all sets Γ, Δ of propositions, $\Gamma \vdash_{\mathfrak{L}_1} \Delta$ entails $\Gamma \vdash_{\mathfrak{L}_2} \Delta$ and

2. every situation of $\boldsymbol{A}$ that is normal in $\mathfrak{L}_2$ is normal in $\mathfrak{L}_1$.

It is important to notice that these inclusions go the opposite directions. The more entailments one has, the fewer normal situations. This reversal is intimately connected with the issue of non-monotonicity.

We can state the main thesis of this paper succinctly; it will take more space to explain it. *Intuitive reasoning takes place relative to an implicit local logic. Anything that changes one part of the logic (the constraints or the set of normal situations) has the potential to change the other, thereby giving rise to (apparent or genuine) non-monotonicity.* Before illustrating this, let us state the following:

Theorem 1 *The local logics on a given classification form a complete lattice under the ordering $\sqsubseteq$.*

PROOF: From *IF*. QED

This allows us to combine and move between local logics representing different background assumptions.

To bring out one aspect of non-monotonicity, let us assume we have a fixed Boolean classification $\boldsymbol{A}$. For any set T of sequents, the logic generated by the T on $\boldsymbol{A}$, Log($\boldsymbol{A}^T$), has as normal situations all situations that satisfy all sequents in T and as constraints all sequents satisfied by all such situations. Similarly, for any set N of situations of $\boldsymbol{A}$, the logic generated by the situations N on $\boldsymbol{A}$, Log($\boldsymbol{A}_N$), has as constraints all sequents satisfied by all situations in$N$ and has as normal situations all the situations of $\boldsymbol{A}$ satisfying such constraints. It is easy to see that these are indeed local logics, that the constraints of Log($\boldsymbol{A}^T$) contain T and that the normal situations of Log($\boldsymbol{A}_T$) contain N.

Proposition 1 *Let $\boldsymbol{A}$ be a fixed Boolean classification.*

1. *If $T_0 \subseteq T_1$ then* $\text{Log}(\boldsymbol{A}^{T_0}) \subseteq \text{Log}(\boldsymbol{A}^{T_1})$.
2. *If $N_0 \supseteq N_1$ then* $\text{Log}(\boldsymbol{A}_{N_0}) \subseteq \text{Log}(\boldsymbol{A}_{N_1})$.

PROOF: Routine. QED

Example: Let's return to our basic example involving the heating system. The reluctance to infer $p_1, p_2, p_3 \vdash q$ from $p_1, p_2 \vdash q$ is surely related to the fact that premise p_3 suggests that the set of situations implicitly assumed to be normal is larger than was assumed in $p_1, p_2 \vdash q$. This is can be seen as an instance of (1). Let T_0 consist of the sequent $p_1, p_2, p_3 \vdash q$ and T_1 consist of both sequents. By (1), we see that $\text{Log}(\boldsymbol{A}^{T_0}) \subseteq \text{Log}(\boldsymbol{A}^{T_1})$, so that the normal situations for the former form a *superset* of those for the latter. But since $p_1, p_2, p_3 \vdash q$ follows from $p_1, p_2 \vdash q$ by Weakening, the logic generated by T_1 is the same as that generated by the single sequent $p_1, p_2 \vdash q$.

Example: Let $\boldsymbol{A}$ be a classification of, say, local bird sightings, and let N consist of actual sightings to date. Then Log($\boldsymbol{A}_N$) has as constraints all sequents satisfied by all these bird sightings to date, and the normal situations consist of all bird sightings, including future sightings, that satisfy all these constraints, a set that certainly contains N. This logic might well include the constraint BIRD $\vdash$ FLY, a constraint that works fine as long as the situations we encounter are enough like those of N. But if a penguin turns up, it will lie outside the normal situations since it violates BIRD $\vdash$ FLY. This sighting by itself gives rise to a new set $N' \supset N$. The previous result shows that their associated logics are related by $\mathrm{Log}(\boldsymbol{A}_{N'}) \sqsubseteq \mathrm{Log}(\boldsymbol{A}_N)$. There well be fewer constraints that hold in $\mathrm{Log}(\boldsymbol{A}_{N'})$, as indeed there will be in our example, since BIRD $\nvdash$ FLY in this new logic.

While it is all well and good to pin the blame for non-monotonicity on the order inverting duality between propositions and situations under the $\sqsubseteq$-ordering on logics, it is not of much help in understanding the mechanisms people use to reason in such circumstances, the main concern of the non-monotonic logic literature. For example, it does give any hint as to how one gets $p_1, p_2, p_3 \vdash \neg q$ rather than $p_1, p_2, p_3 \vdash q$.

In the next section we examine some mechanisms that gives rise to shifting between local logics on certain special kinds of Boolean classifications. We then turn to a brief look at logic infomorphisms, a central notion in *IF*. Our hope is that by beginning to understand these mechanisms in admittedly very special circumstances, we gain a foothold that might eventually lead us to understand them better in a more general setting.

State spaces

The prevailing methodology in the sciences is that of state spaces. We refer the reader to the first chapter of Casti (3) (hereinafter *RR*) for an excellent introduction to state spaces and their use in the physical sciences, life sciences, and social sciences. The overwhelming success of this methodology makes it surprising how little interaction there has been between logic and state spaces. It seems worth investigating this relationship for at least three reasons. First, the proven utility of the state space approach makes understanding the logical aspects of state space models interesting in its own right. Second, it could be that such an approach will allow one, at least in some instances, to automate inference by means of libraries of numerical routines. Finally, it could be that we can learn something about some aspects of non-monotonicity by seeing how the same kinds of problems are handled in other parts of science.

Definition 2 A *state space* $\boldsymbol{S} = \langle \Omega, S, state \rangle$ consists of a set Ω of *states*, a set S of *situations*, a function $state: S \to \Omega$. If $\sigma = state(s)$ then σ is called the *state* of s.

The intuitive idea is that the state σ of a situation s gives one total information about the situation, relevant to the issues under investigation.

Example: We use my heating system as a running example and so start by describing a state space $\boldsymbol{S}_{hs}$ for this heating system. For situations, take some set of objects without additional mathematical structure. Intuitively, these are to be instances of my complete heating system at various times, including the vents, the thermostat, the furnace, and the ambient air in the room where the thermostat is located, and so forth. We that each state is determined by a combination of the following seven "observables" of the system:

Thermostat setting: some real number σ_1 between 55 and 80.

Room Temperature: (in Fahrenheit) a real number σ_2 between 20 and 110.

Power: $\sigma_3 = 1$ (on) or 0 (off).

Outside vents: $\sigma_4 = 0$ (plugged) or 1 (unplugged).

Operating condition: $\sigma_5 =$ -1 (cooling), 0 (off), or 1 (heating).

Running: $\sigma_6 = 1$ (on) or 0 (not on).

Output air temperature: a real number σ_7 between 20 and 110.

Thus for states we let Ω be the set of vectors $B = \langle \sigma_1, \ldots, \sigma_7 \rangle \in \boldsymbol{R}^7$ satisfying the inequalities listed above and the following equations:

$$\begin{aligned} \sigma_6 &= pos(\sigma_5 \cdot sg(\sigma_1 - \sigma_2)) \cdot \sigma_3 \cdot \sigma_4 \\ \sigma_7 &= \begin{cases} 55 & \text{if } \sigma_5 \cdot \sigma_6 = -1 \\ 80 & \text{if } \sigma_5 \cdot \sigma_6 = +1 \\ \sigma_2 & \text{otherwise} \end{cases} \end{aligned}$$

where

$$\begin{aligned} sg(r) &= \begin{cases} +1 & \text{if } r \geq 2 \\ 0 & \text{if } \mid r \mid < 2 \\ -1 & \text{if } r \leq -2 \end{cases} \\ pos(r) &= \begin{cases} 1 & \text{if } r > 0 \\ 0 & \text{if } r \leq 0 \end{cases} \end{aligned}$$

We let $\boldsymbol{S}_{hs}$ consist of these situations and states, with some total function *state* mapping the situations into the states.

Definition 3 A state space $\boldsymbol{S}_0$ is a *subspace* of a state space $\boldsymbol{S}_1$, written $\boldsymbol{S}_0 \subseteq \boldsymbol{S}_1$ iff the situations of $\boldsymbol{S}_0$ are a subset of the situations of $\boldsymbol{S}_1$, the states of $\boldsymbol{S}_0$ are a subset of the states of $\boldsymbol{S}_1$, and the state function of $\boldsymbol{S}_0$ is the restriction of the state function of $\boldsymbol{S}_1$ to the situations of $\boldsymbol{S}_0$.

Logics from state spaces

In *IF*, we associate a Boolean classification $\mathrm{Evt}(\boldsymbol{S})$ with any state space $\boldsymbol{S}$ as follows. (The notation $\mathrm{Evt}(\boldsymbol{S})$ was chosen because is standard to call the propositions of this classification "events.")

Definition 4 For any state space $\boldsymbol{S}$, the Boolean classification $\mathrm{Evt}(\boldsymbol{S})$ is defined by:

- The situations are those of the state space system $\boldsymbol{S}$.
- The propositions are sets $p \subseteq \Omega$ of states.
- A proposition p is true of a situation s, $s \vDash p$, if $state(s) \in p$.
- $p \wedge q = p \cap q$ and $\neg p = \Omega - p$.

The local logic $\mathrm{Log}(\boldsymbol{S})$ has $\Gamma \vdash \Delta$ iff every state in every proposition in Γ is in at least one proposition in Δ. Every situation is normal, i.e., the logic is sound.

In *IF* the case is made that $\mathrm{Log}(\boldsymbol{S})$ is the local logic implicit in the use of the state space $\boldsymbol{S}$ to reasoning about some system.

Example: The propositions used in our example are represented in $\mathrm{Evt}(\boldsymbol{S}_{hs})$ by:

$$
\begin{array}{lcl}
p_1 & = & \{\sigma \in \Omega_{hs} \mid 65 \leq \sigma_1 \leq 70\} \\
p_2 & = & \{\sigma \in \Omega_{hs} \mid \sigma_2 = 58\} \\
p_3 & = & \{\sigma \in \Omega_{hs} \mid \sigma_3 = 0\} \\
q & = & \{\sigma \in \Omega_{hs} \mid \sigma_6 = 1 \text{ and } \sigma_7 > \sigma_2\}
\end{array}
$$

In what follows we assume we are working relative to a fixed state space system $\boldsymbol{S}$ and its subspaces. We want to define a family of local logics on $\mathrm{Evt}(\boldsymbol{S})$ to model the way we reason using the state space.

IF establishes a natural correspondence between the subspaces of a state space $\boldsymbol{S}$ and logics on the Boolean classification $\mathrm{Evt}(\boldsymbol{S})$.

Definition 5 Let $\boldsymbol{S}$ be a state space.

1. An $\boldsymbol{S}$-*logic* is a logic $\mathfrak{L}$ on the classification $\mathrm{Evt}(\boldsymbol{S})$ such that $\mathrm{Log}(\boldsymbol{S}) \sqsubseteq \mathfrak{L}$. A state σ of $\boldsymbol{S}$ is $\mathfrak{L}$-*consistent* if $\{\sigma\} \nvdash_{\mathfrak{L}}$.

2. Let $\boldsymbol{S}$ be a state space and let $\mathfrak{L}$ be an $\boldsymbol{S}$-logic. The *subspace* $\boldsymbol{S}_{\mathfrak{L}}$ *of* $\boldsymbol{S}$ *determined by* $\mathfrak{L}$ has situations the set $N_{\mathfrak{L}}$ of normal situations of $\mathfrak{L}$ and states the set of $\mathfrak{L}$-consistent states.

We summarize the of the relationship between $\boldsymbol{S}$-logics and subspaces of $\boldsymbol{S}$ in *IF* with the following result.

Theorem 2 *Let $\boldsymbol{S}$ be any state space. The mapping*

$$\mathfrak{L} \mapsto \boldsymbol{S}_{\mathfrak{L}}$$

is an order inverting bijection between the set of $\boldsymbol{S}$-logics and the set of subspaces of $\boldsymbol{S}$.

PROOF: See *IF*. QED

This theorem suggests that the state-space analogue of a logic on a classification is that of a subspace of the given state space. The order-inversion is another reflection of the basic phenomenon we claim is at the heart of non-monotonicity. The situations of the subspace correspond to normal situations of the logic, the state of the subspace correspond to the constraints of the logic. The larger the set of states, the fewer the set of constraints, and so the larger the set of normal situations.

Definition 6 Let $\boldsymbol{S}$ be a state space, $\boldsymbol{S}_0$ a subspace of $\boldsymbol{S}$. $\boldsymbol{S}_0$ is *sound* in $\boldsymbol{S}$ if $\boldsymbol{S}_0$ and $\boldsymbol{S}$ have the same situations.

Proposition 2 *Let $\boldsymbol{S}$ be a state space. An $\boldsymbol{S}$-logic $\mathfrak{L}$ is sound iff the associated state space $\boldsymbol{S}_{\mathfrak{L}}$ is sound.*

PROOF: Obvious. QED

Real-valued state spaces and their logics

The earlier discussion is all pretty abstract. In the remainder of this paper we are going to restrict to "real-valued" state spaces, the kind that are most typically used in science, and examine how background assumptions on the observables get reflected in their associated logics. The work here was inspired by the following claim:

> error or surprise always involves a discrepancy between the objects [situations] *open* to interaction and the abstractions [states] *closed* to those same interactions. In principle, the remedy for closing this gap is equally clear: augment the description by including more observables to account for the unmodeled interactions. (*RR*, p. 25)

Let $\boldsymbol{R}^n$ be Euclidean n-space; that is, its points are n-tuples of real numbers.

Definition 7 A *real-valued state space* is a state space $\boldsymbol{S}$ such that $\Omega \subseteq \boldsymbol{R}^n$, for some natural number n, called the *dimension* of the space. The set $n = \{0, 1, \ldots, n-1\}$ is called the set of *observables* and the projection function $\pi_i(\sigma) = \sigma_i$, the i-th coordinate of σ, is called the i-th *observation function* of the system. If σ is the state of s then σ_i is called the value of the i-th observable on s.

The canonical example of a real-valued state space is that used in classical mechanics, where to study k gravitating bodies one uses a state space of dimension $n = 6k$; each body needs three reals to specify its position and three to specify its momentum. Our heating system example $\boldsymbol{S}_{hs}$ has dimension 7. All the state spaces discussed in *RR* are real-valued.

Background conditions

RR further assumes that the observables are partitioned into two sets:

$$Observables \;=\; J \cup O,$$

where J is the set of *input observables* and O is the set of *output observables.* Each output observable $o \in O$ is assumed to be of the form

$$\sigma_o \;=\; F_o(\vec{\sigma_j})$$

for some function F_o of the input observables σ_j.

It typically happens that the input observables are *independent* in that you can vary the inputs independently. This amounts to the following requirement: if σ is some state, $i \in J$ is some input observable, and r is a value of σ'_i for some state σ', then there is a state σ'' such that $\sigma''_i = r$ and $\sigma''_j = \sigma_j$ for all $j \neq i$. We will not assume this in general but will need to invoke it at one point.

Besides dividing the observables into input and output, it is also customary to partition the inputs into two, $J = I \cup P$. The input observables in I are called the *explicit inputs* of the system and those in P the *parameters* of the system. Intuitively, the parameters are those inputs that are held fixed in any given computation or discussion.

Up to this point, the discussion follows *RR* fairly closely. In order to relate this to logic, though, we must now make some new moves.

Definition 8

1. A *background condition* B is a function with domain some set P of input observables, taking real numbers as values. It is required that $B(i)$ be a number in the range of the i-th observation function of the state space. The domain P is called the set of *parameters* of B.

2. A *state σ satisfies B* if $\sigma_i = B(i)$ for each parameter of P. A *situation s satisfies B* if $state(s)$ satisfies B.

3. The set of background conditions is partially ordered by inclusion: $B_1 \leq B_2$ iff the domain of B_1 is a subset of that of B_2 and the two functions agree on this domain.

Every non-empty set of background conditions has a great lower bound under the ordering $\leq$. Also, if a state or situation satisfies B then it clearly satisfies every $B_0 \leq B$. The empty function is the least background condition: it imposes no conditions on states.

In working with a given state space, one assumes one has a fixed background condition B, that one is concerned only with situations that satisfy this background condition, and hence that all computations and inferences take place relative to that background condition. We can make this precise as follows.

Given a set Q of input observables, define $\sigma \equiv_Q \sigma'$ if $\sigma_i = \sigma'_i$ for all $i \notin Q$. This is an equivalence relation on states. We say that a proposition $p \subseteq \Omega$ is *silent on Q* if for all states $\sigma, \sigma' \in \Omega$, if $\sigma \equiv_Q \sigma'$ and $\sigma \in p$ then $\sigma' \in p$. We say that p is silent on an input observable i if it is silent on $\{i\}$. Finally, a proposition p is *silent on B* if p is silent on the set P of parameters of B. If one is given a premise or purported conclusion that is *not* silent on an observable i, it is safe to assume i is *not* a parameter of the system. Put another other way, *if we are reasoning about an observable i, then i must either an explicit input or an output of the system.*

Example: In our heating system example $\boldsymbol{S}_{hs}$, is natural to take to take σ_1, σ_2 as inputs, σ_3, σ_4 and σ_5 as parameters, and σ_6 and σ_7 as outputs. The inputs are clearly independent. The propositions p_1, p_2 and q are indeed silent on the parameters, as can be seen by their definitions. The proposition p_3, by contrast, is not silent on P: it says that $\sigma_3 = 0$. The natural background condition B in the winter is $\sigma_3 = \sigma_4 = \sigma_5 = 1$, representing the case where the power switch is on, the vents are not plugged, and the setting is on "heat." (The is our informal way of indicating the background condition B with domain $\{3, 4, 5\}$ and constant value 1.) In the summer the default background condition is $\sigma_3 = \sigma_4 = 1$ and $\sigma_5 = -1$.

Definition 9 (Background weakening) Given a background condition B and a set Γ of propositions, the *weakening of B by Γ*, written $B \upharpoonright \Gamma$ is the greatest lower bound (in the $\leq$ ordering) of all $B_0 \leq B$ such that every proposition $p \in \Gamma$ is silent on B_0.

The function $B \upharpoonright p$ models the intuitive idea of dropping background assumptions of B on the parameters that are critical to the content of the proposition p. If the input observables of our state space are independent, then this definition is guaranteed to behave the way one would hope.

Theorem 3 *Assume the input observables of the state space $\boldsymbol{S}$ are independent. Then each proposition in Γ is silent on $B \restriction \Gamma$, and this is the greatest such background condition $\leq B$.*

PROOF: This follows easily from the following lemma, of independent use in computing $B \restriction \Gamma$. QED

Lemma 3 *If the input observables of the state space $\boldsymbol{S}$ are independent, then a proposition p is silent on Q iff it is silent on each $i \in Q$. Hence, under these conditions, there is a largest set Q of input observables such that p is silent on Q.*

PROOF: The left to right half of the first claim is trivial. The converse claim is proved by induction on the size of Q. The second claim follows from the first by taking Q to consist of all the input observables i such that p is silent on i. QED

Given a background condition B with parameters P, the lemma lets us compute $B \restriction p$ as simply the restriction of B to the set of input observables $i \in P$ such that p is silent on i.

Example: The inputs of our example $\boldsymbol{S}_{hs}$ are independent so the lemma and theorem apply. The proposition p_3 is silent on each observable except for σ_3. Hence if B is $\{\sigma_3 = \sigma_4 = \sigma_5 = 1\}$ then weakening B by p_3 gives $B \restriction p_3 = \{\sigma_4 = \sigma_5 = 1\}$.

Relativizing to a background condition

Each background condition B determines a subspace $\boldsymbol{S}_B \subseteq \boldsymbol{S}$. The states of $\boldsymbol{S}_B$ are those $\sigma \in \Omega$ that satisfy the background condition B and the situations of $\boldsymbol{S}_B$ are those situations that satisfy B. When we restrict attention to $\boldsymbol{S}_B$, the output equations typically are simplified since the parameters become constants in the equations. In our example, the output equations simplify to

$$\begin{aligned} \sigma_6 &= pos(sg(\sigma_1 - \sigma_2)) \\ \sigma_7 &= \begin{cases} 80 & \text{if } \sigma_6 = +1, \\ \sigma_2 & \text{otherwise.} \end{cases} \end{aligned}$$

Returning to the general case where we have a background condition B on a real-valued space $\boldsymbol{S}$, we recall that since $\boldsymbol{S}_B \subseteq \boldsymbol{S}$, the duality between subspaces of $\boldsymbol{S}$ and $\boldsymbol{S}$-logics gives us an $\boldsymbol{S}$-logic $\mathrm{Log}(\boldsymbol{S}_B)$ with $\mathrm{Log}(\boldsymbol{S}) \sqsubseteq \mathrm{Log}(\boldsymbol{S}_B)$. We call this the *local logic* $\mathrm{Log}(\boldsymbol{S}_B)$ *supported by the background condition* B.

Proposition 4 For each background condition B, the local logic $\mathrm{Log}(\boldsymbol{S}_B)$ supported by B is the logic on $\mathrm{Evt}(\boldsymbol{S})$ given by:

- The $\mathrm{Log}(\boldsymbol{S}_B)$-consistent states (in the sense of Definition 5) σ are those satisfying B.
- If Γ, Δ are sets of propositions of $\boldsymbol{S}$, then $\Gamma \vdash_B \Delta$ iff for every state σ satisfying B, if $\sigma \in p$ for all $p \in \Gamma$ then $\sigma \in q$ for some $q \in \Delta$.
- The normal situations are those situations satisfying B.

PROOF: The proof is routine. QED

This result shows that $\mathrm{Log}(\boldsymbol{S}_B)$ codifies the intuition behind taking the background condition B for granted. Since $\mathrm{Log}(\boldsymbol{S}) \sqsubseteq \mathrm{Log}(\boldsymbol{S}_B)$, we know that $\mathrm{Log}(\boldsymbol{S}_B)$ typically has more constraints but fewer normal situations than $\mathrm{Log}(\boldsymbol{S})$, as one would expect.

Corollary 5 *Let B be a background condition.*

1. *The logic* $\mathrm{Log}(\boldsymbol{S}_B)$ *is sound iff every situation satisfies B.*
2. *The logic* $\mathrm{Log}(\boldsymbol{S}_B)$ *is complete iff every state that satisfies B is the state of some situation.*

The logic $\mathrm{Log}(\boldsymbol{S}_B)$ is not in general sound, since there may well be situations not satisfying the background condition B. For instance, in our working example routine calculation shows that the sequent $p_1, p_2 \vdash q$ is a constraint in the logic $\mathrm{Log}(\boldsymbol{S}_{\sigma_3=\sigma_4=\sigma_5=1})$ with background condition $\sigma_3 = \sigma_4 = \sigma_5 = 1$, but not in the full logic $\mathrm{Log}(\boldsymbol{S})$.

Corollary 6 *If* $B_1 \leq B_2$ *then* $\mathrm{Log}(\boldsymbol{S}_{B_1}) \sqsubseteq \mathrm{Log}(\boldsymbol{S}_{B_2})$.

As remarked earlier, in reasoning about a system we expect to be given information that is silent on the parameters of the system, since when we are given information that is not silent on some input observable, it is no longer a parameter but an explicit input of the system. In particular, if we are given explicit information about the value of some observable, these observables cannot be parameters, which means that the background is weakened according to the new information.

Example: The information in p_3 is about the observable σ_3, the power to the system. Not only is p_3 not silent on the background condition B of the first sequent, it directly conflicts with it. Thus, the natural understanding of claim that $p_1, p_2, p_3 \vdash q$ is as taking place relative to its weakening $B \lceil p_3$ according to p_3. Relative to the local logic $\mathrm{Log}(\boldsymbol{S}_{B\lceil p_3})$, this is not a valid constraint. Indeed, in this logic another routine calculation shows that $p_1, p_2, p_3 \vdash \neg q$, as desired.

We summarize the above discussion by putting forward a pragmatic model of the way people intuitively reason against background conditions. First:

Definition 10 Γ *strictly entails* Δ relative to the background condition B, written $\Gamma \Rightarrow_B \Delta$, if the following four conditions hold:

1. $\Gamma \vdash_B \Delta$
2. that all propositions in $\Gamma \cup \Delta$ are silent on B.
3. $\bigcap \Gamma \neq \emptyset$
4. $\bigcup \Delta \neq \Omega$

The first two conditions have been extensively discussed above. The third and fourth are not important for this discussion, but we include them for the sake of completeness because they do capture intuitions about the way people reason.[1]

Our observations can now be put as follows:

1. The consequence relation $\Gamma \Rightarrow_B \Delta$ is a better model of human reasoning against the background condition B than is $\Gamma \vdash_B \Delta$.
2. The relation $\Gamma \Rightarrow_B \Delta$ is monotonic in Γ and Δ, but only as long as you add propositions that are silent on B.
3. If one is given a proposition p that is not silent on B, the natural thing to do is to weaken the background condition B by p thereby obtaining $B \lceil p$.
4. But $\Gamma \Rightarrow_B \Delta$ does not entail $\Gamma, p \Rightarrow_{B\lceil p} \Delta$ or $\Gamma \Rightarrow_{B\lceil p} \Delta, p$.

[1] The third is that the information in the sequent is consistent in that there should be some possible state of the system compatible with everything in Γ. The last condition is that the information in sequent is non-vacuous in that not every possible state of the system should satisfy Δ.

Logic infomorphisms

A reasonable objection to our presentation so far would be to point out that we have systematically exploited an ambiguity by using "p_1, p_2, p_3" and "q" for both symbolic expressions (like the English sentences where they were introduced) as well as for the "corresponding" (state space) propositions, which are, after all, sets of states, not exactly the sort of thing people reason with in ordinary life. We need to discuss this relationship between the symbolic expressions and propositions modeled as sets of states. To do so we must invoke another notion from *IF*.

Given classifications $\boldsymbol{A}$ and $\boldsymbol{B}$ we work with pairs $f = \langle f^{\wedge}, f^{\vee} \rangle$ of functions, of which $f^{\wedge}$ is a function from the propositions of one of these classifications to the propositions of the other, and $f^{\vee}$ is a function from the situations of one of these classifications to the situations of the other. We say that f is a *contravariant* pair from $\boldsymbol{A}$ to $\boldsymbol{B}$, and write $f: \boldsymbol{A} \rightleftarrows \boldsymbol{B}$, if $f^{\wedge}: \text{typ}(\boldsymbol{A}) \to \text{typ}(\boldsymbol{B})$ and $f^{\vee}: \text{tok}(\boldsymbol{B}) \to \text{tok}(\boldsymbol{A})$.

Definition 11 An *infomorphism* $f: \boldsymbol{A} \rightleftarrows \boldsymbol{B}$ from $\boldsymbol{A}$ to $\boldsymbol{B}$ is a contravariant pair of functions $f = \langle f^{\wedge}, f^{\vee} \rangle$ satisfying the condition:

$$f^{\vee}(b) \models_A \alpha \text{ iff } b \models_B f^{\wedge}(\alpha)$$

for each situation b of $\boldsymbol{B}$ and each proposition α of $\boldsymbol{A}$.[2] A *logic infomorphism* $f: \mathfrak{L}_1 \rightleftarrows \mathfrak{L}_2$ consists of a contravariant pair $f = \langle f^{\wedge}, f^{\vee} \rangle$ of functions that is an infomorphism of the underlying classifications, $f^{\wedge}$ maps constraints of $\mathfrak{L}_1$ to constraints of $\mathfrak{L}_2$, and $f^{\vee}$ maps normal situations of $\mathfrak{L}_2$ to normal situations of $\mathfrak{L}_1$.

To see what this has to do with our problem, let us return to our example and set up a symbolic Boolean classification $\boldsymbol{A}$. The propositions of $\boldsymbol{A}$ are symbolic expressions $\varphi, \psi, \eta, \ldots$, either the English expressions used in the introduction and Boolean combinations of them, or some sort of more formal counterparts of the sort used in an elementary logic course. We use $\varphi_1, \varphi_2, \varphi_3$, and ψ for the premises and conclusions of our classification. The situations of $\boldsymbol{A}$ are instances of my heating system. Let $\boldsymbol{B}$ be the classification $\text{Evt}(\boldsymbol{S})$ associated with the state space $\boldsymbol{S}$ as above. The map $\varphi_i \mapsto p_i$, and $\psi \mapsto q$, extended in the natural way to the Boolean combinations, defines a map $f^{\wedge}$ from the propositions of $\boldsymbol{A}$ to the propositions of

[2]Infomorphisms of classifications are a generalization of the usual notion of an interpretation in logic as well as that of a continuous function in topology. They are also known as Chu transformations in the computer science literature. See, for example, the extensive work of Vaughn Pratt and his colleagues. See, for example, Pratt [9]. A complete list of these papers can be found on the world wide web at http://Boole.stanford.edu/chuguide.html.

Log($\boldsymbol{S}$), while the identity map on situations can be thought of as a map of the situations of Log($\boldsymbol{S}$). This gives us a classification infomorphism from $\boldsymbol{A}$ to the classification Evt($\boldsymbol{S}$),

$$f: \boldsymbol{A} \rightleftarrows \boldsymbol{B}.$$

In *IF*, we show that any infomorphism $f: \boldsymbol{A} \rightleftarrows \boldsymbol{B}$ and local logic $\mathfrak{L}_2$ on $\boldsymbol{B}$ gives rise to a $\sqsubseteq$-largest local logic $\mathfrak{L}_1$ on $\boldsymbol{A}$ such that f is a logic infomorphism from $\mathfrak{L}_1$ to $\mathfrak{L}_2$; this logic is called the *inverse image* of $\mathfrak{L}_2$ by f and is denoted by $f^{-1}[\mathfrak{L}_2]$.[3] This gives us a host of logics on the classification $\boldsymbol{A}$. There is the logic $f^{-1}[\mathrm{Log}(\boldsymbol{S})]$, but also, for each subspace of $\boldsymbol{S}$, there is the inverse image of the logic associated with this subspace. In particular, for each background condition B, there is a local logic $f^{-1}[\mathrm{Log}(\boldsymbol{S}_B)]$. Let us write these logics as $\mathfrak{L}$ and $\mathfrak{L}_B$ respectively, and $\Gamma \vdash \Delta$ and $\Gamma \vdash_B \Delta$ for their entailment relations.

The logic $\mathfrak{L}$ is sound, that is, all its situations are normal. The constraints of this logic are just those sequents of $\boldsymbol{A}$ whose validity is insured by our state space model. The logics of the form $\mathfrak{L}_B$, however, are not sound; the normal situations of $\mathfrak{L}_B$ are those that satisfy the background condition B. For example, the normal situations of $\mathfrak{L}_{\sigma_3=\sigma_4=\sigma_5=1}$ are those in which the power is on, vents are unblocked, and the controls are in heating mode.

While we have done all this for our toy example, the entire discussion clearly generalizes to any sort of symbolic Boolean classification $\boldsymbol{A}$. This puts us in a position to define a version of our three-place pragmatic notion of entailment $\Gamma \Rightarrow_B \Delta$ where Γ and Δ are sets of symbolic propositions of $\boldsymbol{A}$.

Definition 12 If Γ and Δ are sets of symbolic propositions of $\boldsymbol{A}$, then Γ *strictly entails* Δ relative to B, written $\Gamma \Rightarrow_B \Delta$, iff $f^\wedge[\Gamma] \Rightarrow_B f^\wedge[\Delta]$.

This relation has the same inferential properties of the similarly notated relation in terms of which it is defined. It has one additional nice feature, though.

The ineffability principle In the model of the previous section, a background condition B could be packaged into a proposition q_B, namely,

$$q_B \quad = \quad \{\sigma \in \Omega \mid \sigma \text{ satisfies } B\}.$$

This proposition could, in principle, be made explicit as an additional premises of an inference. But this proposition will typically lie outside of the range of $f^\wedge$. Put differently, *there is in general no way to capture*

[3] There is also a result in the other direction, but we do not need that here.

the background condition B *by a symbolic proposition of* $\boldsymbol{A}$. I call this the "ineffability principle" since it seems to model the fact that it is seldom if ever possible to say exactly what background assumptions are in force when we reason in ordinary life.

Turtles all the way down

Having seen the relationship between state spaces and local logics, both symbolic and otherwise, let us return to state spaces.

As a start, let us note that if $\boldsymbol{S}$ is a real-valued state space of dimension n and B is a background condition on k parameters, then $\boldsymbol{S}_B$ is isomorphic to a real-valued state space $\boldsymbol{S}^*{}_B$ of dimension $m = n - k$. The isomorphism is the identity on situations while on states it projects $\langle \vec{\sigma_i}, \vec{\sigma_p}, \vec{\sigma_o} \rangle$ to $\langle \vec{\sigma_i}, \vec{\sigma_o} \rangle$. (In our example, the new space would have as states those 4-tuples $\langle \sigma_1, \sigma_2, \sigma_3, \sigma_4 \rangle$ such that $\langle \sigma_1, \sigma_2, 1, 1, 1, \sigma_3, \sigma_4 \rangle \in \Omega$.) The move from the n dimensional $\boldsymbol{S}$ to the m dimensional $\boldsymbol{S}^*{}_B$ results from setting k parameters as dictated by B.

Now let's look at it the other way around. Suppose we had begun with an m-dimensional state space $\boldsymbol{S}$ and later learned of k additional parameters that we had not taken into account. Letting $n = m + k$, we would then see our space $\boldsymbol{S}$ as (isomorphic to) a subspace of an n-dimensional state space $\boldsymbol{S}'$. Ignoring the isomorphism, we can think of this as $\boldsymbol{S} \subseteq \boldsymbol{S}'$. Then, taking the associated local logics, and recalling the order reversal that takes place, we have $\text{Log}(\boldsymbol{S}') \sqsubseteq \text{Log}(\boldsymbol{S})$. That is, the logic associated with the new state space is weaker in that it has fewer constraints, but more reliable, in that it has more normal situations, which is just what we have seen.

Example: Suppose instead of the information p_3, we had been faced with the new information

> (p_4) The gas line has been broken and there is no gas getting to the furnace.

We would like to get $p_1, p_2, p_4 \vdash \neg q$, but we don't. In fact, the proposition p_4 does not even make sense in our state space, since the gas pressure and gas line have not been taken into account in our 7-dimensional state space $\boldsymbol{S}_{hs}$. To take these into account, we would need to see $\boldsymbol{S}$ as isomorphic to a subspace of a 9-dimensional state space $\boldsymbol{S}'$, one where the additional observables are the state of the gas line and the gas pressure. Then as we have seen, we would have $\text{Log}(\boldsymbol{S}') \sqsubseteq \text{Log}(\boldsymbol{S})$.

One should not think that there is ever an "ultimate" completely perfect state space. In general, it seems it is almost always possible in real-life systems to further refine a state space model of some system by introducing more observables. That, of course, is what we mean by saying it is turtles all the way down.

The example makes an additional point, though. Once we introduce the new observables, we have also implicitly changed the set of situations under consideration. Our situations were just instances of my heating system, something located entirely in my house. But the gas lines run from the house out under the streets of Bloomington. Our situations have greatly expanded. Each of our old situations s is part of a situation $s' = f^{\vee}(s)$ in S'. In other words, our isomorphism is no longer the identity on situations. When we look at this in terms of the associated classifications $\mathrm{Evt}(\boldsymbol{S}')$ and $\mathrm{Evt}(\boldsymbol{S})$, what we have is an infomorphism$f\colon \mathrm{Evt}(\boldsymbol{S}') \rightleftarrows \mathrm{Evt}(\boldsymbol{S})$ that is not the identity on situations. Rather, it takes each situation to a richer situation. And again, there typically seems to be no end to this. As we reason in greater detail, there seems no end to the richness that may have to be considered. (This is not a new observation, it has been known to those working in non-monotonic logic for a decade or so. The point here is just to see how this realization fits into the picture presented here.)

What is the theoretical limit of this process? In a sense, it is possible worlds semantics. There we have only one situation, the entire actual world, and the "set" Ω of a logically possible ways this world might have been. Every actual situation gets mapped to this world, and every proposition gets interpreted as a set of possible ways the world could be. The constraints are just those dictated by logic itself. This is why we prefer to call our tokens situations, not possible worlds.

Conjectures and Conclusions

In spite of their ubiquitous use in science and applied mathematics, the use of state space as a model for human reasoning has, as far as I know, been largely ignored. But there are a couple of reasons why such a move might prove fruitful.

Feasibility In those applications where the equations

$$\sigma_o = F_o(\vec{\sigma_i})$$

are given by equations that really can be computed, the state space/local logic approach might give us an interesting alternative to traditional theorem proving methods, one where fixed a numerical calculations could be used in place of more symbolic approaches. This is the antithesis of Pat Hayes thesis in his famous manifesto [4]. The proposal here is to *exploit* the sciences in modeling common sense reasoning rather than replace it by a parallel symbolic framework.

In this regard, Tarski's Decision Procedure for the reals (Tarski [12]) suggests itself as potentially a useful tool.[4] As long as the output functions F_o, the propositions in Γ, Δ, and the background condition B are first-order definable over the field of reals, as they are in our example, Tarski's decision procedure gives us a mechanical way to determine whether or not $\Gamma \vdash_B \Delta$.

Logic and cognition Much more speculatively, the proposal suggests a way out of a box logic has been put into by some of its detractors. Within recent the cognitive science literature, logic is often seen as irrevocably wed to what is perceived as an outdated symbol processing model of cognition.[5] From there it is but a short step to the conclusion that the study of logic is irrelevant for cognitive science. This step is often taken in spite of the fact that human reasoning is a cognitive activity and so must be part of cognitive science. Perhaps the use of state spaces might allow a marriage of logic with continuous methods like those used dynamical systems and so provide a toe-hold for those who envision a distinctively different model human reasoning (see chapter 10 of Barwise and Perry [1], for example). The (admittedly wild) idea is that the input and output of reasoning could be symbolic, at least sometimes, while reasoning itself might be more modeled by state space equations, with the two linked together by means of something like infomorphisms.

Modifying background conditions The above suggestions are, admittedly, pie in the sky. What we have accomplished here is at best much more modest: a computationally feasible mechanism $B \upharpoonright \Gamma$ for modeling the modification of background ground conditions B given new information Γ. I believe the nice properties of this mechanism suggest that states spaces and their local logics are potentially a useful tool for getting an interestingly different handle on the vexing problems of exceptions and the non-monotonicity of reasoning.

Apology

I feel the need to apologize for the fact that the results here have not been related to the substantial literature on non-monotonic reasoning in A.I. that

[4] See Rabin 10 for a brief exposition of Paul Cohen's improved proof of Tarski's Theorem.

[5] I am thinking here of some of view of logic as it is seen in, for example, Port and van Gelder [8].

has developed in the sixteen years since the early work of McCarthy [7], Reiter [11] and others. My excuse, feeble though it is, is that I don't understand the relationship. This stems in part out of ignorance of the recent work in the non-monotonic community, in part because the ideas presented here are quite recent.

There should be relationships since the intuitive idea of shifting context that motivates the approach taken here can been seen to underlie other formalizations of nonmonotonic reasoning such as default logic, nonmonotonic modal logics and semantics of logic programs. For example, Marek and Truszczynski [5, 6] have developed a logic using a default proof theory where the inference that can be performed are context dependent. The context C for a default theory is determined by the set of conclusions derivable from the default theory given the context C, the circularity being cashed out in by means of a fixed point equation. In their theory, context plays a role only when testing justifications of the default rules. It is not at all clear to me how to go about comparing their ideas with the approach taken here, since there is nothing analogous to default rules in the current proposal. The rules of inference of a fixed local logic are entirely classical. The non-monotonic mechanisms proposed here have to do with how one shifts between these local logics.

Added in proof (26 October 1998): Over the past few months I have been working with Maricarmen Martinez to implement a reasoning system based in part on the ideas in this paper. We are implementing the system in *Mathematica* and are finding it quite interesting. Several ideas not in the paper have come up. A key idea is that of moving information from types to background restrictions in order to make computations more efficient. A second is in using "basic" sets of reals to encode restrictions, where the basic sets form the smallest algebra of sets of reals containing the closed intervals.

References

1. Jon Barwise and John Perry, *Situations and Attitudes*, Bradford Books, MIT Press, 1983.

2. [*IF*] Jon Barwise and Jerry Seligman, *Information Flow in Distributed Systems*, Cambridge University Press, Tracts in Theoretical Computer Science, to appear.

3. [*RR*] John Casti, *Reality Rules, vol. I*, Wiley Interscience, 1992

4. Pat Hayes, "Naive Physics I: Ontology for Liquids," In *Formal Theories of the Commonsense World*, ed. R. Hobbs and R. Moore, Ablex Press, 71-108

5. W. Marek and M. Truszczynski. "Relating autoepistemic and default logics," In *Proceedings of the 1st International Conference on Principles of Knowledge Representation and Reasoning,* Toronto, Canada, 1989, 276-288

6. W. Marek and M. Truszczynski, *Nonmonotonic Logic: Context-Dependent Reasoning,* Springer-Verlag, Berlin, 1993.

7. John McCarthy, "Circumscription – a form of non-monotonic reasoning," *Artificial Intelligence,* 1980, 27-39

8. Robert Port and Timothy van Gelder, *Mind as Motion: Explorations in the Dynamics of Cognition*, Bradford Books, M.I.T. Press, 1995

9. Vaughn Pratt, "Chu spaces: Complementarity and uncertainly in rational mechanics," Course notes, TEMPUS summer school, 35pp, Budapest, 1994

10. Michael Rabin, "Decidable theories," in *Handbook of Mathematical Logic*, ed. by Jon Barwise, North-Holland, 1974, 595-629

11. Ray Reiter, "A logic for default reasoning," *Artificial Intelligence,* 1980, 81-132

12. Alfred Tarski, "A decision method for elementary algebra and geometry," 2nd revised edition, Berkeley and Los Angeles, 1951

96301
21-44

Presupposition Accommodation: A Plea for Common Sense

David Beaver

Linguistics Department
Stanford University

1 Introduction[1]

Life is short. There is not enough time to explain everything. As speakers, or writers, we are forced to make assumptions. It is common to be advised to fix in one's mind a picture of the audience, that is, to make an advance decision as to what the audience can be expected to know. Often, especially given limitations of time for speaking or space for writing, one is forced to take much for granted. As a result, cases of presupposition failure, the situation occurring when the speaker or writer takes for granted something of which the hearer or reader is not previously aware, are surely common. Somehow, hearers and readers cope, and usually without complaining.

The author, in most *genres*, assumes that the text will be read linearly, and further assumes, optimistically, that readers will gather information throughout the reading process. So what the author has is not a fixed picture of the common ground with the intended readership, but a rather rough cut and idealized movie of how this common ground should develop. Each frame in the movie approximates what is common between relevant aspects of (1) the author's beliefs at time of writing, and (2) the readers' beliefs as they reach some point in the text. At risk of straining the cinematic metaphor somewhat, it could be said that the text itself is analogous to a script, but with detailed screenplay and directorial instructions omitted. Barring a major scientific breakthrough, the corresponding film will never be put on general release, so precisely how the writer intends the information state of idealized readers to evolve as they read is never made public in all its technicolor glory. In this paper, I will describe in brief a

Logic, Language, and Computation, Vol. II, edited by Lawrence S. Moss, Jonathan Ginzburg, and Maarten de Rijke.

[1]I would like to express my thanks to the organisers of the ITALLC conference, and to the anonymous referees of this paper. The paper extends work previously published as [Bea94c].

model of how readers' information states do evolve, as they construct their own movies on the basis of the script.

The theory to be developed in this paper can be thought of as providing a formal characterization of what Lewis [Lew79] called *accommodation.* But the model will differ markedly from existing proposals, in that most writers have taken accommodation to be a repair strategy, something that happens when the interpretation process goes wrong. Lewis seems to picture accommodation as a covert adjustment of what he calls the *conversational score*, a sort of creative accounting needed to make conversational ends meet. Van der Sandt's accommodation [vdS92], to take a more recent example, is a sophisticated cut-and-paste operation on Discourse Representation Structures (c.f. [Kam81]).

Accommodation, then, has been viewed as an essentially non-monotonic operation, overwriting our previous record of what had happened in a discourse to fit with new demands. The view espoused in this paper will be quite the contrary: accommodation will be analyzed as a *monotonic* operation, in the sense that it does not replace or destructively revise our information about a speaker or author, but further instantiates our knowledge, reducing the range of possibilities for what the speaker was assuming.

The main claim I will make is that when we accommodate, we look not only at the record of what has been said, but also behind what has been said, and consider explicitly what the author might have intended and what the author might have expected. The property of monotonicity will arise as a direct consequence of incorporating the reader's uncertainty about the authors assumptions. This type of reasoning under uncertainty will involve what is in the A.I. literature sometimes described as *common sense reasoning.* That is, the reasoning processes described will involve not only absolute knowledge of what the speaker believes or doesn't believe, but also knowledge of what is most *plausibly* believed. It will be assumed that agents are able to partially order alternative (partial) models of what another agent believes according to the plausibility of those models.

2 Theories of Presupposition and Accommodation

In this section, existing theories of presupposition will come under discussion. The reader is warned that these theories will be painted with a broad brush: for more detailed exegeses see [So89, Bea97]. Linguistic examples will be introduced which suggest that a more sophisticated notion of accommodation is needed than is found in any previous theories.

We will begin with what may be termed *semantic* theories, a large and heterogeneous class of theories which do not incorporate any notion of accommodation at all. A semantic theory of presupposition is one in which presupposition is accounted for in terms of truth and falsity conditions. Typically, in such theories truth conditions are calculated compositionally, independently of speaker intention or discourse context.

Consider the following two sentences:

1 If Jane takes a bath, Bill will be annoyed that there is no more hot water.

2 If Jane wants a bath, Bill will be annoyed that there is no more hot water.

The predicate "annoyed that" is factive, and in these examples triggers the presupposition that there is no more hot water. Theorists proposing semantic accounts of presupposition may differ according to whether they predict that these triggered presuppositions become full presuppositions of the two sentences.[2] However, neither of the sentences "Jane wants a bath " nor "Jane takes a bath" is logically related to the sentence "there is no hot water", by which I mean that there are no entailment relations between these sentences. From a purely semantic point of view, there is thus no relevant difference between (1) and (2). So, in one crucial respect, I think proponents of semantic accounts will be forced to agree with each other: whatever the form of the analysis given for example (1), the same form of analysis must be used for (2).

An utterance of (1) does not suggest to me that there actually is no more hot water, but only that if Jane takes a bath, there will be no more hot water. On the other hand, (2) suggests strongly that there is no more hot water. I do not believe that any purely semantic theory of presupposition can account for this contrast.

By far the lion's share of the existing literature on presupposition concerns semantic approaches, usually depending on some sort of multi-valued logic. None of these approaches is compatible with the claim that I will advance in this paper that the contrast between (1) and (2) is not naturally explained in terms of truth conditions but in terms of plausibility.

For those who find a contrast between the examples, the reason will presumably be that they are used to a situation where water supplies are limited. Thus it seems plausible to them that there is a causal relation between Jane taking a bath and the exhaustion of the hot-water supply. On the other hand no similarly plausible explanation suggests itself for why the hot-water supply should be dependent on Mary's *desire* for a bath alone,

[2]For instance, a semantic theory based on the Strong Kleene interpretation of the conditional would yield conditionalized presuppositions for both examples, i.e. "If Jane takes a bath then there is no more hot water" and "If Jane wants a bath then there is no more hot water" respectively. See [Bea97] for discussion.

and the conclusion is reached on hearing (2) that the hot-water shortage is an absolute, independent of the antecedent of the conditional.

Let us now move on to consider accounts of presupposition which incorporate a notion of accommodation. The approach developed here contrasts sharply with the formalization of Lewis' accommodation which [Zee92] attributes to Heim, and with the DRT-based theory of presupposition accommodation developed by van der Sandt. Both the Zeevat-Heim and van der Sandt accounts provide notions of accommodation which might be termed *structural*: one might caricature them as involving *move-α* at the level of logical form. Let me explain. Heim [Hei83] characterizes accommodation as follows:

> Suppose [a sentence] S is uttered in a context c which doesn't admit it.... simply amend the context c to a richer context c', one which admits S and is otherwise like c, and then proceed to compute c' [updated with] S instead of c [updated with] S.

But how is the context to be amended? [Hei83] is unclear on this point. On Zeevat's interpretation of Heim's theory, if S is uttered in a context c which doesn't admit it, then there must be a specific unsatisfied proposition p which is presupposed by S, and the amendment consists in adding this proposition to c.[3]. Heim says that accommodation will apply equally if S occurs in an embedded context. Thus if c admits R but does not admit a sentence "if R then S ", and S presupposes the proposition α, then one way of proceeding is to add α to c, and then update this context with "if R then S". This is the alternative that she refers to as *global accommodation*: in effect the presupposition α is moved (and conjoined) to the front of the sentence, to yield "α and if R then S ", and an update is then performed with the resulting formula. Heim also suggests that a presupposition may be added to some of the intermediate contexts involved in calculating an update, what she calls *local accommodation*. This has much the same effect as allowing alternative landing sites for α at the level of logical form. Thus if update with "if R then S " fails, she would offer three alternative ways of updating, corresponding to the sentences: "α and if R then S ", "if (α and R) then S ", and "if R then (α and S) ". In van der Sandt's account, which I will not describe in detail here, the *move-α* flavor is even more obvious than in Heim's theory. Whereas in the Heim-Zeevat account α is essentially

[3] Until recently I shared Zeevat's understanding of [Hei83]. However, Heim (p.c.) denies that the she intended accommodation to be understood as the addition of a specific conventionally indicated proposition such as in Zeevat's interpretation, and has pointed out that her original text is quite non-committal. So what in this paper is referred to as the *Zeevat-Heim account* is probably not an account to which Heim herself would fully subscribe.

seen as a unit of propositional information, in van der Sandt's account α is found at a pseudo-syntactic, almost LF-like level of representational form, namely, as mentioned above, the Discourse Representation Structures of Kamp's DRT.

It can now be stated what I mean by a *structural* account of accommodation. A structural account is one in which any given presuppositional construction has a single conventionally presupposed proposition, and where accommodation associated with a sentence containing that construction consists of adding this proposition to some relevant context. Here the relevant contexts are the initial context, and some set of intermediate contexts which can be specified for each sentence type. The problem that I see for a purely *structural* account of accommodation is as follows: it is not possible to predict on structural grounds alone exactly what should be accommodated. In general, the exact accommodated material can only be calculated with reference to the way in which world knowledge and plausibility criteria interact with the meaning of a given sentence. Let us consider another example:

3 If the North Korean ambassador turned up, then it is amazing that both the North and South Korean ambassadors are here.

If I overheard (3) at one of the high level diplomatic receptions to which I am regularly not invited, I would conclude that the South Korean ambassador was present. It is this inference — call it the SK-inference — that will now come under discussion.

It seems that the SK-inference must be related to the factive "amazed that" in the consequent of (3), which triggers the presupposition that both North and South Korean ambassadors are present. However, a speaker uttering (3) could not be taking for granted that both ambassadors were present, since the possibility of the North Korean ambassador having arrived is precisely what is under consideration. So, global accommodation of the triggered presupposition cannot be the appropriate explanation of the SK-inference.

What are the other possible readings of (3) in the van der Sandt or Heim-Zeevat theories? Accommodation in the context corresponding to the antecedent of the conditional in (3) would produce an interpretation which could be glossed as "if both ambassadors are here then that is amazing." Accommodation in the context in which the trigger is found, i.e. in the consequent of the conditional, produces an interpretation "if the NK ambassador came, then both ambassadors are here, and that is amazing." Of these, the latter form of accommodation comes closest to the intuitively correct interpretation. But, crucially, an account of the SK-inference is still missing.

One is tempted to formulate an analysis of (3) whereby the triggered presupposition is somehow split into two before accommodation kicks in. But such an analysis would be *ad hoc*, since there is no obvious general principle which would license splitting of the presupposition in this case, but not in others. I would advocate an alternative line of explanation for the SK-inference. Firstly, the form of sentence (3) leads to a conditional presupposition, roughly "if the NK ambassador came, then both ambassadors are here". Secondly, world knowledge about what the speaker is likely to take for granted leads to strengthening of this presupposition. the mechanisms behind both of these steps will be detailed shortly.

If I am correct in my analysis of (3), then the failure of structural accounts of presupposition on this and related examples arises from two separate weaknesses. Firstly, such theories provide no way of accounting for conditionalized presuppositions, and secondly they provide no way of strengthening those conditionalized presuppositions where appropriate.

For the moment, the details of the analysis I will advocate are not important. What is important is the fact that structural theories can provide no account of the SK-inference. The reason is simple: what the hearer accommodates is not the same proposition as is presupposed by any trigger in the sentence. Certainly, the accommodated information is related to what is triggered. But, in structural theories there is only one way that accommodated material can be related to triggered material: structural identity.

Here is another case where a conditional reading might be advocated:

4 If Spaceman Spiff lands on Planet X, he'll notice that he weighs more than on Earth.

The factive "notice that" in this case triggers the presupposition that Spiff's weight is higher than it would be on Earth. Structural accounts of accommodation suggest that this proposition should be globally accommodated. However, it is questionable whether this result is appropriate. It is not normal to conclude from (4) that Spiff's weight is definitely higher than it would be on Earth. Indeed, it seems natural for (4) to be uttered under conditions where Spiff is hanging about in space, and completely weight-less. Can non-global accommodation save the structural account? Accommodation into the antecedent produces something like "If Spaceman Spiff weighs more than on Earth and he lands on Planet X, he'll notice he weighs more than on Earth." I do not think this is a possible meaning of (4). Accommodation into the consequent appears to improve, yielding (after charitable adjustment of tense) "If Spaceman Spiff lands on Planet X, then he'll weigh more than on Earth and will notice that he weighs more than on Earth." This provides a reasonable meaning for (4), and offers hope that if only some way could be found of removing the two incorrect readings, the structural account might be saved. Unfortunately, a very slight variation on (4) produces an example where the structural account produces four incorrect (or, at the very least, non-preferred) readings, and completely fails to yield the preferred reading:

5 It is unlikely that if Spaceman Spiff lands on Planet X, he'll notice that he weighs more than on Earth.

The preferred reading of this sentence is still one involving the conditional implication, i.e. if he lands on Planet X, Spiff's weight will be higher than it is on Earth, and quite natural assumptions about the dynamics of the "it is unlikely" construction would lead to the model presented here making the same presuppositional predictions for this example as for (4). But in this case the structural account no longer yields the right reading after accommodation into the consequent of the conditional. This would yield "It is unlikely that, if Spaceman Spiff lands on Planet X, he'll weigh more than on Earth and notice that he weighs more.", which does not imply that if he lands on Planet X, Spiff's weight will be higher than it is on Earth. On the contrary, there is even a slight suggestion from this sentence that if he lands on Planet X his weight probably will not be higher than it is on Earth, which is clearly inappropriate.

There are versions of (4) in which the inability the van der Sandt and Heim-Zeevat accounts to produce the appropriate conditional presupposition are even more obvious:

6 If Spaceman Spiff is in our solar system, he'll land on Planet X, and will notice that he weighs more than on Earth.

7 If Spaceman Spiff lands on Planet X and notices that he weighs more than on Earth, he'll radio home about it.

Both of these examples seem to retain the conditional presupposition: the most likely conclusion of a reader remains that Planet X is the sort of place where one is particularly heavy, and if Spiff lands there, he will be heavy. Consider (6). Global accommodation of Spiff's weight being too high still seems wrong, and similarly accommodation into the antecedent of the conditional. But now even local accommodation fails to produce an appropriate conditional. A reading is derived along the lines of: "if Spiff is in our system, he'll weigh a lot, he'll land on planet X, and will notice that he weighs a lot." Thus the conclusion is that Spiff's weight is dependent on his being in the speaker's solar system. This, I suppose, is a possibility, but it is not the most plausible interpretation. Concerning (7), where the presupposition trigger is found in the antecedent of a conditional, structural accounts predict that only global accommodation and accommodation in that antecedent are possible. Thus there is absolutely no way to produce a reading where Spiff's being heavy is conditionalized. Global accommodation is as wrong as ever, and accommodation in the antecedent produces an interpretation of the sentence something like: "if Spiff's weight is high and he lands on our planet and he notices that his weight is high, then he'll leave." This is not a plausible interpretation of (7). These results pose a serious problem for the van der Sandt and Heim-Zeevat accounts.

What of the cancellation theory of presupposition proposed by Gazdar [Gaz79]? The basic idea is that when a sentence containing a presupposition trigger is uttered, by default the presupposed proposition will be projected. But if the presupposition is inconsistent with the previous context, with what is asserted by the sentence, with implicatures of the sentence, or with other presuppositions, then it is canceled. But the following following wonderful observation from [Hei83] shows that the cancellation account could easily be characterized as a structural theory of accommodation:

> Note that by stipulating a *ceteris paribus* preference for global over local accommodation, we recapture the effect of [Gazdar's] assumption that presupposition cancellation occurs only under the threat of inconsistency.

Under Gazdar's assumption that presuppositions are also entailed by their triggers, his theory will only give one of two readings available in the accommodation accounts[4]. Gazdar's theory always gives one of two

[4]As for whether presuppositions are entailed by their triggers, that remains a moot point. [Bea97] argues that regarding most classes of presupposition trigger, failing to make this assumption would cause severe problems for the cancellation account. The most obvious difficulty that would arise if presuppositions were not taken to be entailed would be the lack of any explanation as to why presuppositions (at least for the vast majority of presupposition triggers) cannot be canceled when they are not embedded within a special context such as that created by negation. For instance, both "Cheese is good for you and Mary knows that it is bad for you" and "Cheese is good for you and it bothers Mary that cheese is bad for you" are odd. If the presuppositions of 'knows that' and 'it bothers Mary that' are also entailments, then the oddity is explained because the sentences are inconsistent. On the other hand, without the assumption of entailment, there is no obvious explanation, because Gazdar's account predicts that inconsistency simply produces cancellation of presuppositions with no further effects. The same applies to other cancellation accounts such as [Mer92, vdS88, Soa82]. Dropping the assumption that presuppositions are entailed by their triggers, though it would not help with the treatment of Spaceman Spiff type examples being discussed here, would result in a model that could produce readings not found in the accommodation accounts. In that case, cancellation of a non-entailed presupposition would produce an interpretation in which the presupposition played no part whatsoever. This is impossible in the accommodation accounts, for in these theories the presupposition always ends up somewhere in the meaning, even if only within an embedded context.

readings available in the accommodation accounts. If the presupposition is not canceled, then we derive the global accommodation reading, and if the presupposition is canceled, the assumption that presuppositions are also entailed by their triggers means that we obtain the same affect as local accommodation would produce. Clearly the above criticisms of structural theories of accommodation also apply to Gazdar's superficially quite different account, for, as has been explained, neither global nor local accommodation produces the right result in the cases discussed.

In fact the prognosis for Gazdar's theory is somewhat worse than for the van der Sandt and Heim-Zeevat accounts. The problem is that cancellation only occurs in the face of possible inconsistency, but in none of the Spiff examples (5–7) is there any entailment or implicature to threaten inconsistency with the presupposition that Spiff is heavier than he would be on Earth. Unless the context of utterance directly contradicts this proposition, Gazdar's model will predict that utterances of all of (4–6) presuppose that Spiff is heavier than he would be on Earth. A similar point can be made concerning all the earlier examples. For instance with regard to (1) and (2), on the assumption that the utterance context does not expressly state that there is hot water, Gazdar's model predicts that both these examples have the presupposition that there is no hot water. No account is given of the difference between these examples, this difference being too subtle for Gazdar's "if it don't fit, trash it" theory of presupposition to pick up.

To sum up this section, it has been argued that all purely semantic theories of presupposition, and all purely structural accommodation or cancellation based theories of presupposition are doomed to failure, for they lack common sense. The non-structural alternative to be proposed will rest on a relation of plausibility being defined across contexts (or, equivalently, across closed theories). It will now be shown how the ordering defined by this relation might underly a model of presuppositional accommodation within a dynamic account of utterance interpretation.

3 The Writer's View of the Common Ground

To return to the metaphor with which I began, in directing their own films readers second-guess the intentions of the original writer-director. But to understand how readers work out the writer-director's intentions, it is firstly necessary to know more of the craft of the writer-director. In the coming section I will elaborate on how, working from an assumption as to initial conditions, the author envisages the evolution of the common ground. It will be helpful to adopt some of the formalism of recent dynamic semantics. I will build particularly on ideas of Stalnaker [Sta74], Karttunen [Kar74] and

Heim [Hei83], and use formal techniques related to those discussed by Groenendijk and Stokhof, e.g. [GS91], and Veltman [Vel91]. I will present what will here be called Presupposition Logic, a simple propositional system with a dynamic semantics and dynamic notion of semantic entailment. Further discussion and motivation can be found in [Bea92, Bea93b, Bea94, Bea95].

Presupposition Logic provides a model of how a speaker or author envisages the common ground evolving. This evolution is iterative, since the common ground at any instant provides the context in which a given chunk of text is interpreted, and it is the effect of this interpretation which determines what the common ground will be prior to interpretation of the next chunk. It is no longer controversial to assert that the interpretation process relies on such iteration, but there remains some question as to the course-grainedness of the iteration. For instance in the work of Gazdar [Gaz79] (and also related proposals such as Mercer's [Mer92]) it is whole sentences which produce a change in the context of interpretation. However in Karttunen-Heim style treatments of presupposition such as that introduced in this paper, as well as in treatments of anaphora due to Heim [Hei82], Kamp [Kam81] and Groenendijk and Stokhof [GS91], a finer grained iteration is involved, with sub-sentential constituents producing their own effects on local contexts of interpretation.

We begin by assuming some set of atomic proposition symbols. A model is a pair $\langle W, I \rangle$, where W is a set of worlds and I is an interpretation function mapping each atomic proposition symbol to a subset of W. The Context Change Potential (to borrow Heim's terminology) of a formula ϕ, written $[\![\phi]\!]$, is a set of pairs of input and output contexts, where a context is the writer's view of the common ground. Following Stalnaker, a context is thought of as a set of possible worlds, the set containing all and only those worlds compatible with the information supposed to be common. I will write $\sigma[\![\phi]\!]\tau$ to indicate that the pair $\langle \sigma, \tau \rangle$ is in the $[\![\phi]\!]$ relation, i.e. an input context σ updated with the information in ϕ can produce an output context τ.

Defining contexts as subsets of W introduces a natural lattice structure with union and intersection as meet and join, and this lattice provides a model-theoretic characterization of the amount of information in a given context. A minimal context (with respect to some model) may be defined as the set of all worlds (in that model): this is the state of blissful ignorance in which no information about the world is available. Similarly a maximally informative non-contradictory context would be a singleton set: the available information rules out all except one world. However, it is possible to add even more information to such a context, and in case this information contradicts previous information we will arrive at a context containing no worlds, a truly maximal but contradictory context.

Definition D1, below, gives an update semantics for the language of propositional logic. The first clause says that the result of updating a context with an atomic proposition is an output containing only those worlds in the input which are in the extension of the proposition. A context can be updated with a conjunction of two formulae (the second clause) just in case it can firstly be updated with the left conjunct to produce an intermediary context (v in the definition), and this context can be updated with the right-hand conjunct to produce the final output (τ). The third clause says that a context can be updated with the negation of a formula just in case there is some state that can be obtained by updating the context with the negated formula itself, in which case the result of updating with the whole formula is the set of worlds in the input which are not present in the update with the negated formula. In other words, the effect of updating with the negation of a formula is to remove all information compatible with the formula. The final two clauses define implication and disjunction by (carefully selected) standard equivalences.

Definition D1 **(Update Semantics for Propositional Logic)** *For all models $\mathcal{M} = \langle W, I \rangle$ and information states σ, τ, the relation $[\![.]\!]^{\mathcal{M}}$ (superscript omitted where unambiguous) is given recursively by:*

$$\sigma[\![p_{\text{atomic}}]\!]\tau \quad \textit{iff} \quad \tau = \{w \in \sigma \mid w \in I(p)\} \tag{1}$$

$$\sigma[\![\phi \wedge \psi]\!]\tau \quad \textit{iff} \quad \exists v \;\; \sigma[\![\phi]\!]v[\![\psi]\!]\tau \tag{2}$$

$$\sigma[\![\neg\phi]\!]\tau \quad \textit{iff} \quad \exists v \;\; \sigma[\![\phi]\!]v \;\wedge\; \tau = \sigma \backslash v \tag{3}$$

$$\sigma[\![\phi \rightarrow \psi]\!]\tau \quad \textit{iff} \quad \sigma[\![\neg(\phi \wedge \neg\psi)]\!]\tau \tag{4}$$

$$\sigma[\![\phi \vee \psi]\!]\tau \quad \textit{iff} \quad \sigma[\![\neg(\neg\phi \wedge \neg\psi)]\!]\tau \tag{5}$$

A context σ satisfies a formula ϕ ($\sigma \models \phi$) if updating adds no new information, producing an output identical to the input. One formula entails another ($\phi \models \psi$) if any update with the first produces a context in which the second is satisfied:

Definition D2 **(Satisfaction and Dynamic Entailment)**

$$\sigma \models \phi \quad \textit{iff} \quad \sigma[\![\phi]\!]\sigma$$

$$\phi \models \psi \quad \textit{iff} \quad \forall \sigma, \tau \;\; (\sigma[\![\phi]\!]\tau \Rightarrow \tau \models \psi)$$

Over the standard propositional language this notion of entailment is classical. However, we will now extend the language with a presupposition operator, written ∂. The resulting logic will be non-classical. For example, commuting conjunctions will no longer uniformly preserve validity. The intuition behind the following definition is that a formula $\partial\phi$ ("the presupposition that ϕ") places a constraint on the input context, only allowing update to continue if the presupposed proposition is already satisfied.

Definition D3 **(Presupposition Logic)** *Presupposition Logic is defined over the language of Propositional Logic with an extra unary operator* ∂. *It has the dynamic notion of semantic entailment above, and semantics consisting of the update semantics for atomic propositions and standard connectives combined with the following interpretation for* ∂*-formulae:*

$$\sigma[\![\partial\phi]\!]\tau \quad \textit{iff} \quad \sigma \models \phi$$

As an example, the sentence "Mary realizes that John is sleepy" might be said to correspond to a formula in Presupposition Logic of the form $\partial p \ \wedge \ q$, where p is atomic proposition that John is sleepy, and q is an atomic proposition that Mary has come to believe that John is sleepy. In this paper I will not be concerned with the details of how such logical forms may be derived from natural language, or with the question of whether it is reasonable to use a representation in which presuppositions are divided explicitly from assertions using the ∂-operator. However, both these issues are dealt with elsewhere: see [Bea93a, Bea93b, Bea95] where two sorted versions of classical type theory are used to provide a Presupposition Logic style semantics for a fragment of natural language.

Suppose a formula contains a presuppositional sub-formula. What will the presuppositions of the whole formula be? This is the presupposition projection problem of Langendoen and Savin [LS71], except applied to Presupposition Logic rather than natural language. A formula $\partial\phi$, "the presupposition that ϕ", defines an update if and only if ϕ is satisfied, so it is natural to say that in general a formula presupposes all those formulae that must be satisfied by the input context in order for there to be an update. We say that a context σ admits a formula ϕ (written $\sigma \rhd \phi$) if and only if it is possible to update σ with ϕ, this being a formalization of Karttunen's notion of admittance in [Kar74]. In that case one formula ϕ presupposes another ψ (written $\phi \gg \psi$) just in case every context that admits the first satisfies the second. Note that admittance provides a counterpart to so-called *presupposition failure*, what happens when updating cannot continue because presuppositions are not satisfied.

Definition D4 **(Admits ($\rhd$) and Presupposes ($\gg$))**

$$\begin{array}{lll} \sigma \rhd \psi & \textit{iff} & \exists\tau\ \sigma[\![\phi]\!]\tau \\ \phi \gg \psi & \textit{iff} & \forall\sigma\ \sigma \rhd \phi \Rightarrow \sigma \models \psi \end{array}$$

It is clear that the definitions for the semantics of the ∂-operator and the meta-logical $\gg$ relation are closely related. For example we have that for any ϕ, $\partial\phi \gg \phi$. Indeed, $\gg$ could have been equivalently defined in terms of ∂, defining $\phi \gg \psi$ *iff* for some χ, $[\![\phi]\!] = [\![\partial\psi \wedge \chi]\!]$.

We can now study projection in Presupposition Logic. As detailed in F1, the system behaves just as anyone familiar with Karttunen's 1974 system would expect. In particular, note that a formula may fail to carry a presupposition of one of its component sub-formulae, but instead carry a logically weaker conditionalized variant.

Fact F1
If $\phi \gg \psi$ *then:*

$$\begin{array}{rcl} \neg\phi & \gg & \psi \\ \phi \wedge \chi & \gg & \psi \\ \phi \rightarrow \chi & \gg & \psi \\ \phi \vee \chi & \gg & \psi \\ \chi \wedge \phi & \gg & \chi \rightarrow \psi \\ \chi \rightarrow \phi & \gg & \chi \rightarrow \psi \\ \chi \vee \phi & \gg & (\neg\chi) \rightarrow \psi \end{array}$$

4 The Naive Reader

Readers who updated their own information state according to principles like those behind the semantics of Presupposition Logic would be stymied whenever some information was presupposed which they did not have. In this situation the reader would lack any further means of updating. So Presupposition Logic, as it stands, does not provide a good model of the evolution of the information state of a hearer or reader. But suppose that you only had a Presupposition Logic-like semantics to help you understand a text. How would you use that semantics to glean information?

If an infinite number of monkeys with typewriters were given time, some of them might produce this text. What if an infinite group of monkeys schooled in the writer-director approach to writing described above were given some rhetorical goal but no description of the intended audience? They might choose the initial common ground randomly, although after that each monkey's view of the common ground at a particular point in the text would be fully determined by what they typed. Now although as a reader you do not know what initial common ground has been assumed when you read a text, you can reason for any particular choice of initial conditions how the common ground would evolve. And this leads me to a suggestion for how a dynamic semantics like that given for Presupposition Logic could be used to understand a text without presupposition failure being problematic.

Begin the reading process by imagining an infinite number of monkeys (or as many as you can manage) with an infinite number of assumed initial common grounds, your goal being to find out which monkey wrote the text. As you read, separately update each of these contexts. At various stages presuppositional constructions will be encountered, and these are what sort out the wheat from the chaff. For whenever something is presupposed which has not been explicitly introduced earlier in the text, a number of monkeys drop out of contention, and there remain only those monkeys for which the assumed common ground corresponding to that point in the text satisfies the presupposition. In general, this process may not tell you exactly which monkey was responsible, but it will at least limit the options, and it will simultaneously tell you quite a lot about what information you were intended to have after reading the text.

The many-monkeys strategy can easily be formalized in terms of the semantics of Presupposition Logic. A reader's information state is identified with a set of contexts — I will use the term *information set* — and is thus a subset of the powerset of worlds. A state can be updated with a formula by updating each of the member contexts separately, so producing the following definition of the update of a state I with a formula ϕ:

Definition D5 **(Updating Information Sets)**

$$I + \phi = \{\tau \mid \exists \sigma \in I\ \sigma[\![\phi]\!]\tau\}$$

By definition, let us say that an information set satisfies a formula only if its member contexts satisfy the formula:

Definition D6 **(Satisfaction by an Information Set)**

$$I \models \phi \quad \textit{iff} \quad \forall \sigma \in I\ \sigma \models \phi$$

The earlier notion of entailment could easily be defined in terms of the new notion of information, as the following fact demonstrates. (Here $P(W)$ is the powerset of the set of worlds W.)

Fact F2

$$\begin{aligned} \phi \models \psi \quad & \textit{iff} \quad (P(W) + \phi) \models \psi \\ & \textit{iff} \quad \forall I\ (I + \phi) \models \psi \end{aligned}$$

5 The Sophisticated Reader

The naive reader might imagine an infinite number of monkeys, and use only information from the text to help find out which monkey is the author. But other information is available, if not of an absolute character. We cannot initially say of any given proposal as to the assumed common ground that it is impossible, and to this extent it is necessary to consider all possibilities. But we can say that some proposals are relatively more plausible than others. The sophisticated reader considers what assumption the author is *likely* to have made as to the initial common ground.

The assumptions of an author, whatever they are, determine a Presupposition Logic context, a set of worlds. A reader's knowledge of which assumptions are most plausible determines an ordering over these contexts, what I will call a *plausibility ordering*. A plausibility ordering relative to some model is a reflexive, transitive binary relation over a subset of the powerset of the set of worlds. For an ordering π, $\sigma \geq_\pi \tau$ is written for $\langle \sigma, \tau \rangle \in \pi$, and $\sigma >_\pi \tau$ is taken to mean that both $\sigma \geq_\pi \tau$ and $\tau \not\geq_\pi \sigma$. An ordering π can be updated with a new formula by considering every pair in the ordering, and updating each element of the pair separately according to the principles of Presupposition Logic. The following definition is obtained:

Definition D7 **(Updating Plausibility Orderings)**

$$\pi + \phi \quad = \quad \{\langle \sigma', \tau' \rangle \mid \exists \langle \sigma, \tau \rangle \in \pi \; \sigma[\![\phi]\!]\sigma' \;\wedge\; \tau[\![\phi]\!]\tau'\}$$

Under this definition, certain contexts may drop out of contention in the update process, just as with the naive updating process considered earlier. An example may clarify. Suppose that $[\![\phi]\!] = \{\langle \sigma, \sigma' \rangle, \langle \upsilon, \upsilon' \rangle\}$, and that we wished to update the ordering $\pi = \{\langle \sigma, \sigma \rangle, \langle \tau, \tau \rangle, \langle \upsilon, \upsilon \rangle, \langle \sigma, \tau \rangle, \langle \sigma, \upsilon \rangle, \langle \tau, \upsilon \rangle\}$ with ϕ. Returning to the earlier metaphor, the reader is considering three different *movies* that the writer-director might have intended, and at the current point in the text the candidates for the correct frame are σ, τ and υ, with a plausibility ordering $\sigma \geq_\pi \tau \geq_\pi \upsilon$. The reader should now verify that $\pi + \phi = \{\langle \sigma', \sigma' \rangle, \langle \upsilon', \upsilon' \rangle, \langle \sigma', \upsilon' \rangle\}$. Observe that since τ cannot be updated with ϕ, there is no next frame from the film containing τ in the new ordering, and we are left with only two candidate films, with current frames σ' and υ' ordered $\sigma' \geq_{\pi+\phi} \upsilon'$. So the fact that frames σ and υ were in a certain ordering relation means that the next frames in those films are in the corresponding ordering relation after update. A more sophisticated model would perhaps allow juggling of orderings in the update process, to allow for what Gricean conversational analysis might tell us about the author's knowledge and intentions.

Before considering how we might make use of plausibility orderings, let us see how the earlier notion of entailment could be defined in terms of them. The *domain* of an ordering π, written $\star\pi$, can be defined as the set of contexts which are at least as plausible as themselves in the ordering, and this allows retrieval from a plausibility ordering of a corresponding information set. This in turn permits the definition of a notion of satisfaction of a formula by a plausibility ordering in terms of the earlier notion of satisfaction by an information set:

Definition D8 **(Domain of ordering and 'Ordinary' Satisfaction)**

$$\begin{array}{rcl} \star\pi & = & \{\sigma \mid \sigma \geq_\pi \sigma\} \\ \pi \models \phi & \textit{iff} & \star\pi \models \phi \end{array}$$

Given such a notion of satisfaction, it should be clear that it would be straightforward to define a notion of entailment equivalent to that given earlier. However, it is also possible to define alternative notions of entailment relative to any given plausibility ordering. Let us say that the set of *preferred contexts* in an ordering π, written $\uparrow \pi$, is the set of all contexts which are at least as plausible as any context in the ordering. Then we can say that an ordering π *preferentially satisfies* a formula ϕ, written $\pi \rhd \phi$, if the set of preferred contexts in π satisfies ϕ. Preferential satisfaction is a weaker notion than satisfaction, in that an ordering may preferentially satisfy more formulae than it satisfies. We may now say that a formula ϕ preferentially entails a formula ψ relative to an ordering π, written $\phi \rhd_\pi \psi$, if updating π with ϕ produces an ordering which preferentially satisfies ψ. Here are the formal definitions:

Definition D9 **(Preferential Satisfaction and Entailment)**

$$\begin{array}{rcl} \uparrow \pi & = & \{\sigma \mid \forall \tau \in \star\pi \; \sigma \geq \tau\} \\ \pi \rhd \phi & \textit{iff} & \uparrow \pi \models \phi \\ \phi \rhd_\pi \psi & \textit{iff} & \pi + \phi \rhd \psi \end{array}$$

In the case of a trivial ordering consisting of the cross-product of the powerset of worlds $\pi_0 = \mathcal{P} \times \mathcal{P}$, for which every set of worlds is at least as plausible as every other set of worlds, this notion collapses into the earlier entailment:

Fact F3

$$\phi \models \psi \quad \textit{iff} \quad \phi \rhd_{\pi_0} \psi$$

Recall the following earlier cited pair of examples:

1 If Jane takes a bath, Bill will be annoyed that there is no more hot water.

2 If Jane wants a bath, Bill will be annoyed that there is no more hot water.

On the earlier analysis, (1) is compatible with the standard CCP prediction of a conditional reading, indicating that if Jane takes a bath, there will be no more hot water. On the other hand (2) leads to a stronger conclusion not predicted by the standard CCP model, namely that there actually is no more hot water. The current revision of the CCP model will predict the contrast provided the following plausibility assumptions hold:

- At least one alternative in which it is established that there is no hot water is more plausible than all alternatives in which it is not known whether there is hot water, but in which it is known that if Jane wants a bath then there will be no hot water.

- An alternative in which it is not known whether or not there is hot water but in which it is established that if Jane has a bath then there will be no more hot water must be at least as plausible as all alternatives where it is definitely established that there is no hot water.

The contrast between (1) and (2) results from our ability to find a common-sensical explanation of the lack of hot water in terms of somebody having taken a bath, as against our inability to fully explain a lack of hot water in terms of somebody simply wanting a bath. The simple assumption that there is a finite amount of relevant hot water — about a bathful — is sufficient to allow justification of there being no more hot water in situations where Jane has just taken a bath. However, the same simple assumption would not suffice in the case of (2), and a number of other assumptions would be needed, such as the assumption that if Jane wants a bath then she will definitely take one. Thus it is the relative plausibility of assumptions not explicitly mentioned in the text of the example sentences that determines what is implicated.

Let us see how some of this analysis of (1) and (2) may be crudely formalized. For expository purposes, I will ignore many obviously relevant issues, such as temporal connections between antecedent and consequent clauses in the conditionals. Let us represent "Jane takes a bath" as "JTB", "Jane wants a bath" as JWB", "there is no hot water" as NHW, and "Bill will be annoyed that there is no more hot water" as ∂NHW $\wedge$ BA. Now suppose that our common sense knowledge of the relative plausibility of different assumptions is encoded in a plausibility ordering π. The two conditions required of π may be formalized as follows:

- $\exists \sigma \in \star\pi \; \sigma \models \text{NHW}$ and
  $\forall \tau \in \star\pi$ if $(\tau \not\models \text{NHW}$ and $\tau \models \text{JWB} \rightarrow \text{NHW})$ then $\sigma >_\pi \tau$

- $\exists \sigma \in \star\pi \; \sigma \not\models \text{NHW}$ and $\sigma \models \text{JTB} \rightarrow \text{NHW}$ and
  $\forall \tau \in \star\pi$ if $\tau \models \text{NHW}$ then $\sigma \geq_\pi \tau$

If π conforms to these requirements, then we have the following preferential entailments:

$$\begin{array}{rcl} \text{JWB} \rightarrow (\partial\text{NHW} \wedge \text{BA}) & \rhd_\pi & \text{NHW} \\ \text{JTB} \rightarrow (\partial\text{NHW} \wedge \text{BA}) & \not\rhd_\pi & \text{NHW} \\ \text{JTB} \rightarrow (\partial\text{NHW} \wedge \text{BA}) & \rhd_\pi & \text{JTB} \rightarrow \text{NHW} \end{array}$$

In other words, with respect to π, whilst example (2) preferentially entails that there is no more hot water, example (1) preferentially entails not that there is no more hot water, but that if Jane takes a bath then there is no more hot water.

Some further comment on this analysis of examples (1) and (2) is appropriate here, for an apparent difficulty with the analysis can in fact be quite instructive as regards revealing more about the nature of presupposition accommodation. Regarding (1), there is the possibility either of accommodating that *if Jane takes a bath there is no hot water* or accommodating that *there is no hot water* period. The weaker conditional appears to be preferred. On the other hand, the treatment of (2) seems to involve the reverse situation, with a logically stronger (i.e. non-conditionalized) proposition being accommodated. How can it be that a sometimes a logically stronger proposition is considered more plausible than a logically weaker one, and sometimes *vice versa*[5]?

The enigma arises from a misidentification of the accommodation process with the addition of single propositions, and a misidentification of the ordering over epistemic alternatives as an ordering over propositions. The plausibility ordering should not be thought of as an ordering over propositions, but over logically closed sets of propositions, that is an ordering over *theories*.

I am not committed to any claim about the conditional *if Jane takes a bath there is no hot water* being more plausible than the proposition *there is no hot water*, and I am not committed to *there is no hot water* being more plausible than *if Jane wants a bath there is no hot water*. Rather, I would claim that there must be a relevant closed theory containing the first conditional which is more plausibly taken to be the common ground than every relevant closed theory containing the simple proposition. Similarly, there must be a theory containing the simple proposition which is more plausible than every theory which contains the second conditional but not the simple proposition. There is no reason to assume that any relation of

[5]This objection, that plausibility appears oddly disconnected from entailment, was first made to me by Henk Zeevat. The same objection can be found in [Ge96]. Note that some other objections of Geurts have been discussed in [Bea94b].

inclusion (the *theory* level counterpart of propositional entailment) holds between these various theories.

So contexts could profitably be thought of as logically closed theories, and what is accommodated does not hang on the relative plausibility of alternative propositions, but on the relative plausibility of alternative theories. And this, indeed, was a theme implicit in my informal analyses of the other examples discussed in section 2 of this paper. The Spaceman Spiff examples, for instance, were explained not in terms of the plausibility of one or other presupposed proposition, but in terms of the relative plausibility of alternative theories concerning the dependence of Spiff's weight on external conditions and the nature of those conditions.

6 Discussion

My aim has been to show how the process of text understanding is related to our common sense knowledge of the world and of other agents, and not to account for the source of that knowledge. No theory has been provided of how agents perform the feat of ordering alternative belief models by plausibility, of why, for instance, the conditions imposed on π to explain the contrast between (1) and (2) arise. Such an analysis would certainly go beyond the intended scope of this paper.

If one takes a narrow view of linguistics, then one can see this paper as establishing firstly a negative conclusion: there are limits on how much of the data concerning presupposition can be naturally explained by linguistic argumentation alone, and previous theorists have sometimes over-stepped those limits.

Looked at programmatically, what this means is that we are now faced with an interdisciplinary task in which linguists and other cognitive scientists must cooperate to provide integrated models of text understanding and common sense reasoning. Does the model presented in this paper, which as I have said does not claim to give any account of common sense reasoning, take us any closer to a practical basis for such collaborative effort? It seems to me that, in this respect, one shortcoming of the model I have presented is that it does not provide an interface with any existing theory of common sense reasoning. Ideally, in so far as the problems of text understanding are really reducible to separate problems of a linguistic and of an inferential nature, the linguist should perhaps be providing a 'plug-in module' which could be bolted on top of an arbitrary theory of general reasoning. I have certainly not achieved that. But it is at least possible to see how plausibility orderings could be related to other cognitive models.

Most obviously, we might ask how plausibility orderings relate to what are sometimes called *Bayesian* models of epistemic states on which there is a large literature (see e.g. [Gar88]). In a Bayesian theory epistemic states are modeled as functions from sentences of some formal language to probabilities, with certain additional constraints on what counts as an appropriate function. In fact there is a fairly obvious way to progress towards a restatement of the model I have presented which does not assume a relative ordering of plausibility over sets of worlds, but instead utilizes an absolute assignment of probability to sentences. We could define a plausibility ordering π_p in terms of a Bayesian epistemic probability function p:

$$\begin{aligned} \pi_p \quad = \quad & \{\langle \sigma, \tau \rangle \in \mathcal{P}(\mathcal{W}) \times \mathcal{P}(\mathcal{W}) \mid \\ & p(\bigwedge\{\phi \mid \sigma \models \phi\}) \geq p(\bigwedge\{\phi \mid \tau \models \phi\}) \} \end{aligned}$$

Defining plausibility orderings in this way raises many questions. For instance, what are the properties of plausibility orderings thus defined, and, conversely, which properties of probability functions would lead to attractive properties for plausibility orderings. Which properties of plausibility orderings π_p are preserved across updates on orderings as defined in the current paper? Studying this last question might in turn lead to suggestions as to how the notion of update might be modified.

More generally, I do not claim that the probability-based definition provides any deeper insight into the nature of common sense reasoning I have invoked in this paper, but at least it taps into a long and venerable tradition of analyses of reasoning under uncertainty.

And here I should perhaps add that interpreting the orderings I have used in terms of probability is not the only possibility. One could also, for instance, draw a profitable analogy between the treatment of presupposition in this paper and the common-sense driven analyses of discourse and temporal relations in the work of Asher, Lascarides and Oberlander (see e.g. [LA93]) who use a special purpose non-monotonic logic. One might take various temporal expressions (e.g. tensed verbs) as presupposing some temporal reference with accommodation then being necessary to build the most plausible link between previously introduced temporal discourse referents and the new reference point.

Hobbs and co-workers (see e.g.[HSAM93]) have presented models of the interpretation process based on *weighted abduction*. In the process of text comprehension, utterances are understood to provide only partial information as to utterance meaning, and some information must be abduced. The abducible information is assigned different weights according to both world knowledge and knowledge of grammar. So far as I know Hobbs is not committed to any direct interpretation of these weights in terms of probability: for instance one imagines that the weights could also in principle depend on processing cost. Now it should be clear that the epistemic alternatives in a plausibility ordering correspond loosely to abducible sets of propositions, with my notion of relative plausibility corresponding to Hobbs' use of absolute weight. Given this analogy, it would be of interest to further explore the possibility of recasting the model I have proposed in terms of abduction, or to attempt to build a computer implementation based on the model of presupposition I have assumed but in which an abductive inference engine served the function of the plausibility ordering.

Having mentioned processing cost, one is reminded of proposals to treat presuppositions in terms of Sperber and Wilson's *relevance theory* [SW84]. Sperber and Wilson have applied a psychologically motivated notion of the cost of an inference to explain a wide range of linguistic phenomena including presupposition. To what extent should the data I have discussed be explained in terms of plausibility and to what extent in terms of processing cost? This is a complicated empirical question, and, given the difficulties of providing psychologically plausible formal models of processing, it also brings with it methodological questions. But I would not wish to deny Sperber and Wilson's claim that processing factors are at work, and it might at least be of philosophical interest to consider reinterpreting plausibility orderings as indicating not relative plausibility but relative cost of different inferences.

To sum up, in this paper I have claimed that when we accommodate, we look not only at the record of what has been said, but also look behind what has been said, and consider what the author might have intended and what the author might have expected. This sort of reasoning may be called *reconstructed reasoning* in the sense that it involves a reconstruction of the speaker's assumptions and intentions. In the context of existing theories of presupposition, the claim that an empirically adequate account of presupposition must take reconstructed reasoning into account appears radical. Yet the model proposed is intended to be *conservative*, in that an existing line of research (i.e. the Karttunen/Stalnaker/Lewis/Heim dynamic treatment of presupposition) is used as the basis of a model which incorporates this sort of reasoning process. Furthermore, as I have tried to show in this closing section, my proposal to model accommodation in terms of general inference processes which are dependent in large part on non-linguistic information is not made in a vacuum: a number of other researchers have come to similar conclusions with respect to both presupposition and other phenomena. I am very much in agreement with Richmond Thomason, who seeks a notion of accommodation that is yet more sophisticated than the one I have proposed here, since it takes into account not only the beliefs, but also the communicative goals of other agents: "Concentrating on accommodation means shifting to reconstructed reasoning that underlies utterances. And it suggests that certain reasoning processes, such as intention recognition and cooperation are central. Successful accommodation requires that we first recognize someone's intention to achieve a goal, and then establish goals of our own that will assist in achieving this goal." [Tho90]

References

[Bea92] Beaver, D., 1992. *The Kinematics of Presupposition* Proceedings of the Eighth Amsterdam Colloquium, ILLC, University of Amsterdam.

[Bea93a] Beaver, D., 1993. "Kinematic Montague Grammar", in Kamp, H. (ed.), *Presupposition*, DYANA-2 deliverable R2.2a, University of Amsterdam.

[Bea93b] Beaver, D., 1993. *What Comes First in Dynamic Semantics*, ILLC report LP-93-15, University of Amsterdam.

[Bea94] Beaver, D., 1994. "When Variables Don't Vary Enough", in Harvey, M. and Santelmann, L. (eds.), *SALT 4*, Cornell.

[Bea94b] Beaver, D., 1994. "Accommodating Topics", in van der Sandt, R. and Bosch, P. (eds.), *The Proceedings of the IBM/Journal of Semantics Conference on Focus, Vol.3*, IBM Heidelberg, pp.439–448.

[Bea94c] Beaver, D., 1994. *An Infinite Number of Monkeys*, Acta Linguistica Hungarica, Vol. 42 (3–4), pp. 253–270.

[Bea95] Beaver, D., 1995. *Presupposition and Assertion in Dynamic Semantics*, PhD Dissertation, University of Edinburgh. To appear, Studies in Logic, Language and Information, CSLI, Stanford.

[Bea97] Beaver, D., 1997. "Presupposition", in van Benthem, J. and ter Meulen, A. (eds.), *The Handbook of Logic and Linguistics*, Elsevier, pp. 939–1008.

[Gar88] Gardenförs, P., 1988. *Knowledge in Flux: Modeling the Dynamics of Epistemic States*, M.I.T. Press, Cambridge, Mass.

[Gaz79] Gazdar, G., 1979. *Pragmatics: Implicature, Presupposition and Logical Form*, Academic Press, New York.

[Ge96] Geurts, B., 1996. *Local Satisfaction Guaranteed: A Presupposition Theory and its Problems*, Linguistics and Philosophy 19(3), 259–294.

[GS91] Groenendijk, J. and Stokhof, M., 1991. *Dynamic Predicate Logic*, Linguistics and Philosophy 14(1), 39–100.

[Hei82] Heim, I., 1982. *On the Semantics of Definite and Indefinite Noun Phrases*, PhD dissertation, University of Amherst.

[Hei83] Heim, I., 1983. "On the Projection Problem for Presuppositions", in Barlow, M., Flickinger, D. and Westcoat, M. (eds.), *Second Annual West Coast Conference on Formal Linguistics*, University of Stanford, pp.114–126.

[HSAM93] Hobbs, J., Stickel, M., Appelt, D. and Martin, P., 1993. *Interpretation as Abduction*, Artificial Intelligence 63, 69–142.

[Kam81] Kamp, H., 1981. "A Theory of Truth and semantic Representation", in Groenendijk, J., Janssen, T. and Stokhof, M. (eds.) *Formal Methods in the Study of Language.*

[Kar74] Karttunen, L., 1974. *Presuppositions and Linguistic Context*, Theoretical Linguistics 1.

[KP79] Karttunen, L., and S. Peters, 1979. "Conventional Implicatures in Montague Grammar", in Oh, C. and Dineen, D. (eds.), *Syntax and Semantics 11: Presupposition*, Academic Press, NY.

[LS71] Langendoen, D. and H. Savin, 1971. "The Projection Problem for Presuppositions", in Fillmore, C. and Langendoen, D. (eds.), *Studies in Linguistic Semantics*, Holt, Reinhardt and Winston, New York.

[LA93] Lascarides, A. and Asher, N., 1993. *Temporal Interpretation, Discourse Relations and Commonsense Entailment*, Linguistics and Philosophy 16:5, pp 437–495.

[Lew79] Lewis, D., 1979. *Scorekeeping in a Language Game*, Journal of Philosophical Logic 8, pp. 339–359. Also appears in Bäuerle, R., Egli, U. and von Stechow, A. (eds.), *Semantics from Different Points of View*, Berlin.

[Mer92] Mercer, R.,1992. *Default Logic: Towards a Common Logical Semantics for Presupposition and Entailment*, Journal of Semantics 9:3.

[vdS88] van der Sandt, R., 1988. *Context and Presupposition*, Croom Helm, London.

[vdS92] van der Sandt, R., 1992. *Presupposition Projection as Anaphora Resolution*, Journal of Semantics 9:4.

[Soa82] Soames, S., 1982. *How Presuppositions Are Inherited: a Solution to the Projection Problem*, Linguistic Inquiry 13:483–545.

[So89] Soames, S., 1989. "Presupposition", in D. Gabbay and F. Guenther (eds.) *Handbook of Philosophical Logic*, vol. IV, Reidel, Dordrecht, pp. 553–616.

[SW84] Sperber, D. and Wilson, D., 1984. *Relevance: Communication and Cognition*, Basil Blackwell, Oxford; Harvard University Press, Cambridge, Massachusetts.

[Sta74] Stalnaker, R., 1974. "Pragmatic Presuppositions", in Munitz, M. and Unger, P. (eds.) *Semantics and Philosophy*, NYP.

[Tho90] Thomason, R., 1990. "Accommodation, Meaning and Implicature: Interdisciplinary Foundations for Pragmatics", in Cohen, P., Morgan, J. and Pollack, E. (eds.), *Intensions in Communication*, Bradford Books, MIT Press, Cambridge, Mass., pp. 325–363.

[Vel91] Veltman, F., 1991. "Defaults in Update Semantics", in DYANA deliverable R2.5.C. (To appear in: Journal of Philosophical Logic).

[Zee92] Zeevat, H., 1992. *Presupposition in Update Semantics*, Journal of Semantics 9:4, pp. 379–412.

A Dynamic Syntax-Semantics Interface

Tsutomu Fujinami

Abstract The relation between syntax and semantics of natural language can be regarded as a constraint. With the ideas from Channel Theory ([Barwise1993], [Barwise and Seligman1994]), the way that an utterance represents a situation can be captured as a linguistic channel. To study the operational aspects of such a channel, we construct it as a system of communicating processes by turning to the π-calculus ([Milner et al.1992]). We show how a concurrent bottom-up chart parser can be encoded in the calculus and how a semantic object similar to those employed in Situation Theoretic Discourse Representation Theory ([Cooper1993]) can be created as the result of interactions between processes encoding the parser.

1 Introduction

The syntax-semantics interface is a relation between syntactic information and semantic information of natural language. The most familiar approach to specifying the interface may be to assume a one-to-one correspondence between syntax and semantics as is taken in Montague grammar. In this tradition, each syntactic element is assigned its counterpart in semantics. Another approach is to provide a set of rules by which items of syntactic information are converted into those of semantic information. Discourse Representation Theory ([Kamp and Reyle1993]), for example, employs a set of construction rules to transform syntactic structures into Discourse Representation Structures. The other approach, to which we are more sympathetic, is to regard the relation as a *constraint*.

The trouble we recognise in the first approach is that it is not flexible enough to capture the context dependency between syntax and semantics because the relation is hard-coded by the strict one-to-one correspondence. In the second approach, we may specify context-dependent relations by extending the antecedent of rules, but we are still left wondering what the ontological status of these rules might be. In Situation

Logic, Language, and Computation, Vol. II, edited by Lawrence S. Moss, Jonathan Ginzburg, and Maarten de Rijke.

Semantics ([Barwise and Perry1983]), on the other hand, the relation plays a central role and is even regarded as part of meaning. The notion of constraint helps us explain how information can be conveyed from one to another and has been further developed in Channel Theory ([Barwise1993], [Barwise and Seligman1994]).

Channel Theory enables us to refine the constraint approach. With the ideas from the theory, we can depict the way an utterance represents a situation as Figure 1:

1. An utterance is classified as an instance of a sentence type through parsing.
2. By constraint, the sentence type is related to a situation type.
3. A situation is classified as an instance of the situation type through evaluation.
4. The utterance signals the situation.
5. The signal is classified as an instance of the constraint.

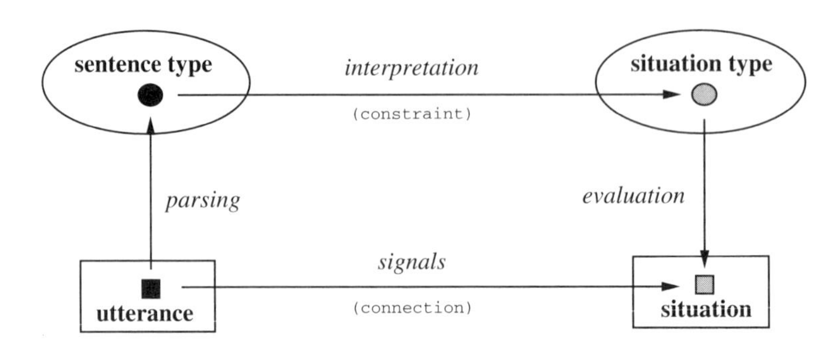

Figure 1: Linguistic channel

We may call the syntax-semantics interface as depicted in the figure *linguistic channel.* In this view the notion of *context* is refined to the signalling relation between the circumstances in which the sentence is uttered and the situation that the sentence is about. One can understand that the relation forms the core part of language understanding by examining a simple sentence such as *"Could you pass me the salt?"* The hearer needs to get access to the utterance situation to anchor *'you'* to the addressee and *'me'* to the speaker. He also needs to get access to the situation they are talking about to anchor *'the salt'* to a particular container of salt.

Given the picture, we are interested in how a signal can be recognised as an instance of a particular constraint by the hearer. We observe that the hearer can extract items of information from various sources in parallel while classifying a signal. We therefore argue that the way a signal is classified must be flexible so that the items of information can be utilised anytime. To implement the idea, we propose to construct the linguistic channel as a system of communicating processes ([Milner et al.1992]). Through the construction, we expect to obtain a cognitive model of language understanding, an efficient constraint satisfaction mechanism, and an operational interpretation of channels.

In what follows, we show how sentences can be parsed as the result of interactions between processes. The resulting representation for the meaning, too, is encoded as a system of processes. We first explain in the next section what the system of communicating processes is. Throughout the paper, we employ a graphic representation for processes and avoid to go into the detail so that we can concentrate on the essential points. Section 3 presents a bottom-up chart parser and Section 4 shows how the syntax-semantics interface can be modelled as a process. We finally conclude the article by relating the idea to other work. The work presented here constitutes part of the author's dissertation ([Fujinami1996]), which gives the reader more detail of the construction.

2 Interaction graph

We present a graphic representation of processes, avoiding to show any algebraic formulas throughout the paper. The graphic representation is proposed by ([Fujinami1995]) as *interaction graph*, which is a variant of Milner's π-nets ([Milner1994]). The reader is referred to these papers for more detail.

Basic actions There are two sorts of primitive action, *export* and *import*. The arc with white head in (1)a below depicts an exporting action of datum, b, via the port a. The other arc with black head in the figure depicts an importing action of a datum via a to replace it for x. The port part is indicated with shadow. We call the upper part of the node *buffer* because a datum can be stored. Not all nodes play both roles of port and buffer. If a node only plays a role of buffer, we do not shadow its under part. By chemical metaphor ([Berry and Boudol1992]), we conceive of these arcs as a molecule floating in solution. The horizontal line is meant to be the surface. These two molecules can interact with each other to substitute b for x through a. After the interaction, they will evaporate into the air as shown in (1)b, leaving no other molecules for interactions.

(1) a.

b.

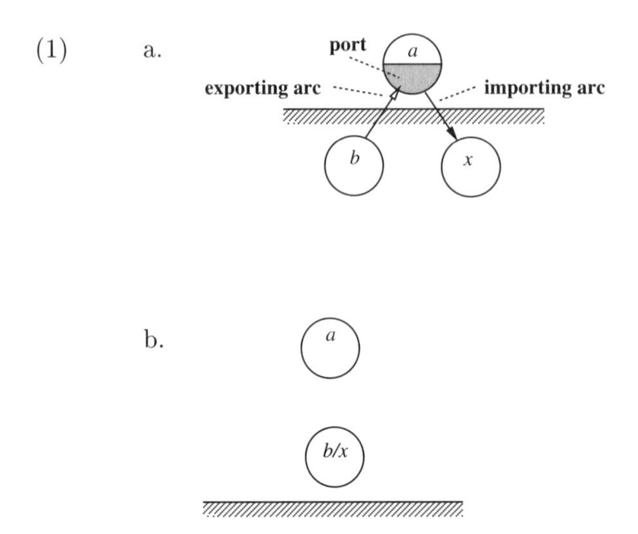

Mobility We may call the port *location* in the sense that the port can be seen as the point of communication. For more flexibility, we allow a port name to be exchanged. Let us consider another example of communicating processes (2). Initially, the exporting action of c via b is located at b, and the importing action to y via x is located at x as shown in (2)–1. Upon the interaction via a, x is substituted by b as shown in (2)–2. Then, the molecules still available establish a direct connection as shown in (2)–3 as they are now located at the same port, b. Subsequently, they will interact with each other as shown in (2)–4.

(2)

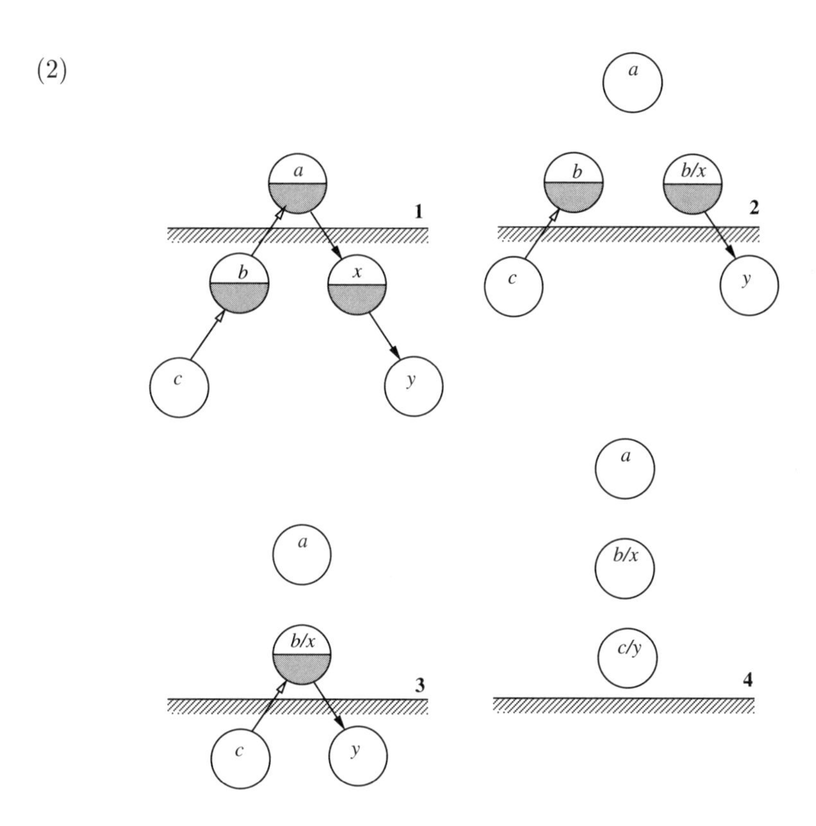

Non-determinacy Two exporting arcs may compete to get access to a port. Suppose there are two exporting actions, one to export e via d and the other to export f via d. They compete for the access to d to replace z as shown in (3)a. In this case, there are two possible interactions, but we cannot predict which interaction will actually take place. The left part of (3)b depicts the case where the first action wins, while the right part of the figure depicts the case where the second action wins. We assume that the lost action will be removed.

(3) a.

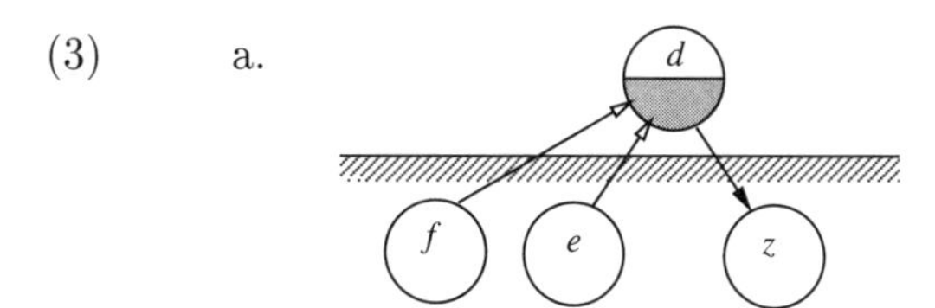

b.

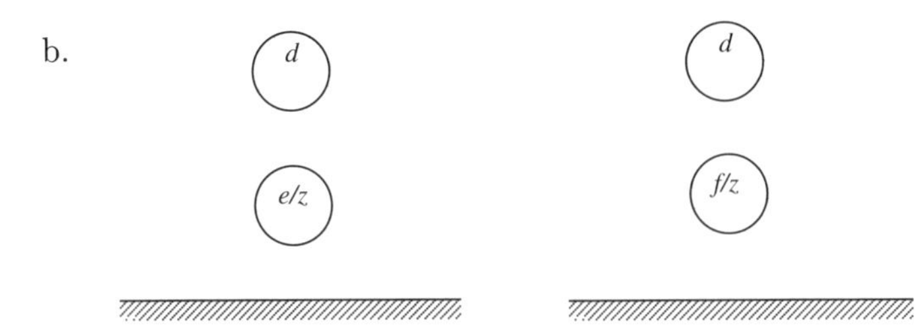

Match There is of course a way to control non-deterministic interactions. Actions can be constrained with a *match* formula referring to substitution environments. Specified a match formula, an action cannot be executed until the condition is satisfied. For example, suppose in the previous example (3)a that the exporting action of f via d is constrained using the match formula, $b = x$, and the other exporting action of e via d is constrained using the match formula, $c = x$, as shown in the left of (4). We assume that x had been substituted by b by now. Then, only the condition, $b = x$, is met as the formula turns into $b = b$ under the substitution environment. Thus, only the exporting action of f via d becomes executable.

(4)

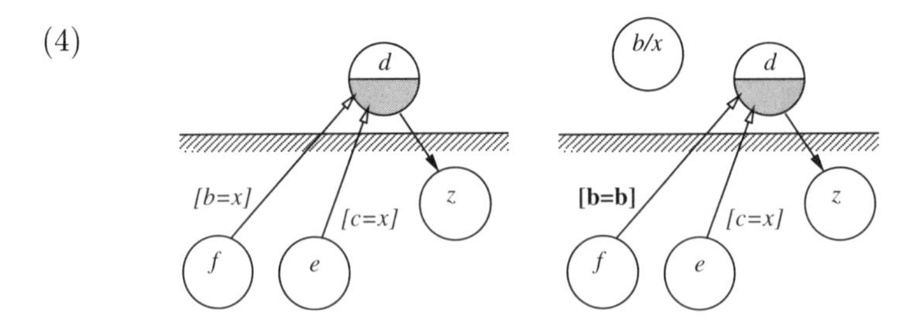

Bound names So far we have regarded all variables as free, but there is a distinction of free/bound variables. In fact, all variables to be substituted are bound. When we depict a system, we enclose bound variables into a box indicating the scope and put the symbols in the upper part. In the example below, the two occurrences of x are regarded to be a distinctive name:

(5)

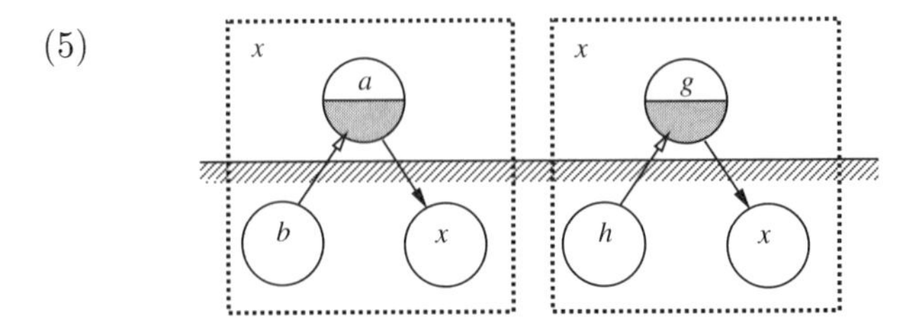

The scope of bound variables can metaphorically be seen as forming a membrane by which other molecules are blocked. The bound variables can also be understood as a name recognisable only within the membrane. Extending the metaphor further, we can think of constants local to the membrane. In the example below, the two occurrences of b are regarded to be a distinctive name:

(6)

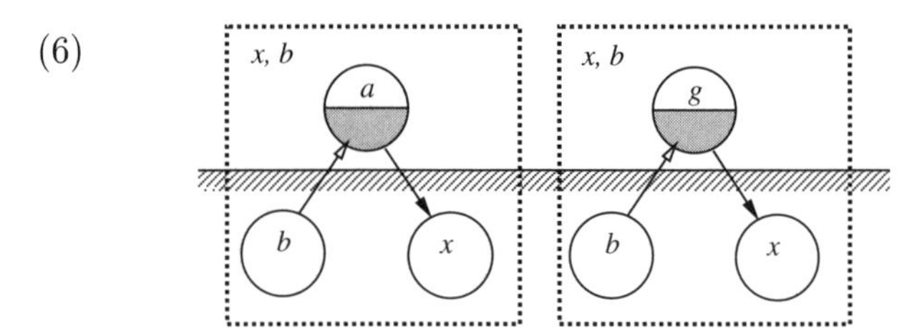

Replication All the actions we have so far discussed are consumed once they are executed. In some cases, however, we would like systems to be replicated as many times as needed. To define such a mechanism, we introduce a special box to our graphic representation as shown in (7)a. The system within the bold box can be replicated many times as shown in (7)b.

(7) a. b.

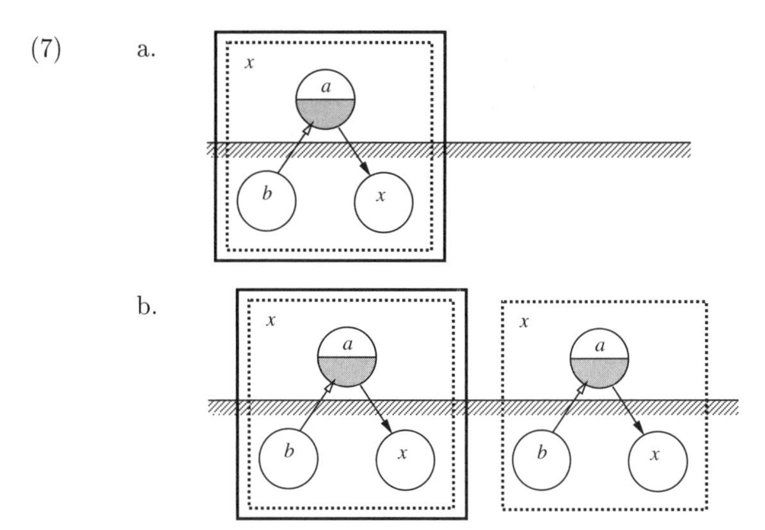

3 A concurrent bottom-up chart parser

Having explained in what computational mechanism we model the syntax-semantics interface, we present how a concurrent bottom-up chart parser can be encoded as a system of processes. We start this section by sketching how a sentence can be parsed as a result of interactions between processes. We show then how feature structures can be encoded and demonstrate how a sentence can be parsed in our approach.

3.1 Parsing through communication

The idea is to parse a sentence as a result of interactions between communicating processes. We assign to each word a sub-system of processes encoding the lexical information and the potential behaviour upon interaction. Each sub-system is connected with its left and right neighbours by a wire. The wire can be seen as *node* and the sub-system as *edge* in chart. Below (8) is a configuration of sub-systems to parse a sentence, *"a man walks."*

(8)

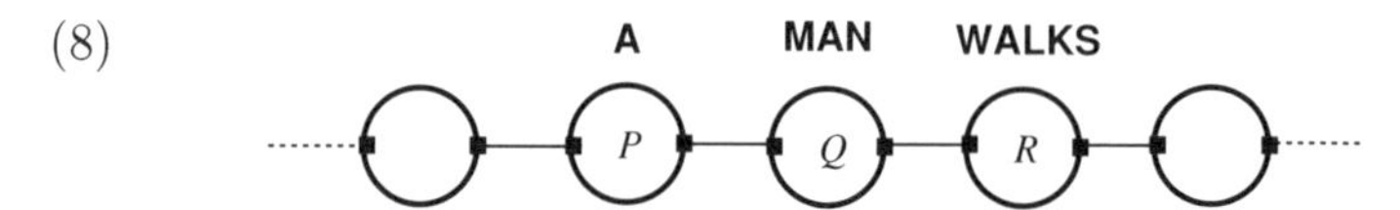

To parse the sentence, the sub-system Q assigned to *'man'* first extracts items of information from its left sub-system P assigned to *'a'*. If the items extracted satisfy Q's requirement, e.g. the category of *'a'* is determiner and the number is singular, Q creates another sub-system, Q', encoding the items of the information for the noun phrase, *'a man'*, as shown in (9):

(9)

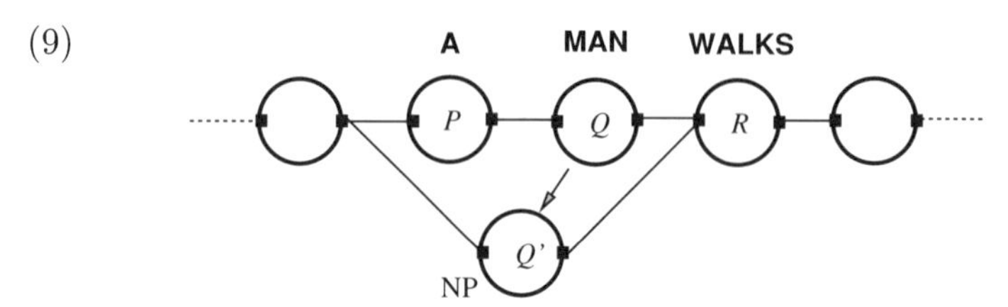

When created, the sub-system Q' is given the connection to R assigned to *'walks'*, through which the sub-system R extracts items of information of Q', e.g. if the category is noun phrase and the number is singular. When satisfied, the sub-system R creates another sub-system R' encoding the items of information for the sentence as shown in (10):

(10)

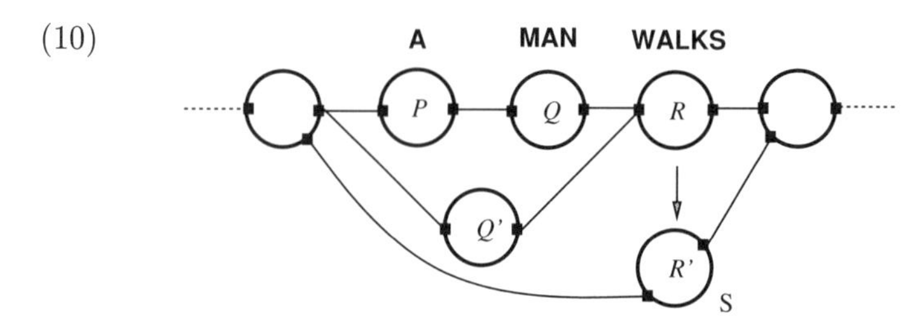

3.2 Feature structures as a system of processes

As an example, we show how the lexical information for *'walk'* presented as a feature structure (11) below can be encoded as a system of processes. For simplicity, we only consider part of the information, i.e. the category is verb, the phonetic form is 'walks', and as for agreement the person is third and the number singular.

(11)

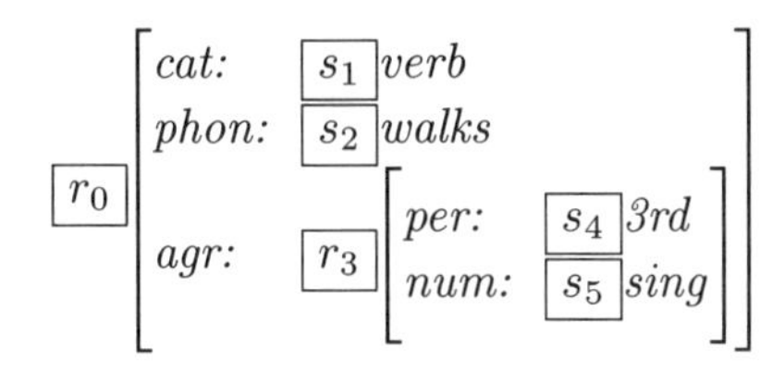

To encode the feature structure, we explicitly index every item of information. With these indices, we can model the structure as a system of processes such that:

1. At r_0 the sub-system receives a pair of *request* and *return address.*
2. If the request matches *cat*, to the return address it sends back s_1, at which the item of information, *verb*, is available.
3. If the request matches *phon*, it sends back s_2, at which the item of information, *walks*, is available.
4. If the request matches *agr*, it sends back r_3, at which it receives another pair of request and return address for further inquiry.

The sub-system is encoded as shown in Figure 2. The indices, r_0, r_3, s_1, s_2, s_4, and s_5, are local to the whole sub-system. The parameters, x and *ret*, for receiving *request* and *return address*, appear both in the outer box and in the inner box, but they are not identical as they are within different scopes. The white bar attached two importing arcs to x and *ret* indicates a polyadic importing action. The items of information can be replicated a number of times as they are enclosed in the bold box.

3.3 Parsing as reaction

We demonstrate how the sub-system for *'man'* may create another sub-system for the noun phrase, *'a man'*. Figure 3 depicts the system composed of the sub-system for *'a'* and that for *'man'*. The left part depicts the sub-system for *'a'* and the right that for *'man'*. Since the encoding of feature

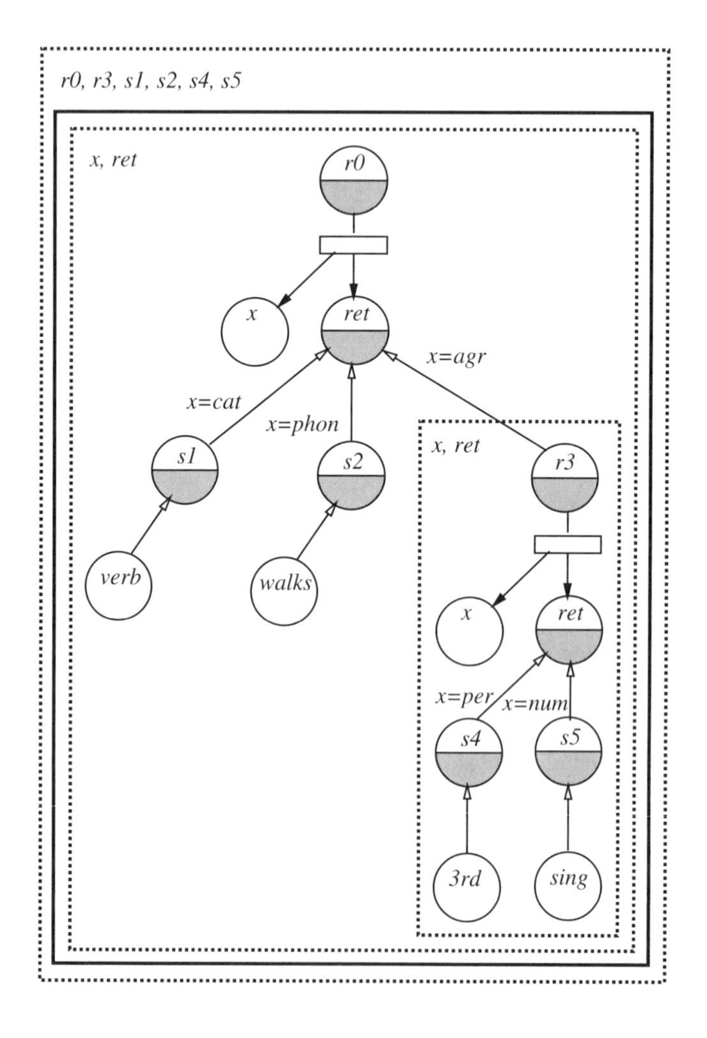

Figure 2: The sub-system encoding the information of *'walks'*.

information has already been explained in the above, we suppress the part encoding the feature information for *'man'* for simplicity. The sub-system for *'man'* is equipped with a module to examine the lexical category of its left neighbour. Note that the module can be executed only once. Suppose that a copy of the feature information for *'a'* is already generated. We assume that the examining module and the sub-system for *'a'* are already connected through some other processes. The figure also includes a black box to be invoked once the condition, $w = det$, is satisfied. The black box encodes the sub-system for the noun phrase, *'a man'*. In what follows, we often suppress the port name for simplicity where it does not matter, leaving the node unlabelled.

Starting from the state, the examining module interacts with the sub-system for *'a'* through the topmost unlabelled port. Upon the interaction, r_0 substitutes t and the scope of these bound variables are subsequently merged. When the two sub-systems are connected at r_0, it is possible for the sub-system for *'man'* to emit *cat* and u to the other for *'a'*. These names substitute for x and *ret*, respectively. At this point, only the exporting arc constrained using the match formula, $x = cat$, becomes executable given the substitution environment. The variable, v, can therefore only be substituted by s_1, thus w, too, can only be substituted by *det*. The figure 4 depicts the system at this stage, where the unmatched arcs have already removed. As the result of these interactions, the part indicated with the black box is now ready for execution, the box that encodes the items of information for the noun phrase, *'a man'*.

4 The dynamic syntax-semantics interface

We complete our construction of linguistic channel (Figure 1) as systems of processes. Before showing how a semantic representation can be constructed, we explain how we encode semantic representations as a system of processes. The semantic objects considered here are very limited and the encoding is simplified. The reader is referred to another paper by the author ([Fujinami1995]) or to his dissertation for more detail. We show then how the syntax-semantics interface can be described as a process.

4.1 Semantic representations as processes

The meaning of the sentence, *"a man walks"*, can be represented as (12) below. This is a graphic notation proposed by ([Barwise and Cooper1993]) for the situation-theoretic object, *infon abstract*, and means that there is an individual parameterised as X whose properties are to be a man and to walk. The parameter X is indexed with r, which we call *roll*.

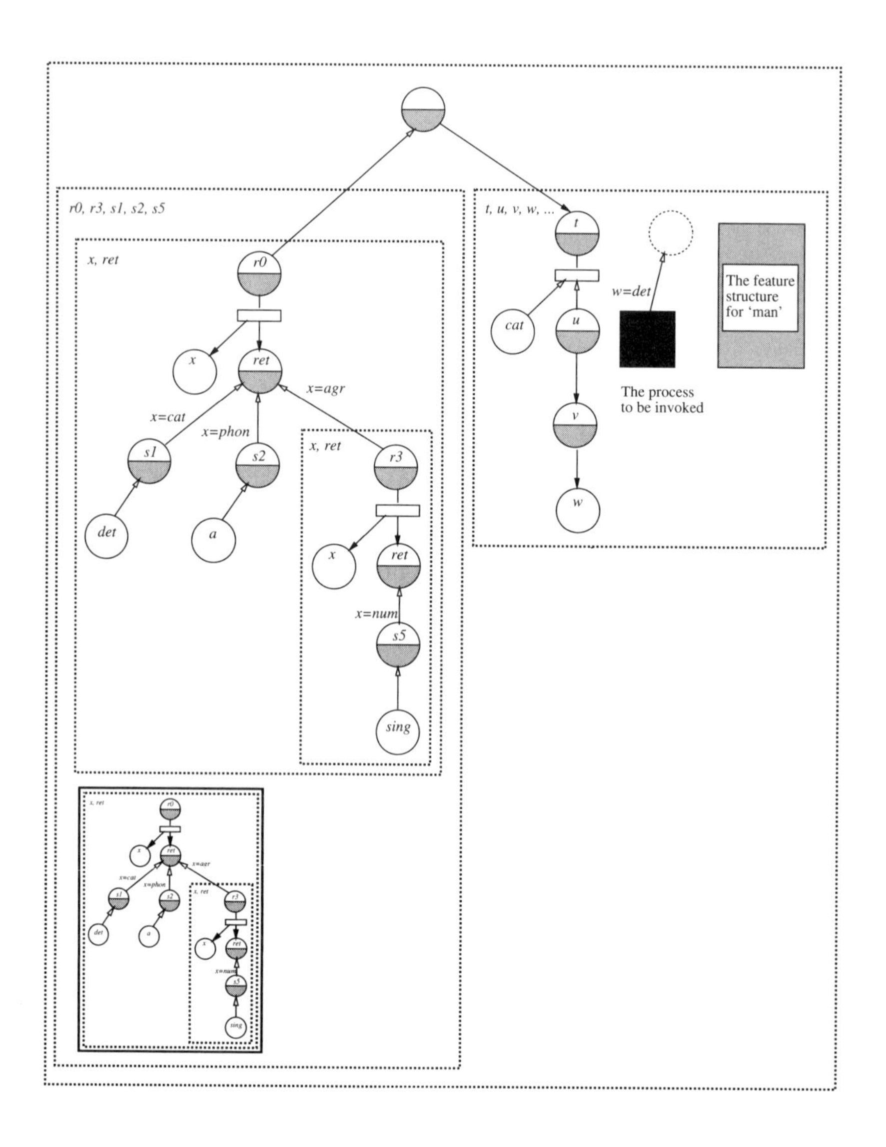

Figure 3: The system composed of the two sub-systems for *'a'* and *'man'*

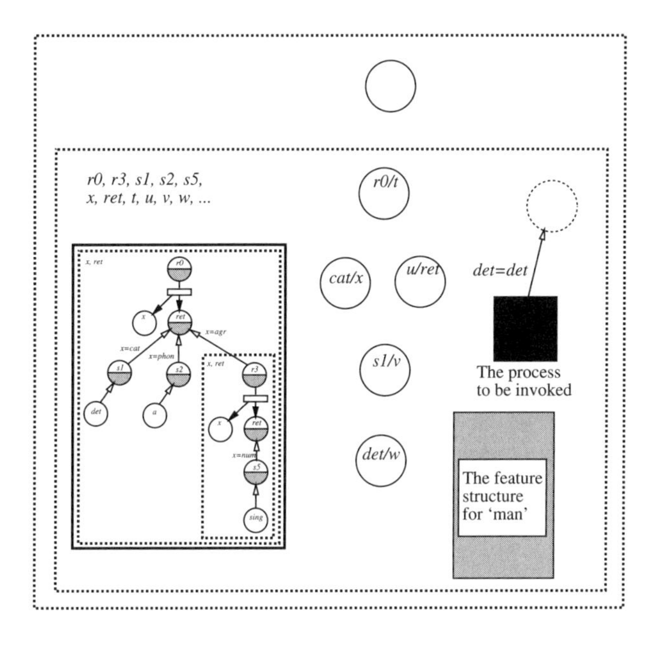

Figure 4: The state where the sub-system for *'a man'* becomes ready

(12) 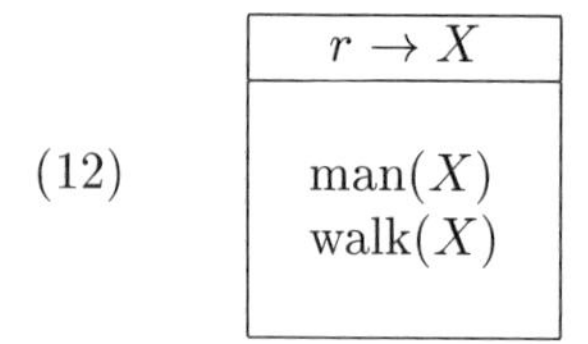

In our graphic representation, we represent the infon abstract (12) as a system of processes as shown in Figure 5. In this figure, we use r as role and x as parameter. To indicate r and x are bound within the system, we enclose the whole system into the dotted box, putting r and x in upper part. The parts enclosed with the bold box means that these parts are available as many times as requested. As a whole, the figure depicts a system of processes such that x will be replaced by some other individual imported through the role r, which is accessible either via man' or via $walk'$.

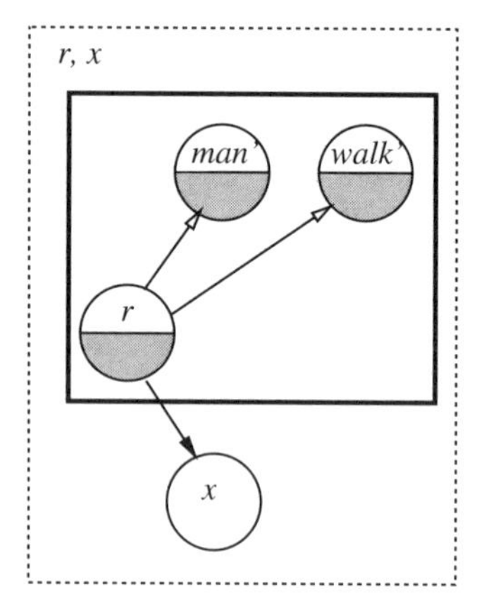

Figure 5: A system encoding the infon abstract (12)

4.2 Representation through interactions

To construct the system as the result of interactions, we add to each lexical item a sub-system encoding parts of the meaning. Figure 6 depicts the sub-systems connected with each other at certain locations. The leftmost part depicts the meaning of 'a' of the sentence. The sub-system contains the part importing some individual through r to replace it for x. The sub-system is abstracted over the role r with an index u, through which it is connected to its right neighbour. The middle part depicts the sub-system encoding the meaning of 'man'. The part to be integrated into the resulting system is abstracted over s with an index v, through which it is connected to its left neighbour. Since we have already explained how each sub-system can be connected with each other, we leave the locations of interactions unlabelled for simplicity. The rightmost part depicts the sub-system encoding the meaning of 'walks'. The part is also abstracted over t with an index w, which is connected to its left neighbour at the unlabelled location.

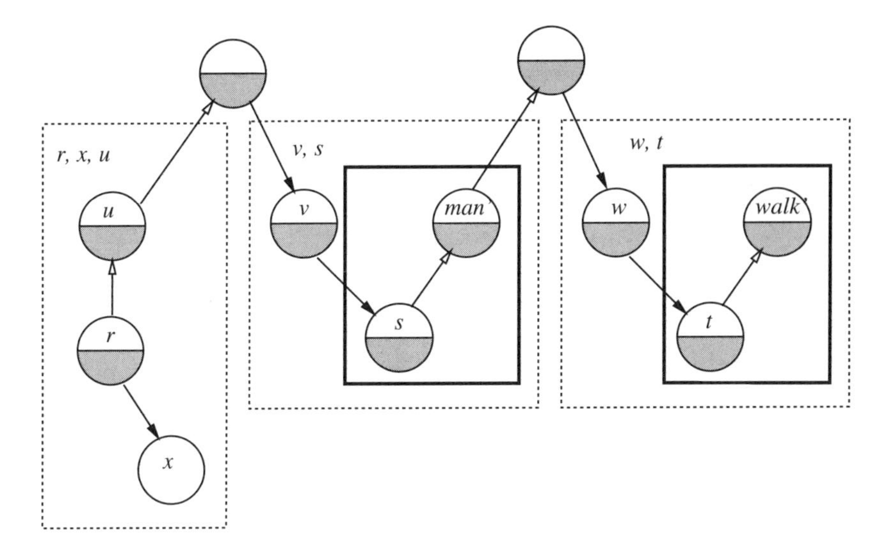

Figure 6: The sub-systems encoding parts of meaning

Given the configuration, we follow briefly the steps from the initial state towards the end result. Firstly, the left two sub-systems in the figure interact with each other when the two words, *'a'* and *'man'*, are recognised as forming the noun phrase as shown in the first row in Figure 7. We assume that the two sub-systems are connected in the same way as the sub-systems encoding feature structures get into touch with each other. The point here is that the connection between the two sub-systems encoding parts of meaning are established as the result of interactions in syntax part. In other words, the way the parts for semantic representations are combined with is determined by the way the parts for syntax is organised. As the result of interaction between the sub-systems encoding syntactic information for *'a'* and *'man'*, the two sub-systems encoding the parts of meaning is connected with each other. Then, u replaces v and the scope of bound names are extruded accordingly. Upon the interaction via u, the created sub-system is further evolving into another sub-system encoding the meaning of 'a man' when r replaces s as shown in the second row of Figure 7.

To continue the interaction, the sub-systems encoding the parts of the meaning for *'a man'* and *'walks'* make a copy of their parts as is shown in the third row of Figure 7. The two sub-systems interact with each other as the two words are recognised as forming a sentence. Upon the interaction, w is replaced by man', and further t is replaced by r as shown in the bottom row of Figure 7. The result is the same as the system shown in Figure 5. By removing the man'/w node and conjoining the nodes bearing the same name, r, we obtain the system encoding the meaning of the sentence, *"A man walks"*.

5 Related work

5.1 Situation Theoretic Grammar

We discuss how the proposed formalisation of feature structures as processes is related to our notion of integrating multiple information sources, as emphasised in the beginning. Why do we want to describe everything as process? To answer the question, we have to trace back our motivation into Cooper's work on Situation-Theoretic Grammar (STG) ([Cooper1991]). The characteristic point of STG is that it applies the same framework primarily developed for studying semantics to studying syntax, too. The application becomes possible by regarding utterances as a situation supporting various linguistic facts. For example, let u_1 be an utterance of *'man'*. The utterance regarded as a situation can support many linguistic facts, among them are the fact that its category is noun and another that it is characterised phonetically as *'man'*. The situation can be expressed in Extended Kamp Notation(EKN) ([Barwise and Cooper1993]) as follows:

(13)

u_1
cat(u_1, noun) use-of(u_1, 'man')

STG can be studied in itself, but what most interests us is its connection with Situation Semantics. For the sake of discussion, we consider the same example, *"A man walks"*. The representation (12) is an infon abstracted over the parameter X indexed with the role, r. Presented the representation, one is entitled to ask what the role is. Technically, we can leave the question out by assuming unlimited supply of roles, e.g. using natural numbers as role indices. But more philosophically sound answer is desirable and one solution proposed by ([Cooper1993]) is to regard utterances as role

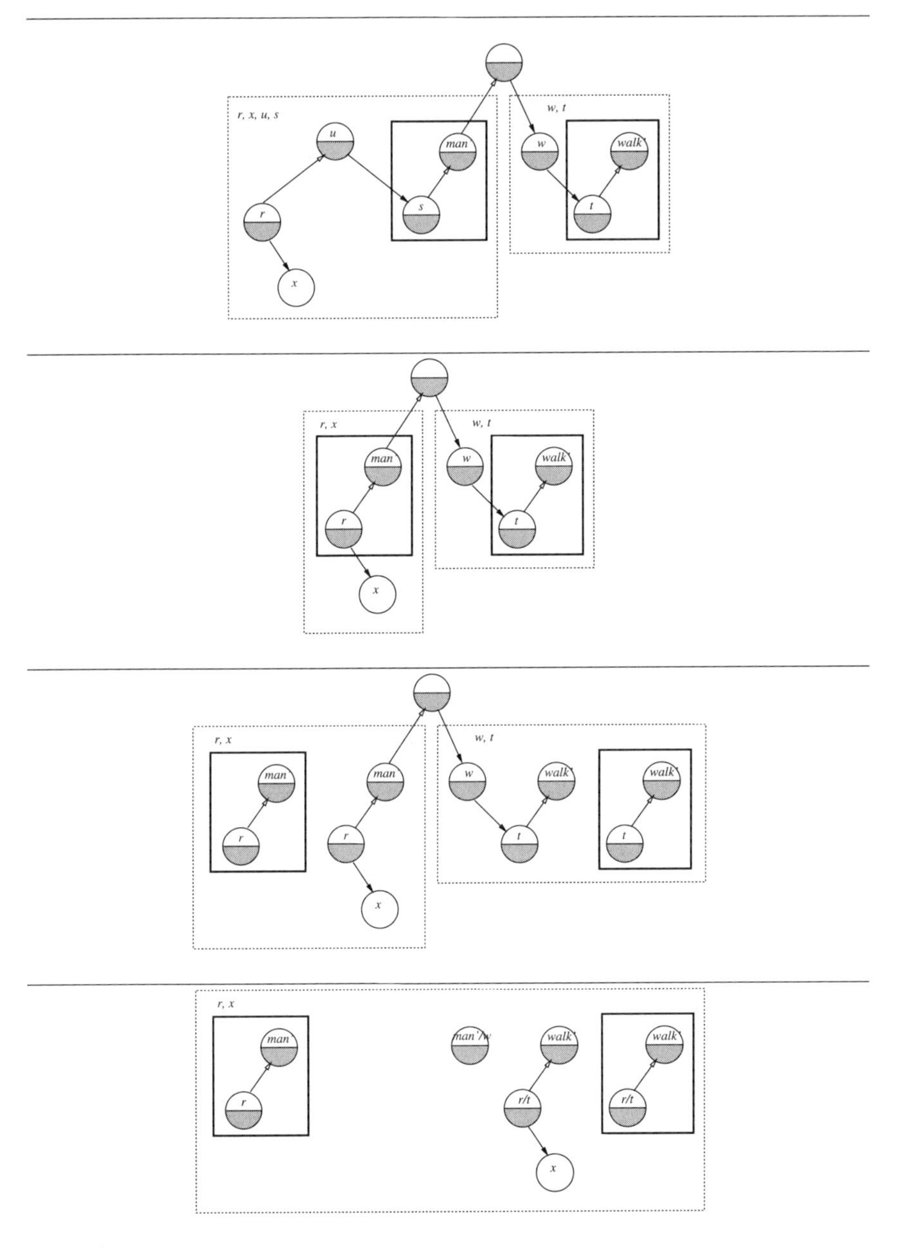

Figure 7: The transitions to construct the semantic representation

indices. In the example above, we can use the utterance u_1 as the role to index the parameter X as below:

(14) a.

$u_1 \rightarrow X$
man(X) walk(X)

b.

u_1
cat(u_1, noun) use-of(u_1, 'man')

As Cooper claims, the technical advantage of this approach is that it avoids us having to make some arbitrary decisions about generating enough unique role indices. More importantly, however, there is also a theoretical advantage obtained by combining the items of information concerning syntax and semantics. One application among others presented by ([Cooper1993]) is to study how utterances contribute to transferring a meaning from the speaker to the addressee in dialogue. The point is that the utterance situation can be shared between dialogue participants, thus, it serves as a common ground by providing them with common indices to parameters that may be represented differently among participants.

We do not go into the study of dialogue phenomena in the approach. What is insightful for us in the use of utterance situation as a role is that the semantic representation (14a) is abstracted by its corresponding syntactic information (14b). We argue that our formalisation of the syntax-semantics interface as a process is a refinement of the approach proposed by ([Cooper1993]) and that our approach is also a faithful implementation of the relational theory of meaning proposed by ([Barwise and Perry1983]). Recall that they argue that the meaning of sentences must be captured in terms of the relation between utterance situation and described situation. Our work can be understood as implementing Situation Semantics in process algebra.

5.2 Dynamic semantics

The most important point advocated by dynamic semantics is to capture the meaning of sentences as the state changes caused by the utterance (see, e.g. ([Kamp and Reyle1993]) and ([Groenendijk and Stokhof1991])). Although approaches to dynamic semantics are divergent, the view to the meaning is shared among them. The difference will only show up by their different definitions and formalisations of states and changes. In this broader perspective, our approach can be seen as a member of dynamic semantics family. Recall that state changes can be depicted schematically as:

$$s_0 \xrightarrow{u_1} s_1 \xrightarrow{u_2} s_2 \cdots s_{n-1} \xrightarrow{u_n} s_n$$

where an utterance, u_1, turns the state, s_0, into another, s_1, and each succeeding utterance, u_i, turns s_{i-1} into s_i until u_n turns s_{n-1} into s_n. Each state, s_i, represents a static meaning such as a discourse representation structure. Notice that the scheme can be seen as a simple process. One may see that at the bottom our approach shares the same view as dynamic semantics.

We have, however, enriched the dynamic model further by combining several processes and allowing them to run in parallel. We have also allowed processes to communicate with each other. The notions of concurrency and communications are what we have added to the dynamics. The benefit of enriching the underlying model is that we can study how particular states can be updated by an utterance within the same framework as the one we adopt to study semantic representations. To show that such an approach is possible to syntax, we have modelled the way a sentence is parsed as a state change caused to a system of communicating processes. By formalising the two domains within the same framework, we can study the dynamics of syntax-semantics interface in great detail.

5.3 Concurrent natural language processing

As for concurrent natural language processing, the most relevant model in spirit to ours is proposed by Bröker, Hahn, and Schacht ([Schacht et al.1994], [Bröeker et al.1994a], [Bröeker et al.1994b]). In their *ParseTalk* model, parsing is modelled as message passing based on Hewitt's actor model ([Agha1986]). Grammar rules are not separated from lexicon, but encoded into each lexical item. Such a lexical item is encoded as an actor, and actors can communicate with each other to parse sentences. Our project shares the intuition in modelling lexical items as autonomous agents, but the level of modelling is different. While their model is defined at a higher level, i.e. actor model, ours is defined at more primitive level, i.e. process algebra. Our advantage is that we can model feature structures and other data structures in the same framework as the model is fine grained, while their model forces to combine different frameworks, e.g. introducing feature structures in separate form. But the difference should not be exaggerated; their aim is to build a practical system while ours to investigate a concurrent theory for NLP. Our model should serve as a basis to analyse and compare various concurrent NLP models such as *ParseTalk* model in broader perspective. Another difference is their emphasis on general knowledge and ambiguity. Class hierarchy is, for instance, employed to disambiguate syntax. That sort of issue was not considered here.

Oz ([Smolka1994b]) is a concurrent constraint programming language developed by Gert Smolka and his group at DFKI(German Research Center for Artificial Intelligence). The language combines ideas from the π-calculus with constraint programming and can find a solution satisfying constraints effectively owing to concurrency. The emphasis is on the design of concurrent constraint language, but applications to NLP are in consideration. One way to encode feature structures as concurrent objects has been proposed by ([Smolka1994a]) and a bottom-up parser is implemented as a demo program. The difference with our project in strategy and objectives should be pointed out, however. While the language is rich in facilities, we start with primitive elements of computation. In our approach the complex behaviour of systems is modelled by combining small building blocks. This is because we use the π-calculus to study concurrent NLP, not to design a particular NLP system. It is therefore inappropriate to discuss the usefulness in implementation in our approach. Our work should, however, be useful for building NLP systems using concurrent languages such as Oz.

6 Conclusion

We have modelled the way an utterance represents a situation in process algebra. We have shown how feature structures and a bottom-up chart parser can be encoded as processes. We have also shown how a semantic representation, which is also encoded as a system of processes, can be created as the result of interactions between processes encoding the parser. We discussed related work and argued that our work could be seen as a refinement of Cooper's Situation Theoretic Grammar and as an implementation of Situation Semantics.

References

[Aczel et al.1993] Aczel, P., Israel, D., Katagiri, Y., and Peters, S. (eds.): 1993, *Situation Theory and its Applications*, Vol. 3, Center for the Study of Language and Informaiton, Stanford, California

[Agha1986] Agha, G. A.: 1986, *Actors : a model of concurrent computation in distributed systems*, The MIT Press series in artificial intelligence, MIT Press, Cambridge, Mass.

[Barwise1993] Barwise, J.: 1993, Constraints, channels, and the flow of information, in [Aczel et al.1993], pp 3–27

[Barwise and Cooper1993] Barwise, J. and Cooper, R.: 1993, Extended Kamp Notation: a graphical notation for situation theory, in [Aczel et al.1993], pp 29–53

[Barwise and Perry1983] Barwise, J. and Perry, J.: 1983, *Situations and Attitudes*, MIT Press, Cambridge, Mass.

[Barwise and Seligman1994] Barwise, J. and Seligman, J.: 1994, The rights and wrongs of natural regularity, in J. Tomberlin (ed.), *Philosophical Perspectives*, Vol. 8, pp 331–365, Ridgeview

[Berry and Boudol1992] Berry, G. and Boudol, G.: 1992, The chemical abstract machine, *Theoretical Computer Science* 96, 217–248

[Bröeker et al.1994a] Bröeker, N., Hahn, U., and Schacht, S.: 1994a, Concurrent lexicalized dependency parsing: The *ParseTalk* model, in *15th International Conference on Computational Linguistics*, Vol. 2, pp 379–385

[Bröeker et al.1994b] Bröeker, N., Strube, M., Schacht, S., and Hahn, U.: 1994b, Coarse-grained parallelism in natural language understanding: parsing as message passing, in *The International Conference on New Methods in Language Processing (NeMLaP)*, pp 182–189, Manchester, U.K.

[Cooper1991] Cooper, R.: 1991, Three lectures on situation theoretic grammar, in M. Filgueiras, L. Damas, N. Moreira, and A. P. Tomás (eds.), *Proceedings of Natural Language Processing, EAIA 90*, No. 476 in Lecture Notes in Artificial Intelligence, pp 101–140, Springer Verlag

[Cooper1993] Cooper, R.: 1993, Integrating different information sources in linguistic interpretation, in *First International Conference on Linguistics at Chosun University*, pp 79–109, Foreign Culture Research Institute, Chosun University, Kwangju, Korea

[Fujinami1995] Fujinami, T.: 1995, A process algebraic approach to situation semantics, in P. Dekker and martin Stokhof (eds.), *Proceedings of the 10th Amsterdam Colloquium*, Vol. 2, pp 263–282, ILLC/Department of Philosophy, University of Amsterdam, The Netherlands

[Fujinami1996] Fujinami, T.: 1996, *A process algebraic approach to computational linguistics*, Ph.D. thesis, Centre for Cognitive Science, University of Edinburgh, Edinburgh

[Groenendijk and Stokhof1991] Groenendijk, J. and Stokhof, M.: 1991, Dynamic predicate logic, *Linguistic and Philosophy* 14(1), 39–100

[Kamp and Reyle1993] Kamp, H. and Reyle, U.: 1993, *From Discourse to Logic: Introduction to Modeltheoretic Semantics of Natural Language, Formal Logic and Discourse Representation Theory*, Dordrecht: Kluwer

[Milner1994] Milner, R.: 1994, Pi-nets: a graphical form of π-calculus, in *Proceedings of ESOP '94*, Vol. 788 of *LNCS*, pp 26–42, Springer Verlag

[Milner et al.1992] Milner, R., Parrow, J., and Walker, D.: 1992, A calculus of mobile processes, parts I and II, *Information and Computation* 100, 1–40 and 41–77

[Schacht et al.1994] Schacht, S., Hahn, U., and Bröeker, N.: 1994, Concurrent lexicalized dependency parsing: A behavioral view on *ParseTalk* events, in *15th International Conference on Computational Linguistics*, Vol. 2, pp 489–493

[Smolka1994a] Smolka, G.: 1994a, A foundation for concurrent constraint programming, in *Constraints in Computational Logics*, Vol. 845 of *LNCS*, pp 50–72, Springer Verlag

[Smolka1994b] Smolka, G.: 1994b, *An Oz Primer*, Technical report, German Research Center for Artificial Intelligence (DFKI), Saarbrücken, Germany

Dynamic Epistemic Logic

Jelle Gerbrandy

ILLC/Department of Philosophy
University of Amsterdam

1 Introduction

This paper is the result of combining two traditions in formal logic: epistemic logic and dynamic semantics.

Dynamic semantics is a branch of formal semantics that is concerned with *change*, and more in particular with change of information. The main idea in dynamic semantics is that the meaning of a syntactic unit—be it a sentence of natural language or a computer program—is best described as the change it brings about in the state of a human being or a computer. The motivation for, and applications of this 'paradigm-shift' can be found in areas such as semantics of programming languages (cf. Harel, 1984), default logic (Veltman, 1996), pragmatics of natural language (Stalnaker, 1972) and of man-computer interaction, theory of anaphora (Groenendijk et al., 1996) and presupposition theory (Beaver, 1995). Van Benthem (1996) provides a survey.

This paper is firmly rooted in this paradigm, but at the same time it is much influenced by another tradition: that of the analysis of epistemic logic in terms of multi-modal Kripke models.

This article contains a semantics and a deduction system for a multi-agent modal language extended with a repertoire of programs that describe information change. The language is designed in such a way that everything that is expressible in the object language can be known or learned by each of the agents. The possible use of this system is twofold: it might be used as a tool for reasoning agents in computer science and it might be used as a logic for formalizing certain parts of pragmatics and discourse theory.[1]

Logic, Language, and Computation, Vol. II, edited by Lawrence S. Moss, Jonathan Ginzburg, and Maarten de Rijke.

[1]As a first step in this direction, Gerbrandy and Groeneveld (1997) show how a logic similar to the one introduced in this paper can be used to formalize puzzles like the Conway paradox or the puzzle of the dirty children.

The paper is organized as follows. The next section contains a short description of classical modal logic and introduces models based on non-well-founded sets as an alternative to Kripke semantics. In the section after that I introduce programs and their interpretation and I give a sound and complete axiomatization of the resulting logic in section 4. The last section is devoted to a comparison with update semantics of Veltman (1996).

Finally, I would like to mention the dissertations of Groeneveld (1995), Jaspars (1994) and de Rijke (1992) and the book by Fagin, Halpern, Moses and Vardi (1995) as precursors and sources of inspiration. The article by Willem Groeneveld and myself (1997) contains ideas that are similar to those presented here.

2 Static Modal Semantics

The classical language of multi-modal logic is the following:

Definition 2.1 Let $\mathcal{A}$ be a set of agents and $\mathcal{P}$ a set of propositional variables. The language of classical modal logic is given by:

$$\Phi ::= p \mid \phi \wedge \psi \mid \neg\phi \mid \Box_a \phi$$

where $p \in \mathcal{P}$ and $a \in \mathcal{A}$.

One way of providing a semantics for this language is in terms of Kripke models. A *pointed Kripke model* is a quadruple $\langle W, \{R_a\}_{a\in\mathcal{A}}, V, w\rangle$, where W is a set of possible worlds, w is a distinguished element of W (the point of evaluation), R_a is a relation on W for each $a \in \mathcal{A}$, V is a valuation function that assigns a subset of W to each propositional variable $p \in \mathcal{P}$.

Intuitively, given a Kripke model and a world w in it, a proposition p is true just in case $w \in V(p)$, and the information of an agent a in w is represented by the set of worlds that are accessible from w via R_a; these worlds are the worlds compatible with a's information in w.

Kripke models have been studied extensively and they provide a very perspicuous semantics for the classical language of epistemic logic. Unfortunately, it turns out that Kripke-models are not very suitable structures for defining operations that correspond to intuitive notions of information change.[2] To avoid this problem, I use a different (but equivalent) representation.

Definition 2.2 Possibilities.
Let $\mathcal{A}$, a set of agents, and $\mathcal{P}$, a set of propositional variables, be given. The class of possibilities is the largest class such that:

[2] Cf. Groeneveld (1995) for a discussion of the problems one encounters.

- A possibility w is a function that assigns to each propositional variable $p \in \mathcal{P}$ a truth value $w(p) \in \{0,1\}$ and to each agent $a \in \mathcal{A}$ an information state $w(a)$.

- An information state σ is a set of possibilities.

A possibility w characterizes which propositions are true and which are false, and it characterizes the information that each of the agents has in the form of an information state σ, that consists of the set of possibilities the agent considers possible in w.

This definition of possibilities should be read to range over the universe of non-well-founded sets in the sense of Aczel (1988).[3] Defining classes as 'the largest class such that....', is a standard form of definition in non-well-founded set theory.

Truth of classical modal sentences in a possibility can be defined in a way analogous to the definition of truth for Kripke models.

Definition 2.3 Truth.
Let w be a possibility.

$$\begin{array}{rll} w \models p & \text{iff} & w(p) = 1 \\ w \models \phi \wedge \psi & \text{iff} & w \models \phi \text{ and } w \models \psi \\ w \models \neg\phi & \text{iff} & w \not\models \phi \\ w \models \Box_a \phi & \text{iff} & \text{for all } v \in w(a) : v \models \phi \end{array}$$

It turns out that using possibilities instead of Kripke-models does not make an essential logical difference: there is a close relation between possibilities and pointed Kripke models.

Definition 2.4 Let $\mathcal{K} = (W, \{R_a\}_{a \in \mathcal{A}}, V, w)$ be a pointed Kripke model.

- A *decoration* of $\mathcal{K}$ is a function d that assigns to each world $v \in W$ a function with $\mathcal{P} \cup \mathcal{A}$ as its domain, for each $p \in \mathcal{P}$ such that $d(v)(p) = 1$ iff $p \in V(v)$, and $d(v)(p) = 0$ otherwise, and $d(v)(a) = \{d(u) \mid vR_a u\}$ for each $a \in \mathcal{A}$.

- If $\mathcal{K} = (W, \{R_a\}_{a \in \mathcal{A}}, V, w)$ is a Kripke model, and d is a decoration of it, $d(w)$ is its *solution*, and $\mathcal{K}$ is a *picture* of $d(w)$.

[3] The class of possibilities can be defined more precisely as the greatest fixed point of the operator Φ such that for each class A: $\Phi(A) = \{f \mid f$ is a function that assigns to each $p \in \mathcal{P}$ an element of $\{0,1\}$, and to each $a \in \mathcal{A}$ a subset of $A\}$. Aczel (1988) uses structures that are similar to possibilities to define a semantics for process algebra, and Barwise and Moss (1996) similar models are used for a semantics for modal logic.

A decoration of a Kripke model assigns to each possible world w in the model a possibility that assigns the same truth-values to the propositional variables as they get in the model at w, and that assigns to each agent a the set of possibilities that are assigned to worlds accessible from w by R_a.

The notions of solution and picture give us a correspondence between Kripke-models and possibilities:

Proposition 2.5

- Each Kripke model has a unique solution, which is a possibility.
- Each possibility has a Kripke model as its picture.
- Two Kripke-models are pictures of the same possibility iff they are bisimilar.

Defining truth of a formula in a Kripke model in the standard way, it holds that:

Proposition 2.6 For each possibility w:

$$w \models \phi \text{ iff } \phi \text{ is true in each picture of } w$$

So a possibility and a picture of it are descriptively equivalent. This means that one can see possibilities as representatives of equivalence classes of Kripke models under bisimulation.

3 Programs

In this section we will define operations on possibilities that correspond to changes in the information states of the agents. The kind of information change we want to model is that of agents getting new information or learning that the information state of some other agent has changed in a certain way. I will introduce 'programs' in the object language that describe such changes. Changes in the 'real world' will not be modeled, and I will ignore other operations of information change such as belief contraction or belief revision.

The programming language is built up as follows. There are programs of the form $?\phi$ for each sentence ϕ. A program of the form $?\phi$ will be interpreted as a test that succeeds in a possibility when ϕ is true and fails otherwise. The language contains a program operator U_a for each agent a. A program of the form $U_a\pi$ corresponds to agent a learning that program π has been executed. Finally, the language contains two operators that combine programs to form a new program: sequencing and disjunction.

A program of the form $\pi; \pi'$ is interpreted as: "first execute π, then π'." Disjunction is interpreted as choice: $\pi \cup \pi'$ corresponds to executing either π or π'.

To connect the programming language to the 'static part' of the language, we add a modal operator $[\pi]$ for each program π. Intuitively, a sentence $[\pi]\psi$ is true in a possibility just in case that after executing the program π in that possibility, ψ must be true. The set of programs is defined simultaneously with the set of sentences in a way that might be familiar from propositional dynamic logic (cf. for example Pratt, 1976 or Goldblatt, 1987).

Definition 3.1 Language.
Given a set of agents $\mathcal{A}$ and a set of propositional variables $\mathcal{P}$, the set of sentences of dynamic epistemic logic is the smallest set given by:

$$\Phi ::= p \mid \phi \wedge \psi \mid \neg\phi \mid \Box_a \phi \mid [\pi]\phi$$

where $a \in \mathcal{A}$, $p \in \mathcal{P}$, and π is any program. The set of programs is the smallest set given by:

$$\Pi ::= ?\phi \mid U_a \pi \mid \pi; \pi' \mid \pi \cup \pi'$$

Programs are interpreted as relations over possibilities: a pair of possibilities (w, v) will be in the denotation of a program π (written as $w[\![\pi]\!]v$) just in case the execution of the program π in possibility w may result in v. I propose the following definition (in the definition, I use the abbreviation $w[a]v$ for the statement that w differs at most from v in the state it assigns to a):

Definition 3.2 Interpretation of programs.

$$\begin{array}{rll} w[\![?\phi]\!]v & \text{iff} & w \models \phi \text{ and } w = v \\ w[\![U_a\pi]\!]v & \text{iff} & w[a]v \text{ and } v(a) = \{v' \mid \exists w' \in w(a) : w'[\![\pi]\!]v'\} \\ w[\![\pi;\pi']\!]v & \text{iff} & \text{there is a } u \text{ such that } w[\![\pi]\!]u[\![\pi']\!]v \\ w[\![\pi \cup \pi']\!]v & \text{iff} & \text{either } w[\![\pi]\!]v \text{ or } w[\![\pi']\!]v \end{array}$$

Furthermore, the definition of truth is extended with the following clause:

$$w \models [\pi]\phi \quad \text{iff} \quad \text{for all } v \text{ if } w[\![\pi]\!]v \text{ then } v \models \phi$$

Programs of the form $U_a\pi$ are to be read as "a learns that π has been executed," or, alternatively, as "a updates her information state with π." This is modeled as follows. Executing a program of the form $U_a\pi$ in a possibility w results in a new possibility v in which only a's information state has changed. The information state of a in v contains all and only those possibilities that are the possible result of an execution of π in one of the possibilities that in a's information state in w. Note that a program of the form $U_a\pi$ is deterministic; in fact, $[\![U_a\pi]\!]$ is always a total function, which means that the update always exists, and the result is unique.

In the case that π is a test of the form $?\phi$, the result of executing $U_a?\phi$ is such that in a's new information state, all possibilities in which ϕ is not true are discarded: the new information state of a contains only possibilities in which ϕ is true. So, one might say that $U_a?\phi$ corresponds to a getting the information that ϕ is the case.

The programming language is constructed in such a way that each program can be executed by each of the agents. This has the effect that any change in the model that we can express as a program in the object language can be 'learned' by each of the agents. In particular, this means that each sentence can be 'learned' by each of the agents, because there is a test $?\phi$ in the programming language for each sentence ϕ.

I will give some examples. A program of the form $U_aU_b?p$ denotes the action that a updates her state with the information that b has updated his information state with $?p$. This corresponds with a getting the information that b has gotten the information that p is the case.

We can also express –without knowing whether p is in fact the case– the action of a learning whether p is true or not by the program $(?p; U_a?p) \cup (?\neg p; U_a?\neg p)$. If we let '$p$' stand for 'the value of the bit is 1' (and $\neg p$ for 'the value is 0'), then this corresponds to the action of a software agent a learning what the value of a certain bit is, and if 'p' means 'it is raining', this program can be used to describe the information change of a philosopher looking out of the window to see whether it is raining or not.

Conscious Updates

A natural assumption to make is that information is *introspective*: if you have certain information, you also know that you have that information, and vice versa, knowing that you have certain information implies that you indeed have that information. In our framework an agent has introspective information in a possibility w just in case she only considers worlds possible in which she has the information she in fact has (i.e. the information she has in w). Formally, this means that for all $v \in w(a)$, it should hold that $v(a) = w(a)$.

Unfortunately, introspection is not preserved under the notion of 'getting information' as we have defined it above: in general, updates of the form $[U_a\pi]$ will take you from introspective states to states in which introspection fails to hold.

An example may help to appreciate the problem. Let w be an introspective possibility and suppose that $w(a)$ contains possibilities where p is true, and where p is not true. So a does not know whether p is true, and the assumption that her information is introspective implies that she is aware of this lack of information.

Consider now the possibility that results from updating w with $U_a?p$, i.e. the unique possibility v such that $w[\![U_a?p]\!]v$. This is a possibility in which a has gotten the information that p: $v \models \Box_a p$. By definition, the update with $U_a?p$ has only eliminated possibilities from the information state $w(a)$, it has not changed them. So it will continue to hold that $v' \models \neg\Box_a p$ in each of the possibilities v' in a's new information state $v(a)$. In other words, an update of w with $U_a?p$ will change a's information state in such a way that she has learned that p is the case, but the information she has about her own information will not change at all.

To solve this problem, one would like that an update of a's information state with π not only changes each of a's possibilities in accord with π, but also changes each of these updated possibilities to the effect that she has learned π. It turns out that it is not very hard to define a notion of update which reflects this. I will refer to such an update as a 'conscious update,' because this update models the idea that if a updates with π, she is conscious of this fact.[4] I use the notation $U_a^*\pi$ for a conscious update of a's information state with π.

Definition 3.3 Conscious update.

$$w[\![U_a^*\pi]\!]v \text{ iff } w[a]v \text{ and } v(a) = \{v' \mid \exists w' \in w(a) : w'[\![\pi]\!][\![U_a^*\pi]\!]v'\}$$

This definition is circular as it stands. Nevertheless, it is not very hard to prove that for each program π there is a unique relation $[\![U_a^*\pi]\!]$ that conforms to the definition.[5] Again, $[\![U_a^*\pi]\!]$ is a total function for each π.

The definition says that consciously updating a's information state in a possibility w with π results in a possibility v that differs only from w in that all possibilities in $w(a)$ are first updated with π, and after that with $U_a^*\pi$. That the interpretation of $U_a^*\pi$ gives the effect of a conscious update is corroborated by the fact that it holds that if a's information in w is introspective, it is introspective after the update of w with $U_a^*?\phi$.

[4]The terminology is from Groeneveld (1995), who introduces a notion of conscious update in a single agent setting.

[5]A direct proof of the correctness of a similar definition can be found in Gerbrandy and Groeneveld (1997). Alternatively, we can see this definition as an application of the *corecursion theorem* (theorem 17.5) of Barwise and Moss (1996). For those who are initiated: the operator Φ defined in footnote 3 is a 'smooth' operator, so the corecursion theorem applies. The function $[U_a\pi]$ acts as the 'pump' of the corecursive definition of $[U_a^*\pi]$.

Group Updates

Common knowledge is a concept that occurs under different names (mutual knowledge, common ground) in the literature. The usual definition is that a sentence ϕ is common knowledge in a group $\mathcal{B}$ just in case each agent in the group knows that ϕ is the case, each agent knows that each of the other agents knows that ϕ, etcetera. As Barwise (1989) shows, a semantics based on non-well-founded sets is quite useful for modeling this concept.

Instead of concentrating on this static notion of mutuality, I will introduce the notion of a 'group update': an update with a program π in a group of agents that has the effect of changing the state of each agent in the group in the way described by π in such a way that each agent in the group is aware of the fact that each agent has executed π, each agent knows that each agent in the group knows that π is executed by each agent in the group, etc. In case π is a test of the form $?\phi$, a common update with $?\phi$ corresponds to the sentence ϕ becoming common knowledge within the group.

To express this in the object language, I add program operators of the form $U^*_{\mathcal{B}}$ for each subset $\mathcal{B}$ of $\mathcal{A}$ to the language. They are interpreted as follows:

Definition 3.4 Group update
For each π and $\mathcal{B} \subseteq \mathcal{A}$:

$$w[\![U^*_{\mathcal{B}}\pi]\!]v \quad \text{iff} \quad w[\mathcal{B}]v \text{ and } \forall a \in \mathcal{B}: \\ v(a) = \{v' \mid \exists w' \in w(a) : w'[\![\pi]\!][\![U^*_{\mathcal{B}}\pi]\!]v'\}$$

Updating a possibility with a program $U^*_{\mathcal{B}}\pi$ results in a possibility v that differs only from w in that for each $a \in \mathcal{B}$, all situations in $w(a)$ are first updated with π, and then with $U^*_{\mathcal{B}}\pi$.

Note that a group update in a group consisting of a single agent boils down to the notion of a conscious update defined above: updating with $U^*_{\{a\}}\pi$ is exactly the same thing as updating with $U^*_a\pi$.

4 Axiomatization

The following set of axioms and rules provides a sound and complete characterization of the set of sentences that are true in all possibilities. (For sake of presentation, I have left out the conscious single agent updates, since they are a special case of the group updates with a group consisting of a single agent. I have also left out axioms for the non-conscious updates introduced in definition 3.2. The axioms for U_a are just as those for $U^*_{\{a\}}$, except for axiom 7, which should be changed into: $\vdash [U_a\pi]\Box_a\psi \leftrightarrow \Box_a[\pi]\psi$.)

Axioms

1 $\vdash \phi$ if ϕ is valid in classical propositional logic.

2 $\vdash \Box_a(\phi \rightarrow \psi) \rightarrow (\Box_a\phi \rightarrow \Box_a\psi)$

3 $\vdash [\pi](\phi \rightarrow \psi) \rightarrow ([\pi]\phi \rightarrow [\pi]\psi)$.

4 $\vdash [?\phi]\psi \leftrightarrow (\phi \rightarrow \psi)$

5 $\vdash \neg[U^*_{\mathcal{B}}\pi]\psi \leftrightarrow [U^*_{\mathcal{B}}\pi]\neg\psi$

6 $\vdash [U^*_{\mathcal{B}}\pi]p \leftrightarrow p$

7 $\vdash [U^*_{\mathcal{B}}\pi]\Box_a\phi \leftrightarrow \Box_a[\pi][U^*_{\mathcal{B}}\pi]\phi$ if $a \in \mathcal{B}$

8 $\vdash [U^*_{\mathcal{B}}\pi]\Box_a\phi \leftrightarrow \Box_a\phi$ if $a \notin \mathcal{B}$

9 $\vdash [\pi;\pi']\phi \leftrightarrow [\pi][\pi']\phi$

10 $\vdash [\pi \cup \pi']\psi \leftrightarrow ([\pi]\psi \wedge [\pi']\psi)$

Rules

MP $\phi, \phi \rightarrow \psi \vdash \psi$

Nec$\Box$ If $\vdash \phi$ then $\vdash \Box_a\phi$

Nec$[\cdot]$ If $\vdash \phi$ then $\vdash [\pi]\phi$

$\Gamma \vdash \phi$ iff there is derivation of ϕ from the premises in Γ using the rules and axioms above.

In addition to the rules and axioms of classical modal logic, the deduction system consists of axioms describing the behavior of the program operators. Axiom 3 and the rule Nec$[\cdot]$ guarantee that the program operators behave as normal modal operators. Axiom 4 says that performing a test $?\phi$ boils down to checking whether ϕ is true. Axiom 5 reflects the fact that U^*_a-updates are total functions: an update with $U^*_a\pi$ always gives a unique result. This means that if it is not the case that a certain sentence is true after an update with a program of the form $U_{\mathcal{B}}\pi$, then it must be the case that the negation of that sentence is true in the updated possibility, and vice versa. Axiom 6 expresses that the update of an information state has no effect on the 'real' world; the same propositional atoms will be true or false before and after an update. Axiom 7 expresses that after a group update with π, an agent in the group knows that ψ just in case that agent already knew that after executing π, an update with $U^*_{\mathcal{B}}\pi$ could only result in a possibility in which ψ were true. Axiom 8 expresses that a group update has no effect on the information of agents outside of that group. The axioms 9 and 10 govern the behavior of sequencing and disjunction respectively.

Proposition 4.1 Soundness
If $\Gamma \vdash \phi$ then $\Gamma \models \phi$.

proof: By a standard induction. By way of illustration, I will show the correctness of axiom 5. In the proof, I make use of the fact that $[\![U^*_{\mathcal{B}}\pi]\!]$ is a total function for each π, and write $w[\![U^*_{\mathcal{B}}\pi]\!]$ for the unique v such that $w[\![U^*_{\mathcal{B}}\pi]\!]v$. The following equivalences hold, if $a \in \mathcal{B}$:

$$\begin{array}{lll} w \models [U^*_{\mathcal{B}}\pi]\Box_a\phi & \text{iff} & w[\![U^*_{\mathcal{B}}\pi]\!] \models \Box_a\phi \\ & \text{iff} & \forall v \in w[\![U^*_{\mathcal{B}}\pi]\!](a) : v \models \phi \\ & \text{iff} & \forall v : \text{ if } \exists w' \in w(a) : w'[\![\pi]\!][\![U^*_{\mathcal{B}}\pi]\!]v \text{ then } v \models \phi \\ & \text{iff} & \forall w' \in w(a)\ \forall v : \text{ if } w'[\![\pi]\!][\![U^*_{\mathcal{B}}\pi]\!]v \text{ then } v \models \phi \\ & \text{iff} & \forall w' \in w(a) : w' \models [\pi][U^*_{\mathcal{B}}\pi]\phi \\ & \text{iff} & w \models \Box_a[\pi][U^*_{\mathcal{B}}\pi]\phi \end{array}$$

Proposition 4.2 Completeness
If $\Gamma \models \phi$, then $\Gamma \vdash \phi$.

proof: The completeness proof is rather long. I give here the main structure; the details are delegated to the appendix.

The proof is a variation on the classical Henkin proof for completeness of modal logic. It is easy to show that each consistent set can be extended to a maximal consistent set (this will be referred to as 'Lindenbaum's Lemma'). We must show that for each consistent set of sentences there is a possibility in which these sentences are true. Completeness then follows by a standard argument.

Let, for each maximal consistent set Σ, w_Σ be that possibility such that $w_\Sigma(p) = 1$ iff $p \in \Sigma$, and for each agent a: $w_\Sigma(a) = \{w_\Gamma \mid \Gamma$ is maximal consistent and if $\Box_a\psi \in \Sigma$, then $\psi \in \Gamma\}$.[6] We prove the usual truth lemma, namely that for each sentence ϕ it holds that $\phi \in \Sigma$ iff $w_\Sigma \models \phi$.

The truth lemma is proven by an induction on the structure of ϕ, in which all cases are standard, except the case where ϕ is of the form $[\pi]\psi$. The proof for this case rests on the following idea. Just as membership in $w_\Sigma(a)$ depends on the formulae of the form $\Box_a\phi$ in Σ, the π-update of w_Σ is closely related to the formulae of the form $[\pi]\psi$ in Σ. This is reflected by the following relation between maximal consistent sets:

$$\Sigma R_\pi \Gamma \quad \text{iff} \quad \Gamma \text{ is a maximal consistent set and if } [\pi]\psi \in \Sigma \text{ then } \psi \in \Gamma$$

I will prove in the appendix, as lemma A.1, that $w_\Sigma[\![\pi]\!]v$ iff there is a Γ such that $v = w_\Gamma$ and $\Sigma R_\pi \Gamma$. The relevant step in the proof of the truth

[6]In the terminology of definition 2.4, the possibility w_Σ is the solution of Σ in the canonical Kripke model of the modal logic K.

lemma then runs as follows:

$$\begin{array}{rcl} w_\Sigma \models [\pi]\psi & \Leftrightarrow & w_\Sigma[\![\pi]\!] \models \psi \\ & \Leftrightarrow & \text{for each } \Gamma: \text{ if } \Sigma R_\pi \Gamma \text{ then } w_\Gamma \models \psi \text{ (by lemma A.1)} \\ & \Leftrightarrow & \psi \in \Gamma \text{ for each } \Gamma \text{ such that } \Sigma R_\pi \Gamma \text{ (by ind. hypothesis)} \\ & \Leftrightarrow & [\pi]\psi \in \Sigma \text{ (by definition of } R_\pi) \end{array}$$

5 Update Semantics

Update semantics, as it is presented in Veltman (1996), has been an important source of inspiration for this paper. It turns out that update semantics can be seen as a special case of the present approach: update semantics can be seen as describing the updates of an information state of a single agent who has fully introspective knowledge.

In update semantics, sentences are interpreted as functions that operate on information states. Information states are sets of classical possible worlds. The relevant definitions are the following:

Definition 5.1

- $\mathcal{L}^{US}$ is the language built up from a set of propositional variables $\mathcal{P}$ and the connectives $\neg, \wedge$ and a unary sentence operator *might* in the obvious way.[7]

- A classical information state s is a set of classical possible worlds, i.e. a set of assignments of truth-values to the propositional variables.

- For each sentence $\phi \in \mathcal{L}^{US}$ and each classical information state s, the update of s with ϕ, $s[\phi]$, is defined as:

$$\begin{array}{rcl} s[p] & = & \{w \in s \mid w(p) = 1\} \\ s[\phi \wedge \psi] & = & s[\phi] \cap s[\psi] \\ s[\neg\phi] & = & s \setminus s[\phi] \\ s[\textit{might } \phi] & = & s \text{ if } s[\phi] \neq \emptyset \\ & = & \emptyset \text{ otherwise} \end{array}$$

- A sentence ϕ is accepted in an information state s, written as $s \models \phi$, just in case $s[\phi]s$. An argument $\phi_1 \ldots \phi_n/\psi$ is valid, written as $\phi_1 \ldots \phi_n \models_{US} \psi$, iff for each s: $s[\phi_1] \ldots [\phi_n] \models \psi$.

[7]In Veltman's paper the language is restricted to those sentences in which *might* occurs only as the outermost operator in a sentence.

There is a close correspondence between updates of information states in update semantics and consciously updating in possibilities in which agents have introspective information. More precisely, we can associate with each introspective possibility w and agent a a classical information state w^a, which consists of the set of classical worlds that correspond to the possibilities in $w(a)$. Vice versa, given a classical world w and an agent a, we can associate with each classical information state s a possibility s^a_w that assigns to each propositional variable the same value as w does and that assigns to a a set containing a possibility s^a_v for each $v \in s$ (a classical information state does not provide us with any information about which agent we are talking about, or what the 'real world' looks like, so we have to supply these parameters ourselves).

More formally:

Definition 5.2

- If w is a possibility and a an agent, then $w^a = \{v \text{ restricted to } \mathcal{P} \mid v \in w(a)\}$.
- If s is a classical information state, w a classical possible world, and a an agent, then s^a_w is a possibility such that $s^a_w(p) = w(p)$ for each $p \in \mathcal{P}$, and $s^a_w(a) = \{s^a_v \mid v \in s\}$.

It is not hard to see that s^a_w is an introspective possibility. The following proposition expresses how US-updates can be viewed as conscious updates of an introspective information state, if one reads $\neg\Box\neg$ for *might*. More precisely, seeing a classical information state as the information state of a certain agent a, updating such an information state with *might* ϕ in update semantics corresponds to updating a's information state with the test $?\neg\Box_a\neg\phi$.

Proposition 5.3 For each $\phi \in \mathcal{L}^{US}$, let ϕ' be just as ϕ but with all occurrences of *might* replaced by $\neg\Box_a\neg$. Then it holds that:

- For all classical information states s and t: $s[\phi]t$ iff $s^a_w[\![U^*_a?\phi']\!]t^a_w$.
- For all possibilities w and v and each a such that a has introspective information in w: $w[\![U^*_a?\phi']\!]v$ iff $w^a[\phi]v^a$.
- $\phi_1 \ldots \phi_n \models_{US} \psi$ iff for all introspective w:

$$w \models [U^*_a?\phi'_1] \ldots [U^*_a?\phi'_n]\Box_a\psi'.$$

What this proposition expresses is that a US-update can be seen as a conscious update of the information state of an agent who has fully introspective information.

6 Conclusions and further research.

In this paper, I have defined a semantics for a logic of information change in a multi-agent epistemic framework, and given a sound and complete axiomatization of the logic. I'll state here, without proof, that the expressive power of dynamic epistemic logic is the same as that of classical modal logic, which implies that the complexity of dynamic logic is not higher than that of classical modal logic. The comparison with the work of Veltman shows that the logic is an extension of a more simple 'one-agent' version of dynamic logic.

Two directions of further research readily suggest themselves. On the logical side, one can extend the logic with extra 'static' modal operators (operators that express that a certain sentence is common knowledge or distributed knowledge, for example). The difficulty here is not so much how such operations should be interpreted semantically, but finding deduction systems that are complete. Another natural extension of the work done here is to add to the 'dynamical' repertoire of the language. There is a lot of work in this area ready to be picked up, such as the work on belief revision, and on dynamic predicate logic (e.g. Groenendijk et al., 1996). Apart from completeness results, the challenge here lies in finding interesting notions of interaction between between the information change of different agents.

Another open question is to find an axiomatization of the identities that hold between programs. The literature on process algebra contains axioms for $+$, $\cup$ and $?\phi$, and the way they interact, the problem is to find equations that describe the behavior of the $U_{\mathcal{B}}$-operator.

The other direction for further research lies in the area of applications of the semantics. The work of Fagin et al. is closely related to the work presented here, and it would be interesting to see whether and how the present semantics could be applied to problems in computer science. Another line of research is to apply the semantics presented to specify certain aspects of information change that occur in dialogue.

A Appendix

In this appendix, I will prove the lemma that was needed in the completeness proof in section 4. The lemma is the following:

Lemma A.1 *Let π be a program, and assume that for each maximal consistent set of sentences Σ and each subprogram of the form $?\phi$ of π it holds that $w_\Sigma \models \phi$ iff $\phi \in \Sigma$. Then it holds for all Σ:*

$$w_\Sigma [\![\pi]\!] v \text{ iff there is a } \Gamma \text{ such that } \Sigma R_\pi \Gamma \text{ and } v = w_\Gamma$$

proof: The proof is by induction on the structure of π.

- Tests: π is of the form $?\phi$.
  It follows by axiom 4 and maximality of Σ that $\Sigma R_{?\phi}\Sigma$ iff $\phi \in \Sigma$. The argument is then quite simple: $w_\Sigma[\![?\phi]\!]v$ iff $v = w_\Sigma$ and $w_\Sigma \models \phi$ iff $\phi \in \Sigma$ and $\Sigma R_{?\phi}\Sigma$.

- Conscious updates: $U^*_{\mathcal{B}}\pi$. This step is proven in lemma A.3.

- Disjunction: $\pi \cup \pi'$.
  Assume $w_\Sigma[\![\pi \cup \pi']\!]v$. Then $w_\Sigma[\![\pi]\!]v$ or $w_\Sigma[\![\pi']\!]v$. By induction hypothesis, there must be a Γ such that $v = w_\Gamma$ and $\Sigma R_\pi \Gamma$ or $\Sigma R_{\pi'}\Gamma$. To show that $\Sigma R_{\pi\cup\pi'}\Gamma$ take any $[\pi \cup \pi']\psi \in \Sigma$. Then $[\pi]\psi \wedge [\pi']\psi \in \Sigma$, and hence, by maximality of Σ, both $[\pi]\psi \in \Sigma$ and $[\pi']\psi \in \Sigma$. But that means, since $\Sigma R_\pi \Gamma$ or $\Sigma R_{\pi'}\Gamma$, that $\psi \in \Gamma$.
  For the other direction, assume that it does not hold that $w_\Sigma[\![\pi \cup \pi']\!]w_\Gamma$. Then neither $w_\Sigma[\![\pi]\!]w_\Gamma$, nor $w_\Sigma[\![\pi']\!]w_\Gamma$. So, by induction hypothesis, there is a ψ such that $[\pi]\psi \in \Sigma$ but $\psi \notin \Gamma$, and there is a ψ' such that $[\pi']\psi' \in \Sigma$ but $\psi' \notin \Gamma$. But then, $[\pi](\psi \vee \psi') \in \Sigma$ and $[\pi'](\psi \vee \psi') \in \Sigma$, whence $[\pi \cup \pi'](\psi \vee \psi') \in \Sigma$. But by maximality of Γ, $\neg\psi \wedge \neg\psi' \in \Gamma$, so it is not the case that $\Sigma R_{\pi\cup\pi'}\Gamma$.

- Sequencing: $\pi; \pi'$.
  Assume that $w_\Sigma[\![\pi;\pi']\!]v$. By induction hypothesis, there must be Δ and Γ such that $w_\Sigma[\![\pi]\!]w_\Delta[\![\pi']\!]w_\Gamma$, and $v = w_\Gamma$. Assume $[\pi;\pi']\psi \in \Sigma$. Then, by induction hypothesis, $[\pi']\psi \in \Delta$, and $\psi \in \Gamma$. Since $[\phi;\phi']\psi$ was arbitrary, it follows that $\Sigma R_{\pi;\pi'}\Gamma$.
  For the other direction, assume that $\Sigma R_{\pi;\pi'}\Gamma$. This means that the set $\{\psi \mid [\pi][\pi']\psi \in \Sigma\}$ is consistent (since it is a subset of Γ), and hence, $\{\psi \mid [\pi]\psi \in \Sigma\}$ is consistent. But then there is a Δ such that $\Sigma R_\pi \Delta$ and $\Delta R_{\pi'}\Gamma$. But then, by induction hypothesis, $w_\Sigma[\![\pi]\!]w_\Delta[\![\pi']\!]w_\Gamma$, and hence, $w_\Sigma[\![\pi;\pi']\!]w_\Gamma$.

Before giving the proof for the second step in the induction, I would like to make a general remark about the method of proof that will be used. We will prove that two possibilities are equal by showing that there exists a bisimulation between them. A relation $\mathcal{R}$ is a bisimulation between possibilities iff for every two possibilities w and v it holds that if $w\mathcal{R}v$, then (1) $w(p) = v(p)$ for each $p \in \mathcal{P}$, and (2) for each $w' \in w(a)$ there is a $v' \in v(a)$ such that $w'\mathcal{R}v'$, and (3) for each $v' \in v(a)$ there is a $w' \in w(a)$ such that $w'\mathcal{R}v'$. It turns out that the following proposition, which is closely related to proposition 2.5, is a consequence of the axiom of anti-foundation:

Proposition A.2 $w = v$ iff there is a bisimulation $\mathcal{R}$ such that $w\mathcal{R}v$.

I will make use of this fact in the proof of the following lemma.

Lemma A.3 *Fix any $U^*_{\mathcal{B}}\pi$ and assume that it holds that $w_\Sigma[\![\pi]\!]v$ iff there is a Γ such that $\Sigma R_\pi \Gamma$ and $v = w_\Gamma$. (This is the induction hypothesis of the previous lemma.) It holds that:*

$$w_\Sigma[\![U^*_{\mathcal{B}}\pi]\!]v \textit{ iff there is a } \Gamma \textit{ such that } v = w_\Gamma \textit{ and } \Sigma R_{U^*_{\mathcal{B}}\pi}\Gamma$$

proof: Note that by axiom 5, the set $\{\psi \mid [U^*_{\mathcal{B}}\pi]\psi \in \Sigma\}$ is maximal consistent if Σ is. That means that there always is a unique Γ such that $\Sigma R_{U^*_{\mathcal{B}}\pi}\Gamma$. This implies that to prove the lemma, it is enough to show that if $w_\Sigma[\![U^*_{\mathcal{B}}\pi]\!]v$, and $\Sigma R_{U^*_{\mathcal{B}}\pi}\Gamma$, then there is a bisimulation between v and w_Γ. By proposition A.2 this shows that v and w_Γ are in fact equal.

Given a program π and a set of agents $\mathcal{B}$, define a relation $\mathcal{R}$ on possibilities by

$w\mathcal{R}v$ iff $w = v$ or there exist maximal consistent sets Σ and Γ such that $w_\Sigma[\![U^*_{\mathcal{B}}\pi]\!]v, \Sigma R_{U^*_{\mathcal{B}}\pi}\Gamma$ and $w = w_\Gamma$

We will show that $\mathcal{R}$ is a bisimulation. Let $w\mathcal{R}v$, and let Σ and Γ be such that $w_\Sigma[\![U^*_{\mathcal{B}}\pi]\!]v$, $\Sigma R_{U^*_{\mathcal{B}}\pi}\Sigma$, and $w = w_\Gamma$ (the case that $w = v$ is easy). We need to show that the three clauses that define a bisimulation hold:

(1)
We first show that $w(p) = v(p)$ for all $p \in \mathcal{P}$.
$v(p) = 1$ iff $w_\Sigma[\![U^*_{\mathcal{B}}\pi]\!](p) = 1$ iff $w_\Sigma(p) = 1$ (by the semantics) iff $p \in \Sigma$ (by the definition of w_Σ) iff $p \in U^*_{\mathcal{B}}\pi(\Sigma)$ (by axiom 6) iff $w_\Gamma(p) = 1$.

(2)
Next we must show that for each $a \in \mathcal{A}$, if $v' \in v(a)$ then there is a $w' \in w_\Gamma(a)$ such that $w'\mathcal{R}v'$. We distinguish two cases: $a \in \mathcal{B}$, and $a \notin \mathcal{B}$.

First assume that $a \notin \mathcal{B}$. It follows by axiom 8 that $\Box_a\psi \in \Sigma$ iff $\Box_a\psi \in \Gamma$, which implies that $w_\Sigma(a) = w_\Gamma(a)$. But by the definition of $[\![U^*_{\mathcal{B}}\pi]\!]$ and the fact that $a \notin \mathcal{B}$, we have $w_\Sigma(a) = v(a)$. That means that $w_\Gamma(a) = v(a)$, which is sufficient, since by definition $\mathcal{R}$ includes the identity relation.

For the case that $a \in \mathcal{B}$, take any $v' \in v(a)$. We need to show that there is a $w_{\Gamma'} \in w_\Gamma(a)$ such that $w_{\Gamma'}\mathcal{R}v'$. The following picture might make matters more clear.

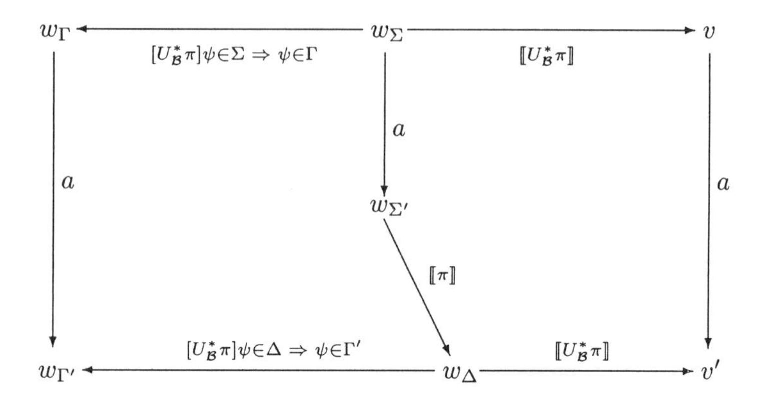

Since $v' \in v(a)$, there must be, by definition of $[\![U^*_{\mathcal{B}}\pi]\!]$, a Σ' such that $w_{\Sigma'} \in w_\Sigma(a)$ and a u such that $w_{\Sigma'}[\![\pi]\!]u[\![U^*_{\mathcal{B}}\pi]\!]v'$. By induction hypothesis, then, there is a Δ such that $\Sigma' R_\pi \Delta$ and $u = w_\Delta$. Take any such Δ and consider the set $\Gamma' = \{\psi \mid [U^*_{\mathcal{B}}\pi]\psi \in \Delta\}$. By the functionality axiom 5, this set is maximal consistent. From the definition of Δ and Γ' it then follows that $w_\Delta[\![U^*_{\mathcal{B}}\pi]\!]v'$ and that $\Delta R_{U^*_{\mathcal{B}}\pi}\Gamma'$, and hence that $w_{\Gamma'}\mathcal{R}v'$.

To show that $w_{\Gamma'} \in w_\Gamma(a)$, take any $\Box_a\psi \in \Gamma$. Then $[U^*_{\mathcal{B}}\pi]\Box_a\psi \in \Sigma$ (by axiom 5), and by axiom 7, $\Box_a[\pi][U^*_{\mathcal{B}}\pi]\psi \in \Sigma$. But then, $[\pi][U^*_{\mathcal{B}}\pi]\psi \in \Sigma'$, so $[U^*_{\mathcal{B}}\pi]\psi \in \Delta$, and hence, $\psi \in \Gamma'$.

(3)
Finally we must show that for each $a \in \mathcal{A}$, if $w' \in w_\Gamma(a)$ then there is a $v' \in v(a)$ such that $w'\mathcal{R}v'$.

If $a \notin \mathcal{B}$, we can use the same argument as in case (2).

For the other case, take any Γ' such that $w_{\Gamma'} \in w(a)$, i.e. such that $\Box_a\psi \in \Gamma \Rightarrow \psi \in \Gamma'$. We need to find a $v' \in v(a)$ such that $w_{\Gamma'}\mathcal{R}v'$.

What we will do is show that there must exist maximal consistent sets Σ' and Δ such that: (i) $\Box_a\psi \in \Sigma \Rightarrow \psi \in \Sigma'$ (ii) $[\pi]\psi \in \Sigma' \Rightarrow \psi \in \Delta$ and (iii) $[U^*_{\mathcal{B}}\pi]\psi \in \Delta \Leftrightarrow \psi \in \Gamma'$.

From (i) it follows that $w_{\Sigma'} \in w_\Sigma(a)$ and from (ii) that $w_{\Sigma'}[\![\pi]\!]w_\Delta$. Since $[\![U^*_{\mathcal{B}}\pi]\!]$ is a total function, there must be a v' such that $w_\Delta[\![U^*_{\mathcal{B}}\pi]\!]v'$, and since $w_\Sigma[\![U^*_{\mathcal{B}}\pi]\!]v$, it follows that $v' \in v(a)$. Finally, it follows from (iii) that $\Delta R_{U^*_{\mathcal{B}}}\Gamma'$ and hence that $w_{\Gamma'}\mathcal{R}v'$. (See the picture above.)

To show the existence of the sets Σ' and Δ, consider first the following set σ (the notation $\langle\pi\rangle$ stands for $\neg[\pi]\neg$):

$$\{\psi \mid \Box_a\psi \in \Sigma\} \cup \{\langle\pi\rangle[U^*_{\mathcal{B}}\pi]\psi \mid \psi \in \Gamma'\} \qquad (\sigma)$$

We show that this set is consistent. For assume it is not. Then there must be $\phi_1 \ldots \phi_n$ such that $\Box_a \phi_i \in \Sigma$ for $i \leq n$ and $\psi_1 \ldots \psi_m$ such that $\psi_i \in \Gamma'$ for $i \leq m$ for which the following holds:

$\phi_1 \ldots \phi_n, \langle\pi\rangle[U^*_{\mathcal{B}}\pi]\psi_1 \ldots \langle\pi\rangle[U^*_{\mathcal{B}}\pi]\psi_m \vdash \bot$. That means that
$\phi_1 \ldots \phi_n \vdash [\pi]\neg[U^*_{\mathcal{B}}\pi]\psi_1 \vee \ldots \vee [\pi]\neg[U^*_{\mathcal{B}}\pi]\psi_m$, whence by axiom 5
$\phi_1 \ldots \phi_n \vdash [\pi][U^*_{\mathcal{B}}\pi]\neg(\psi_1 \wedge \ldots \wedge \psi_m)$, so, using the necessitation rule,
$\Box_a\phi_1 \ldots \Box_a\phi_n \vdash \Box_a[\pi][U^*_{\mathcal{B}}\pi]\neg(\psi_1 \wedge \ldots \wedge \psi_m)$, which means that
$\Sigma \vdash [U^*_{\mathcal{B}}\pi]\Box_a\neg(\psi_1 \wedge \ldots \wedge \psi_m)$. But then
$\Box_a\neg(\psi_1 \wedge \ldots \wedge \psi_m) \in \Gamma$, and thus
$\neg(\psi_1 \wedge \ldots \wedge \psi_m) \in \Gamma'$, in contradiction with the consistency of Γ' and the assumption that $\psi_i \in \Gamma$ for all $i \leq m$.

So, since the set σ is consistent, it has, by Lindenbaum's lemma, a maximal consistent extension Σ'. Take any such Σ', and consider the set δ:

$$\{\psi \mid [\pi]\psi \in \Sigma'\} \cup \{[U^*_{\mathcal{B}}\pi]\psi \mid \psi \in \Gamma'\} \qquad (\delta)$$

We show that δ is consistent as well, by a similar argument. For if it is not, there must be $\phi_1 \ldots \phi_n$ and $\psi_1 \ldots \psi_m$ such that $[\pi]\phi_i \in \Sigma'$ for $i \leq n$ and $\psi_i \in \Gamma'$ for each $i \leq m$ such that:

$\phi_1 \ldots \phi_n, [U^*_{\mathcal{B}}\pi]\psi_1 \ldots [U^*_{\mathcal{B}}\pi]\psi_m \vdash \bot$, so, using axiom 5,
$\phi_1 \ldots \phi_n \vdash \neg[U^*_{\mathcal{B}}\pi](\psi_1 \wedge \ldots \wedge \psi_m)$, and using necessitation
$[\pi]\phi_1 \ldots [\pi]\phi_n \vdash [\pi]\neg[U^*_{\mathcal{B}}\pi](\psi_1 \wedge \ldots \wedge \psi_m)$,
$[\pi]\phi_1 \ldots [\pi]\phi_n, \langle\pi\rangle[U^*_{\mathcal{B}}\pi](\psi_1 \wedge \ldots \wedge \psi_m) \vdash \bot$.
This contradicts the fact that Σ' is consistent, because, since $\psi_i \in \Gamma'$ for each $i \leq m$, it holds that $\psi_1 \wedge \ldots \wedge \psi_m \in \Gamma'$, and hence that $\langle\pi\rangle[U^*_{\mathcal{B}}\pi](\psi_1 \wedge \ldots \wedge \psi_m) \in \Sigma'$, while $[\pi]\phi_i \in \Sigma'$ for each $i \leq n$.

So, the set δ has a maximal consistent extension Δ, by Lindenbaum's lemma. By definition of Δ, properties (i), (ii) and (iii) hold, which completes the proof.

References

[1] Peter Aczel (1988) *Non-well-founded Sets.* CSLI Lecture Notes 14. Stanford.

[2] Jon Barwise (1989) *On the Model Theory of Common Knowledge.* In *The Situation in Logic.* CSLI Lecture notes 17, pages 201–220. CSLI Publications, Stanford.

[3] Jon Barwise and Lawrence S. Moss (1996) *Vicious Circles.* CSLI Publications, Stanford.

[4] David Beaver (1995) *Presupposition and Assertion in Dynamic Semantics.* Center for Cognitive Science, University of Edinburgh.

[5] Maarten de Rijke (1993) *Extending modal logic.* ILLC Dissertation Series 1993-4.

[6] R. Fagin, J.Y. Halpern, Y. Moses and M. Vardi (1995) *Reasoning about Knowledge.* The MIT Press, Cambridge (Mass.).

[7] Jelle Gerbrandy and Willem Groeneveld (1997) *Reasoning about Information Change.* Journal of Logic, Language, and Information 6:147–169. Also available as an ILLC Report LP–1996-10.

[8] Robert Goldblatt (1987) *Logics of Time and Computation.* CSLI Lecture Notes 7. CSLI Publications, Stanford.

[9] Jeroen Groenendijk, Martin Stokhof and Frank Veltman (1996) *Coreference and Modality.* In Shalom Lappin (ed.), *Handbook of Contemporary Semantic Theory,* pages 179–213. Blackwell, Oxford.

[10] Willem Groeneveld (1995) *Logical Investigations into Dynamic Semantics.* ILLC Dissertation Series 1995-18.

[11] D. Harel (1984) *Dynamic Logic.* In D. M. Gabbay and F. Guenthner (eds.), *Handbook of Philosophical Logic, Vol. 2,* pages 497–604. Reidel, Dordrecht.

[12] Jan Jaspars (1994) *Calculi for Constructive Communication.* ILLC Dissertation Series 1994-4, ITK Dissertation Series 1994-1.

[13] Robert C. Stalnaker (1972) *Pragmatics.* In Harman and Davidson (eds.), *Semantics of Natural Language,* pages 380–397. Reidel, Dordrecht.

[14] Johan van Benthem (1996) *Exploring Logical Dynamics.* Studies in Logic, Language and Information. CSLI Publications, Stanford.

[15] Frank Veltman (1996) *Defaults in Update Semantics.* Journal of Philosophical Logic 25:221–261.

85 - 105

Bare Plurals, Situations and Discourse Context

Sheila Glasbey

Department of Applied Computing
University of Dundee

Acknowledgements: I would like to thank Nicholas Asher, Robin Cooper, Manfred Krifka, Carlota Smith, and the Tense and Aspect Group at the Centre for Cognitive Science, Edinburgh University for their helpful comments and suggestions. I am also grateful to the participants of ITALLC 96, the anonymous referees for that conference and for this volume, and to the audience of a seminar I gave at the University of Texas at Austin in September 1996, for many perceptive questions and comments on this paper and related work.

1 Introduction

There appears to be general agreement among English speakers that it is difficult (or even, for some, impossible) to obtain existential readings for the bare plurals 'plates' and 'children' in (1) and (2).[1]

(1) Plates were dirty.

(2) Children are ill.

The difficulty in obtaining existential readings for these examples goes against the predictions made by the principal accounts of bare plurals in the literature. Carlson [7] predicts that stage level (henceforth s-level) predicates always allow existential readings for their bare plural subjects, and classifies predicates of a "temporary" nature such as **dirty** and **ill** as s-level. In the accounts of Wilkinson [22], Kratzer [17] and Diesing [8] (which are

Logic, Language, and Computation, Vol. II, edited by Lawrence S. Moss, Jonathan Ginzburg, and Maarten de Rijke.

[1]By 'existential' I mean a reading corresponding roughly to "some plates" or "some children". The more one thinks about these examples, the easier it often becomes to get an existential reading. This may have something to do with the effort made to "construct an appropriate context"—which would fit in with my proposal below concerning the need for a contextual situation.

sufficiently similar for us to group them together here and refer to them as WKD), these predicates would similarly be classed as s-level, and would similarly lead us to predict existential readings for the bare plurals.

Few accounts of bare plurals appear to notice this reluctance on the part of many s-level predicates to give existential readings.[2] Interestingly, the restriction seems to apply only to **adjectival** predicates. **Verbal** predicates classified as s-level,[3] on the other hand, appear always to allow existential readings.

Adjectival s-level predicates (henceforth ASLPs) which readily give existential readings for their bare plural subjects appear to be in a minority. They include **available**, **present** and (perhaps marginally) '**drunk**'.

What is even more interesting is that it is possible to make the existential reading much more readily available by adding other material or by setting utterances like (1) and (2) in a particular context. Consider, for example:

(3) The hotel inspector filed a bad report on Fawlty Towers. The standard of service, he reported, was disgraceful. Plates were dirty, cutlery was bent and floors were thick with grease.

(4) We must get a doctor. Children are ill.

Notice how much easier it now becomes to get existential readings for 'plates' and 'children'. What seems to be important is the presence of an appropriate context. If the hearer has such a context "in mind" when she comes to process 'Children were ill or 'Plates were dirty' it becomes much easier to get an existential reading. In (3), the context provided by the utterance is something like "the condition of Fawlty Towers" and in (4) it is something like "the situation that the speaker, the hearer and the children in question are in, at the time of the utterance".

What is going on here? Intuitively, one feels that the presence of these contexts has some kind of "localizing effect"—concentrating our attention on a specific group of plates or children. Yet this does not work for any kind of predicate. If I am telling you about my recent holiday on a Pacific island, and I say:

(5) The climatic conditions were awful. Hurricanes were dangerous there.

then you will be unable to obtain an existential reading for 'hurricanes'.[4]

[2]Exceptions are Kiss [16] and Greenberg [14].

[3]Such classification is often done on the basis of whether the predicate is "temporary" or "permanent" in nature. See below for further discussion.

[4]If you can get any interpretation at all, it will be some kind of "restricted generic". See Glasbey [12] for further discussion of such readings.

So, what I tentatively called the **localizing effect** does not seem to work for i-level predicates, but only s-level ones.[5] It is as though ASLPs are **potential** providers of existential readings, but certain conditions involving some kind of "focusing" on a certain location are required to realize such readings.

It is not immediately obvious how to model this localizing effect in a semantics for bare plurals. Let us summarize at this point exactly what is required. We need an account of bare plurals that allows us to explain why existential readings for the subjects of ASLPs are not readily available. Our account will also need to explain why, for a minority of adjectival s-level predicates (like **available** and **present**), existential readings *are* readily available. Furthermore, we need to explain why the presence of certain kinds of discourse contexts make the existential reading much more readily available for many ASLPs. And we need to explain the observation that **verbal** s-level predicates (VSLPs) appear always to allow existential readings, irrespective of context.

I will shortly describe an analysis that makes some progress towards achieving these objectives. First, I will present a a brief summary of some earlier accounts of bare plurals. Next I will briefly introduce an account of bare plurals based on situation theory and channel theory (this is described more fully in Glasbey [12]). I will then show how this account can be used to explain the observations discussed above. In particular, I will show how we can distinguish between cases where the context "supplies a situation" and cases where it does not, which in turn allows us to predict whether or not an existential reading can be obtained.

2 Background

Carlson's (1977) account of bare plurals explains why 'firemen' in:

(6) Firemen are altruistic.

and in:

(7) Firemen hate false alarms.

gets only a **generic** interpretation[6] while 'firemen' in:

(8) Firemen are available.

[5] A much wider range of examples is obviously required to establish this. I have studied a wide range and satisfied myself that this is the case. Further examples are given in Glasbey [13].

[6] Corresponding roughly to "all or most firemen".

gets only an existential interpretation. He does this by distinguishing between two types of predicate, individual level (i-level) and stage level (s-level). The former are predicates of **individuals** and the latter are predicates of **stages** of individuals, where an individual may be an object like **john** or a kind like **firemen**. A stage is a spatiotemporal realization or "slice" of a kind, so that a stage of the kind **firemen**, for example, corresponds to a number of firemen at a particular time and place. An i-level predicate (ILP) gives a "kind" reading for the subject, which corresponds to a generic reading, as we get in (6) and (7). An s-level predicate (SLP) gives a "stage" reading for the subject, and this corresponds to an existential reading as we get in (8).[7]

Thus Carlson [7], as I mentioned in the introduction, predicts that existential readings will always be available for bare plural subjects of SLPs, no matter whether the predicate is verbal or adjectival.

In effect, Carlson's account predicts that any sentence with an ILP (either one that is "basically" i-level one that has been "raised" from s-level to i-level by the generic operator) will give a generic reading for its bare plural subject. The fact that the predicate is i-level forces the subject to be interpreted generically. However, there are exceptions to this, such as:

(9) Hurricanes arise in the South Pacific.

where, although the predicate is i-level (the statement is a generic one so **arise** has been raised to i-level), 'hurricanes' can receive an existential interpretation. Examples such as this one helped to motivate the WKD account. These authors employ a similar i-level/s-level distinction among predicates, but reject Carlson's ontology of kinds and stages. In the WKD account, bare plurals, like other indefinites, introduce variables over individuals (in the ordinary, not the Carlsonian sense). WKD offer a quantificational account of generic sentences involving indefinites. The generic operator, called GEN, is viewed as an unselective quantifier in the manner that Lewis [19] treats adverbs like 'usually'. GEN can thus bind any variables in its scope. Any variables that are not bound by GEN are bound by existential closure, in the manner of Heim [15]. GEN relates two elements of the sentence, the

[7] Carlson also explains how VSLPs give both generic and existential readings for their bare plural subjects. This is because a VSLP may be "raised" to i-level by the generic operator. No such raising is available, according to him, for ASLPs. Thus no generic reading is predicted for (8). This is arguably unsatisfactory, as such a reading does seem to be available, at least for some ASLPs. The WKD account (see below), on the other hand, predicts that the generic reading is always available for bare plural subjects of SLPs, both the verbal and the adjectival ones. This seems wrong, too—consider (1) and (2), for example, where it is very difficult to get generic readings. I will not address the issue of generic readings for ASLPs any further here as it is not central to our concerns.

restrictor (or restrictive clause) and the **matrix** (or nuclear scope). Variables in the restrictor are bound by GEN, and variables that appear only in the matrix are bound by existential closure. Thus the logical structure is: GEN [Res] [Matrix].

WKD then give a syntactic explanation for the readings obtainable for bare plurals. They propose that VP-internal material is mapped into the matrix in the logical representation, while VP-external material goes into the restrictor. The subjects of s-level predicates originate within the VP and move to the specifier of IP at S-structure (for English). Because of this, s-level subjects can map into either the restrictor or the matrix in the logical representation. The subjects of i-level predicates, on the other hand, are base-generated in [SPEC, IP], and remain there at S-structure and LF. I-level subjects must therefore map into the restrictor. As noted above, variables in the restrictor are bound by GEN and thus get a generic reading, while those in the matrix are existentially quantified. Because **arise** is s-level, both generic and existential interpretations are available for 'hurricanes'. Compare:

(10) Hurricanes are dangerous in the South Pacific.

The reason that only a generic reading is available for the subject here is that **dangerous** is i-level, and therefore the subject variable must be mapped into the restrictor, where it receives a generic interpretation.[8]

Thus WKD predicts, like Carlson, that **all** SLPs allow existential readings for their bare plural subjects. As in Carlson's account, no distinction is made between verbal and adjectival predicates in this respect. WKD is therefore unable to explain the data on ASLPs with which we began.

The incorrect predictions made by Carlson and WKD here are noted by Kiss [16] and Greenberg [14]. Kiss observes that while these accounts make the correct predictions for a small set of ASLPs (e.g. **available** and **present**, which appear always to allow existential readings), they falsely predict that existential readings are generally available for the bare plural subjects of ASLPs. Kiss attempts to explain this by arguing that the existential interpretation of a bare plural is allowed in the presence of predicates which express **existence**, and in the presence of VSLPs (which, she maintains, indirectly express existence of the arguments via existential quantification over a Davidsonian event argument). The i-level/s-level distinction is thus not relevant, on her account, to the question of bare plural readings. Kiss can therefore explain why:

---

[8]However, the WKD account is unable to account in a straightforward way for the readings available for bare plural **objects**. I argue in Glasbey [12] that the scrambling mechanism they propose is inadequate to explain the data, and show how the situation theoretic account is able to do this. I will not discuss this matter further here.

(11) Firemen are present.

gives an existential reading—the predicate **present** contains an implicit existential quantifier. In contrast, the predicate **dirty** contains no such quantifier, thus explaining why we can't get an existential reading in (1). In order to explain why we can get an existential reading in

(12) Firemen are drunk.

Kiss has to postulate that **drunk** is a "circumstantial predicate" (after Emonds [9]). Such predicates do not function as primary predicates but behave rather like adverbs modifying the primary predicate **be** (and thus contain an implicit existential quantifier). Space does not allow further discussion here, but see Glasbey [13] for data that shows that there are problems with this explanation. Furthermore, Kiss does not appear to notice the dependency on context, and it is not easy to see how her account could be modified to explain this.

Greenberg notices that many ASLPs do not give the existential reading, but that things are much improved by the addition of a locative modifier.[9] For example, in:

(13) Professors are busy.

we don't readily get an existential interpretation for 'professors'. However, adding 'in this room' makes all the difference:

(14) Professors are busy in this room.

Notice that it is not necessary to add an explicit locative in order to get this effect. The same effect can be achieved by having an implicit locative in the context. In fact, I will argue below that it is not so much the presence of a "place" in the discourse context as that of a "situation" which is responsible for this effect.

We should consider the possibility that the source of Carlson's and WKD's problems lies in precisely how we draw the distinction between s-level and i-level predicates. In WKD, SLPs are clearly identified as "temporary" in nature, and ILPs as "permanent". Some flexibility is allowed in whether a particular predicate is i-level or s-level in a particular context. For example, Kratzer [17] points out that **have brown hair** is normally seen as a permanent property (and thus is i-level), but it is possible to set up contexts where it becomes a temporary property. We can then say things like:

[9]She observes, too, a similar effect in the corresponding Hebrew examples. Furthermore, Bosveld-de Smet [6] notices something very similar in French.

(15) Mary looks much prettier when she has brown hair than when she has bleached it.

McNally [21] points out that such contextual "coercion" of a predicate from i-level to s-level does not, in fact, change the readings that are available for bare plurals. She discusses an 'Alice in Wonderland' scenario where Alice and her animal and plant companions are able to change their heights at will by drinking a magic potion. In these circumstances we can say things like:

(16) When she is tall, Alice can't get through the little wooden door.

However, the fact that we have made **tall** an SLP (in WKD terms) here does *not* mean that we can get existential readings for the subject in sentences like:

(17) Daffodils were tall in Alice's garden yesterday.

even if (16) and (17) are part of the same narrative.[10] McNally proposes that the relevant distinction is between predicates which are spatially located and those which are not.[11] She argues that the former allow existential readings for bare plural subjects, while the latter do not. This enables her to account for the example above. Coercing **tall** into a temporary predicate makes no difference to its spatial locatedness (it is still **location independent**), and location independent predicates do not (for reasons explained in McNally [21]) give existential readings.

It is not clear, however, how McNally's account explains the fact that we can get an existential reading for the subject in:

(19) Firemen were drunk.

Drunk appears to be (spatially) location independent, and trying to argue that it is not so would result in a very unintuitive characterization of this property. It is hard to imagine that if I am drunk in my office, for example,

[10] A few speakers have told me that they can get an existential reading for 'daffodils' here. Others can get an existential reading for 'lilies' in the example:

(18) Mary is a suberb gardener. Lilies are eight feet tall in her garden!

There are other examples, too, which space does not permit me to discuss here, of apparently i-level predicates which allow existential readings in certain contexts. Clearly, further investigation is required here, and I am currently working on this and hope to report preliminary results soon. The analysis I am developing will introduce minor modifications to the account presented here, but the essential analysis will remain substantially the same.

[11] Actually what she proposes is that it is states of affairs rather than predicates in themselves which are (spatially) location-dependent or otherwise. I will gloss over this here as it does not affect the point I are making.

then there is any possibility of ceasing to be drunk by moving out of my office. Similar arguments apply to **angry** and **ill**.

I believe, however, that McNally is right in arguing that the s-level/i-level distinction (or whatever distinction is important for bare plural readings) should not be identified (at least not fully) with the temporary/permanent distinction among predicates. I will indicate below how a distinction among kinds of predicates can be expressed in situation theory which is distinct from the temporary/permanent one, and how this allows us to explain the availability of existential v. generic readings for bare plurals.

3 Bare Plurals: a Situation Theoretic Account

I will give only a brief account of this here. The reader should consult Glasbey [12] and [13] for more detail.

I base my account on the distinction between two kinds of predicate—**relations** and **types**—that can be made in situation theory. I will use the Extended Kamp Notation (EKN) of Barwise and Cooper [3], and the version of situation theory defined in that paper. Propositions in EKN include objects of the form:

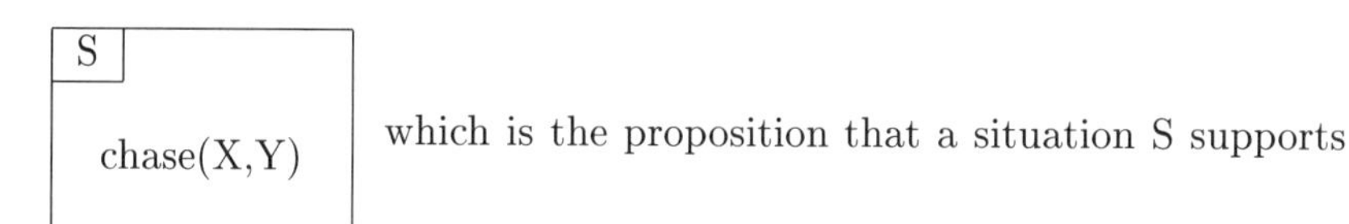

which is the proposition that a situation S supports the infon chase(X,Y). A proposition like this one is called an **Austinian** proposition. A second kind of situation theoretic proposition is exemplified by:

X
dangerous

which is the proposition that an individual X is of the type **dangerous**. No situation is involved here. I call this a **Russellian** proposition.[12]

[12]My use of the term 'Russellian' is slightly different from that of Barwise [2], who uses it to refer to a proposition that a situation supports an infon, where the situation is existentially quantified.

I propose that a certain class of predicates, roughly corresponding to what are called i-level predicates in the literature, correspond to ST **types** (and thus can form only Russellian propositions), while a second class of predicates (roughly corresponding to s-level ones) correspond to ST **relations** (and can form either Austinian propositions or Russellian ones).[13]

In addition, I take bare plurals to denote situation theoretic **types**.[14]

The non-generic reading (i.e., the "single event" reading) of:

(20) Dogs barked.

is represented as:[15]

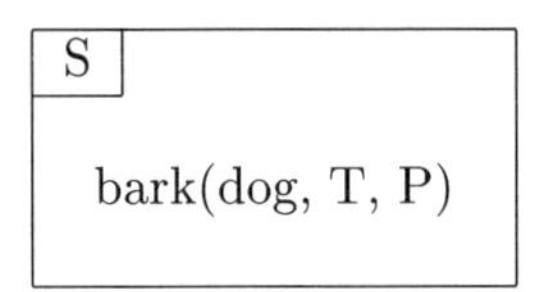

where **dog** is the type of dog objects.[16]

Thus the denotation of bare plurals differs from that of, say, proper names or pronouns, which are normally taken in situation theory to denote individuals.

Now compare:

(21) Hurricanes are dangerous.

which is taken to denote the truth of the proposition:

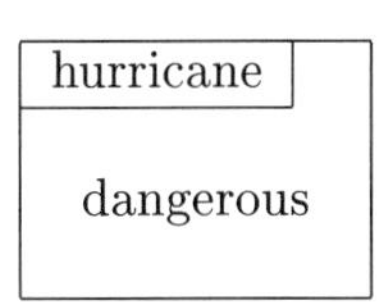

where **hurricane** is the type of hurricane objects. Further information can be derived from such a proposition. I show in Glasbey [12] how we can

[13]This is based on a proposal by Robin Cooper (p.c.), which was in turn inspired by Ladusaw [18].

[14]See McNally [20] for an independent but closely-related proposal that bare plurals in Spanish should be taken to refer to properties.

[15]I ignore tense here, for simplicity. I assume that relations have time (T) and place (P) argument roles (which need not necessarily be filled in the syntax). See Glasbey [10], ch.3 for discussion. I also allow (some) types to have T arguments. This enables me to deal with predicates like **tall** which, as we saw earlier, may in certain circumstances be seen as temporary. I can thus separate out the temporary/permanent nature of a predicate from whether it is a relation or a type, which allows me to address the problems raised by McNally.

[16]Thus a relation like **bark** or **chase**, for example, can take either a parameter corresponding to an individual, or a type, as an argument.

use Channel Theory (Barwise and Seligman [5]) to express the information we can derive about the relation between being a hurricane and being dangerous.

Here, I concentrate on existential readings. Consider again the representation for the event reading of (20).

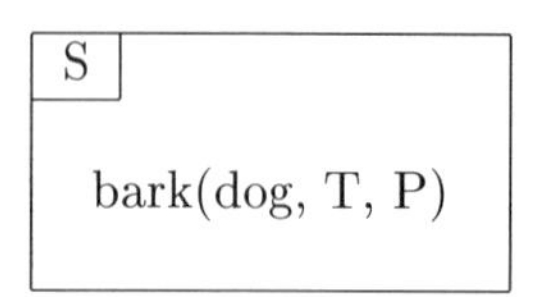

I propose that if a bare plural type is an argument of a relation, forming an infon which is supported by a situation, then an inference can be made to the effect that there are (at least two) instantiations of the bare plural type. Thus the truth of the above proposition allows us to infer that:

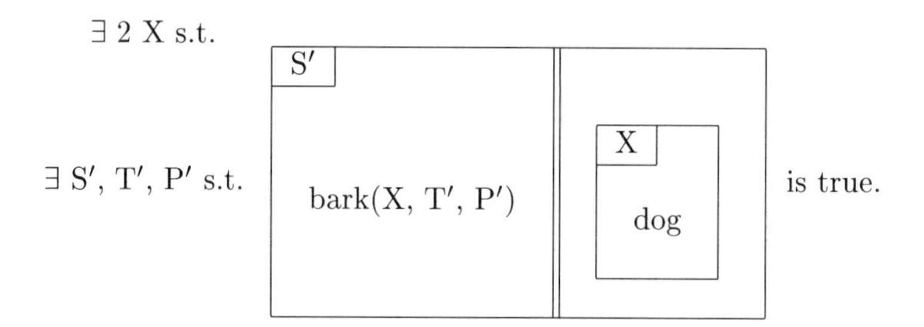

where we require at least two instances of X. That is, we infer that some (at least two) dogs barked.[17] The information about particular instances of

[17] It appears that S and S′ may be different situations, though it seems clear that they must be, in some sense, closely related, as must T and T′, and P and P′. Further work is needed to determine exactly how they are related. Note that S′, T′ and P′ for the different Xs need not necessarily be distinct. There are interesting issues here regarding the distinction between distributive and collective predicates, which need to be explored further. Here, I make the simplifying assumption that the predicate is distributive. Robin Cooper (p.c.) asks whether S and S′ *can* be distinct. Certainly, 'Dogs barked' sounds intuitively as though it describes one situation only. But compare 'I wrote papers last year', which sounds intuitively as though it refers to a number of situations, or can do. This isn't just a feature of object position—notice that 'People climbed Ben Nevis last year' can similarly refer to a number of situations. I am, of course, using some intuitive notion of a situation corresponding to an "event" in order to make these judgments. We often (though not always) have strong intuitions about whether one or more event(s) are involved—presumably using criteria like temporal connectedness to decide. But, of course, in situation theory it is always possible to combine two situations to make a third, raising the question of what implications this might have for temporal connectedness. Perhaps we need to place constraints on situations that are events—such as requiring temporal connectedness. Further work is clearly required.

the type **dog** and their participation in the relevant proposition is thus not information that is **directly described** by (20), but is, rather, information that is obtained **indirectly** by a process of inference. The fact that we infer the relation to hold of "some" rather than "all/most" dogs is a consequence of the fact that **bark** is a relation—the "situatedness" of the information is what leads to what I call the '**existential inference**'.[18]

Thus my account predicts that existential readings will be available for bare plurals in a "situated" context—i.e., where the bare plural is an argument of a predicate that is a **relation** (and can thus participate in an Austinian proposition) as opposed to a **type**.

No existential reading is available for arguments of predicates that are **types** (such as **dangerous**, or the verbal **like**). The reason is that such predicates can only form Russellian propositions, of the form:

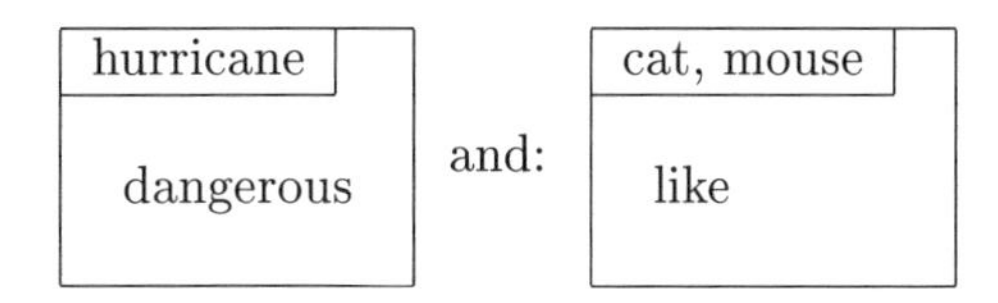

Here, because there is no situation, we cannot make the existential inference, and therefore we cannot get an existential interpretation for hurricanes.

It is useful at this point to consider whether we can find independent evidence to support the division of predicates into relations and types. One such test involves the possible use of predicates in perception complements. Situation semantics has for a long time used the data on **perception complements** as part of the motivation for situations in natural language semantics. Barwise [1] showed that unfortunate consequences arise if we use a proposition as the denotation of, for example 'Mary run' in a naked infinitive perception sentence like:

(22) John saw Mary run.

Barwise proposed that, instead, the perception complement is used to introduce a situation, and this analysis has since become standard in situation semantics. An NI-perception complement must introduce a situation.[19]

[18]My proposal is clearly related to Carlson's (1977) idea that an SLP selects the "stage reading" of a kind—myy notion of the existential inference being quite similar to the idea that **bark** selects a stage of the kind **dogs**. Perhaps my account of bare plurals can be viewed in a certain light as a reworking of Carlson's, within an situation theoretic framework, although of course I differ from him in not using stages.

[19]The converse is not required—i.e., it is not necessary that if something is a situation it can be perceived. See Barwise and Perry [4].

This means that if we can use a predicate to form such a perception complement, then that predicate must correspond to a relation. Thus, the fact that we cannot say:

(23) *John saw Mary love Peter.

whereas:

(24) John saw Mary chase Peter/ Mary drunk.

sounds fine, concurs with the proposal above (motivated by the data on bare plurals) that **love** is a type and **chase** and **drunk** are relations.[20]

I show in Glasbey [12] how we can infer from the truth of the first of these propositions that there exists a channel supporting a constraint holding between the type **hurricane** and the type **dangerous**. I represent the proposition that a channel C supports the above constraint as follows:

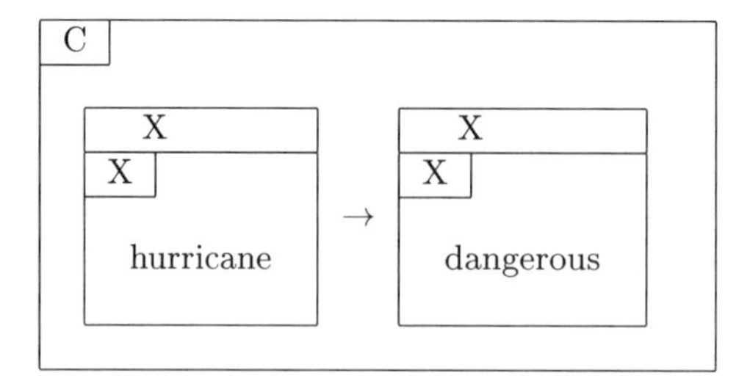

Compare the "event description":

(25) Hurricanes arose in the South Pacific last night.

which we can represent as:[21]

[20]One of the anonymous referees points out that there may be problems with this test, in that sentences like (23) may become acceptable given a suitably epistemically/perceptually equipped perceiver—e.g., 'God saw Mary love Peter'—meaning, presumably, that God saw the whole of Mary's loving of Peter, from beginning to end. I am not convinced that this example is acceptable, although I find the example given by the referee (attributed to S. Neale) 'God saw Bill own three houses' marginally better. But there seems to be evidence for classifying **own** as a relation, anyway, as it allows existential readings for bare plurals in some contexts. Further discussion is not possible here, but this is an interesting issue which would merit further investigation.

[21]B, B′ are "background" situations supporting the relevant information.

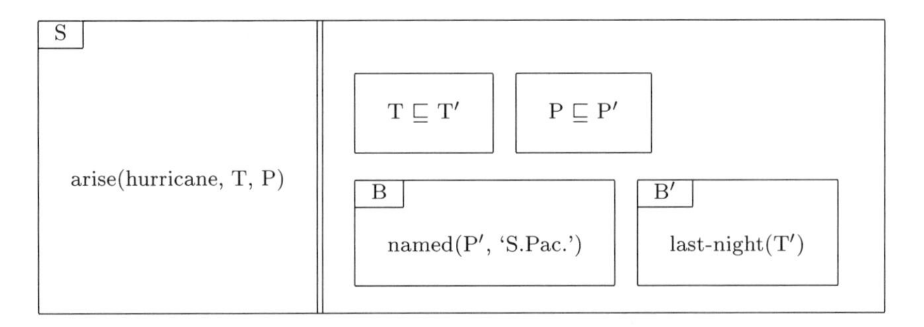

As I explained above, the fact that the type **hurricane** appears as a constituent of an infon, supported by a situation, licenses the existential inference, and thus we get an existential reading.

Of course, some sentences allow both generic and existential interpretations of their bare plural subjects. For example:

(26) Dinosaurs howled.

can either express a constraint between being a dinosaur and having the property of howling, or it can describe what happened on a particular occasion, when an event of some dinosaurs howling occurred. My account allows both these readings, as shown in Glasbey [12]. It also predicts exactly the obtainable readings for more complex generic sentences such as:

(27) Hurricanes arise in the South Pacific.

where both generic and existential readings are available for ‘hurricanes’—yet in each case the utterance is still being used to express a generalization, not to describe a particular event, and:

(28) Hurricanes are dangerous in the South Pacific.

which gives only the generic reading for ‘hurricanes’. It also predicts the observed readings for bare plurals in object position, such as ‘mice’ in:

(29) Cats love mice.

which gives a generic reading only for ‘mice’, and in:

(30) Cats chase mice.

which gives both generic and existential readings for ‘mice’.[22]

[22]The ability to predict correctly the readings for object bare plurals is, I claim, a significant advantage of my account over those of Carlson, WKD and others. See Glasbey [12].

But so far I have not explained the phenomenon with which we began this paper—the lack of existential readings for many ASLPs (or what we can now call **adjectival relations**), and the dependency of these readings on discourse context. My account up to this point appears to predict that existential readings will be available for **all** relations, both verbal and adjectival (just as do the accounts of Carlson and WKD). Let us now turn to this matter.

4 Adjectival Predicates and Situations

As we have seen, some adjectival predicates (such as **available** and **present** appear to give existential readings very readily to their bare plural subjects. Other adjectival predicates (such as **hungry**, **ill**, **dirty** and many others) appear to need much more contextual support in order to give existential readings.

Verbal predicates, on the other hand, appear to divide neatly into those which give existential readings and those that do not. The former seem to correspond closely to verbs expressing temporary states of affairs—(e.g. **run**, **eat**, **chase**, ...)—those that have traditionally been called s-level. The latter seem to correspond to verbs expressing relatively permanent states of affairs—(e.g. **like**, **love**, **hate**, **know**, **believe**, ...)—those that have been called i-level. Context appears to have relatively little effect here.[23]

We can start, then, by saying that "temporary" verbal predicates are **relations** and "permanent" verbal predicates are **types**. Then the account given above can explain why the former give existential readings to subject bare plurals, and the latter do not. But what shall we say about adjectival predicates? The first thing we can do is to classify those adjectival predicates which never allow existential readings (in any context we can manage to create), as types. This leaves us with the rest—the context-dependent ones. Are we to classify them as relations or as types?

We are aided in this decision by the independent criterion of perception complements, mentioned above. Notice that we can say:

(31) John saw Mary happy.

and:

(32) John saw Mary sick.

and:

[23] There are some exceptions, however. See footnote 10.

(33) John saw the kitchen dirty.

which tells us, if we accept the situation semantics analysis of perception complements, that we have situations here as the objects of perception. And, if this is the case, **happy**, **sick** and **dirty** must be relations. That is, we have evidence for classifying these "uncertain" predicates (which sometimes give existential readings and sometimes don't, depending on context) as relations rather than types.

Now, if such adjectives denote relations, why is the existential reading not freely available? Recall that the existential inference, which is what gives us the existential reading, is licensed by the presence of a situation supporting the relevant infon. Now the fact that we have chosen to regard **sick** as a relation means that we can always form an infon such as sick(child) to correspond to 'Children are sick'. Now what if only certain contexts can supply a situation to support the above infon? If there is no situation, then we can't form an Austinian proposition, and therefore we can't make the existential inference. Perhaps we can say that in an empty context (such as we have for (1) and (2)), there is simply no appropriate situation on hand to support the infon. Adding descriptive, scene-setting, "localizing" material as in (3) and (4) has the effect of adding such a situation to the context, thus allowing us to make the existential inference and get an existential reading. Thus what we called the localizing effect in Section 1 can now be seen as a "situating effect" or "adding an appropriate situation to the discourse context".[24]

The above analysis raises the question of why, in the absence of an appropriate situation in the discourse context, we do not simply existentially quantify over the "missing" situation and thereby make the existential reading possible. This happens, of course, in other cases where contexts do not anchor referential parameters in an obvious way. For example, if someone says 'John phoned" and I don't know who 'John' refers to, I will at least get the content "Someone called John phoned".

Our response to this puzzle is to suggest that we need to look more closely at the conditions under which existential quantification over non-anchored parameters is allowed to take place. Clearly it takes place in the example above, where there is no anchor for the parameter corresponding to 'John'. But perhaps it is not always allowed—indeed, one possibility is that such existential quantifying away of parameters is only permitted in what we have called "situated contexts". This might allow us to explain why

[24]Note that the fact that (31–33) are acceptable requires us to say that use of an NI perception complement "forces" there to be an appropriate situation in the discourse context in a way that the use of a bare plural does not. This is the case if these NI perception complements are always acceptable with adjectival relations. Whether or not they are requires further investigation.

such existential quantification over situations is not possible in the cases discussed above. Clearly, a detailed investigation of this matter is required, which we are unable to carry out at this point but hope to undertake in the near future.

Returning to the main thread, we now have to explain why verbal relations *always* allow existential readings. On the account we have just given, this would mean that a verbal relation always introduces a situation into the context—we get the situation "for free" along with the verb, as it were. Why should this be?

It seems intuitively reasonable to identify verbal relations with "events"[25] in a way that we do not necessarily want to identify adjectival relations with events. It is natural to think of utterances of 'John ran' and 'Dogs barked', for example, as introducing some kind of "action" or event corresponding to the act of running, the act of barking or whatever. It does not seem so natural to think of an utterance of 'Fido was present' or 'Fido was hungry' as introducing an event, but rather as adding more information to a situation that is already contextually present (see Glasbey [11] for a proposal that many stative sentences have a "backgrounding" role in that they add further information to situations that are already "present" in the discourse). Now if we think of an event as a (certain kind of) situation, then we can see why it might be that verbal relations always supply a situation to the discourse context. The situation supplied is the one corresponding to the event introduced by the verbal relation. In situation theoretic terms, we can identify this event with the "minimal situation" supporting the relevant infon—the situation supporting this infon and no other facts.[26] For a detailed discussion of events, minimal situations and related notions, see Glasbey [10], chapters 3 and 5.

The proposal that verbal relations necessarily "provide situations" in a way that adjectival ones do not will be developed further below.

Because adjectival relations do not correspond to events in this way, they do not (in general) introduce situations into the discourse context. A situation which can support the infon must already be present in the context—which explains the heavy context-dependency of existential readings for bare plural subjects of most adjectival relations.

Now why is it that some adjectival relations—the "exceptional" ones such as **available** and **present**—do not appear to require contextual support in order to give existential readings, but give existential readings freely in simple sentences such as:

(34) Firemen were present/available.

[25] On an intuitive notion of "event".

[26] This is not quite correct. There may be other facts that we must necessarily infer. See Glasbey [10] for discussion.

I suggest that this is because such predicates are very naturally seen as predicates of situations (in a sense to be explained below), while predicates like **happy** and **dirty** are not. It seems intuitive to think of "a situation in which firemen are available" for example, much more readily than we can think of "a situation in which firemen are happy". We might make this idea more precise by proposing that **available** and **present** both have a "hidden argument" for a situation. Thus we might represent:

(35) Fido was present.

as:[27]

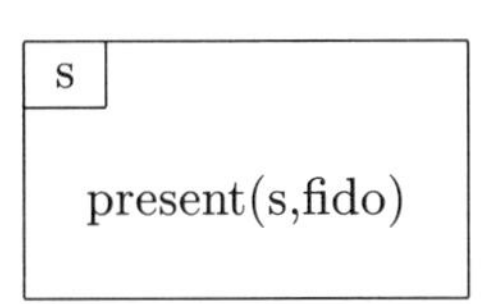

where s is a situation supplied by the relation **present**.

By contrast, we can then say that a predicate like **hungry**, for example, has no such situation argument—the predicate does not "supply its own situation role". We represent:

(36) Fido was hungry.

as:

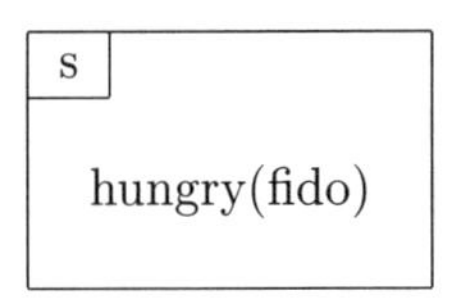

The only way that we can get an existential reading in this case is if the context supplies a situation to support the infon. We saw earlier how some contexts make it much easier to get an existential reading for, say:

(37) Children were hungry/ill.

We need to consider whether the situation that is the first argument of the relation is necessarily the same situation that supports the infon. Perhaps this is more clearly seen in the case of verbal relations, as in:

(38) Fido chased Kitty.

[27]We drop T and P arguments henceforth, in the interests of simplicity.

I suggest that a verbal relation may introduce a situation corresponding to the "minimal event"—the "smallest' situation that supports that infon (see Glasbey [10]). Then we might have:

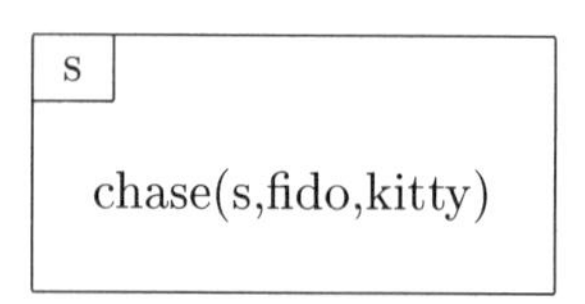

where s is the minimal situation for the infon. But it might also be true that some larger situation s′ (of which s is part) supports that infon, i.e:

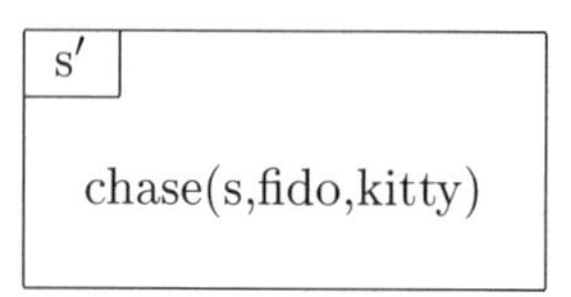

Hence it seems to make sense to think of the situation argument of a verbal relation as corresponding to the minimal situation that supports the infon in question.

Summing up, we appear to have three types of relations:

1. Verbal relations like **bark** and **chase**, which have an event/situation argument, and thus always allow existential readings, irrespective of context.

2. Adjectival relations like **hungry**, which do not have a situation argument, and which only give existential readings if the context supplies a suitable situation to support the infon in question.

3. Adjectival relations like **available** and **present**, which have a situation argument and thus "supply their own situation", with the result that they always allow existential readings, irrespective of context.

A further question to be addressed is whether there are other adjectival predicates that behave like **available** and **present**. Locatives such as **lie along the river bank** readily give existential readings—for example:

(39) Holiday cottages lie along the river bank.

It appears intuitively reasonable here to regard the locative PP as supplying the situation argument. It is interesting that (39) describes a relatively permanent state of affairs—demonstrating once again the irrelavance of the temporary/permanent distinction to the issues under consideration. Other predicates that appear to allow existential readings irrespective of context are **early** and **late**.[28] Consider, for example:

(40) Children were early at the party.

(41) Players were late for the match.

The next obvious question is what do these predicates have in common. Intuitively, something to do with "presence" or "existence" appears to be involved. Whether it is possible to generalize any further about which adjectival predicates supply their own situation role is a matter for further research.

5 Conclusion

In this paper I have drawn attention to the context-dependency of existential readings for bare plural subjects of s-level ("relational") predicates, and proposed an account to explain this. I proposed that the predicates in question correspond to situation-theoretic relations, which combine with suitable arguments to form infons, which, when supported by situations, form Austinian propositions. If the bare plural corresponds to an argument of such an infon, forming part of an Austinian proposition, then the existential reading is licensed, by virtue of what I call the 'existential inference'. Those relations that supply their own situation role (verbal relations and a few adjectival ones like **present** and **available**) allow an existential reading irrespective of context. Those relations which do not supply their own situation role (most of the adjectival predicates) require contextual support, in the form of a situation provided by the context, in order to license the existential inference and the corresponding existential reading.

Further work is required on writing a grammar that will allow the available readings to be derived compositionally from the syntactic form of the sentence together with some representation of the context. Further investigation is also required into some predicates that the present account treats as types, but which occasionally allow existential readings. Such investigation, as noted in the text, is currently underway by the author.

References

[1] Barwise, J. (1981). Scenes and other situations. *Journal of Philosophy*, 78(7):369–397.

[28] Thanks to Carlota Smith for pointing out these examples.

[2] Barwise, J. (1989). *The Situation in Logic.* Number 17 in CSLI Lecture Notes. Center for the Study of Language and Information, Stanford, Ca.

[3] Barwise, J. and Cooper, R. (1993). Extended Kamp Notation: A graphical notation for situation theory. In Aczel, P., Israel, D., Katagiri, Y., and Peters, S., editors, *Situation Theory and its Applications*, volume 3 of *CSLI Lecture Notes*, chapter 2, pages 29–53. Center for the Study of Language and Information, Stanford, Ca.

[4] Barwise, J. and Perry, J. (1983). *Situations and Attitudes.* MIT Press, Cambridge, Mass.

[5] Barwise, J. and Seligman, J. (1994). The rights and wrongs of natural regularity. In Tomberlin, J., editor, *Philosophical Perspectives*, volume 8, pages 331–365. Ridgeview, California.

[6] Bosveld-de Smet, L. (1993). Indefinite subjects in French and stage-level versus individual-level predicates. In de Boer, A., de Jong, J., and Landeweerd, R., editors, *Language and Cognition*, volume 3, pages 29–38. University of Groningen.

[7] Carlson, G. (1977). *Reference to Kinds in English.* PhD thesis, University of Massachusetts, Amherst.

[8] Diesing, M. (1988). Bare plural subjects and the stage/individual contrast. In Krifka, M., editor, *Genericity in Natural Language: Proceedings of the 1988 Tübingen Conference*, pages 107–154. Seminar für Natürlich-Sprachliche Systeme, University of Tübingen, Germany. Report SNS-Bericht 88-42.

[9] Emonds, J. (1976). *A Transformational Approach to English Syntax: Root, Structure Preserving, and Local Transformations.* Academic Press, New York.

[10] Glasbey, S. (1994a). *Event Structure in Natural Language Discourse.* PhD thesis, University of Edinburgh.

[11] Glasbey, S. (1994b). Progressives, events and states. In Dekker, P. and Stokhof, M., editors, *Proceedings of the Ninth Amsterdam Colloquium.* ILLC/Department of Philosophy, University of Amsterdam.

[12] Glasbey, S. (1995a). A situation theoretic interpretation of bare plurals. In J. Ginzburg et al (eds.), *The Tbilisi Symposium on Logic, Language and Computation: Selected Papers* Studies in Logic, Language and Information, 1998, 35–54.

[13] Glasbey, S. (1995b). A situation theoretic interpretation of bare plurals (longer version). Unpublished manuscript, Centre for Cognitive Science, University of Edinburgh.

[14] Greenberg, Y. (1994). Hebrew nominal sentences and the stage/individual level distinction. Master's thesis, Bar Ilan University, Israel.

[15] Heim, I. (1982). *The Semantics of Definite and Indefinite Noun Phrases.* PhD thesis, University of Massachusetts, Amherst, Mass.

[16] Kiss, K. É. (1994). Generic and existential bare plurals and the classification of predicates. In *Working Papers in the Theory of Grammar*, volume 1 of *Theoretical Linguistics Programme.* Budapest University (ELTE), Budapest.

[17] Kratzer, A. (1988). Stage-level and individual-level predicates. In Krifka, M., editor, *Genericity in Natural Language: Proceedings of the 1988 Tübingen Conference*, pages 247–284. Seminar für Natürlich-Sprachliche Systeme, University of Tübingen, Germany. Report SNS-Bericht 88-42.

[18] Ladusaw, W. (1994). Thetic and categorical, stage and individual, weak and strong. In Harvey, M. and Santelmann, L., editors, *Proceedings of SALT IV*, pages 220–229, Ithaca, New York. Cornell University DMLL.

[19] Lewis, D. (1975). Adverbs of quantification. In Keenan, E. L., editor, *Formal Semantics of Natural Language: Papers from a Colloquium Sponsored by King's College Research Centre, Cambridge*, pages 3–15. Cambridge University Press, Cambridge.

[20] McNally, L. (1995a). Bare plurals in Spanish are interpreted as properties. Unpublished ms, Universitat Pompeu Fabra, Spain.

[21] McNally, L. (1995b). Stativity and theticity. Unpublished ms., Centre for Cognitive Science, Ohio State University.

[22] Wilkinson, K. J. (1991). *Studies in the Semantics of Generic Noun Phrases.* PhD thesis, University of Massachusetts, Amherst.

Interleaved Contractions

Wiebe van der Hoek[1] and Maarten de Rijke[2]

[1] Dept. of Computer Science, Utrecht University
P.O. Box 80089, 3508 TB Utrecht, the Netherlands
wiebe@cs.ruu.nl
[2] ILLC, University of Amsterdam
Plant. Muidergracht 24, 1018 TV Amsterdam, the Netherlands
mdr@wins.uva.nl

Abstract We study an approach to concurrent contractions, that is, to simultaneous contractions performed by multiple agents. Using ideas from the semantics of programming we adopt an interleaved approach to reason about concurrent contractions. Although many of the notions from the traditional Gärdenfors approach transfer to this setting, our approach also forces us to depart from the Gärdenfors framework in important ways. We present laws describing rational concurrent contractions, as well as a construction that satisfies these laws.

1 Introduction

In real life concurrent accessing of data is the rule. Multiple agents are working on the same theory, and multiple copies of some data are kept in different locations. Typical examples include scientific research or writing a joint-publication, and practical applications vary from networks of personal computers and workstations sharing some common information to widely distributed applications such as automatic teller machines. The primary advantage of concurrent theory change as opposed to single agent theory change is the ability to share, access and engineer data in an efficient manner. The primary disadvantage is the added complexity required to ensure proper coordination between the agents taking part.

In a multi-agent setting, managing a belief set is a *concurrent* task: not only may several agents *retrieve* information from one and the same source, but it may also be the case that multiple agents have permission to *alter* a

---

Logic, Language, and Computation, Vol. II, edited by Lawrence S. Moss, Jonathan Ginzburg, and Maarten de Rijke.

database (the flight booking procedures are a typical example here). What are sensible strategies for conflict resolution in case inconsistency strikes? The task of maintaining consistency in the setting of multi-agent theory change is more complex than in the single agent case, if only because of the many possibilities that become visible.

This paper is a first report in our study of on concurrent theory change. Its purpose is to demonstrate that concurrent theory change forms an interesting extension of the traditional Gärdenfors style approach towards theory change, one that has many faces and that calls for new tools. Here we will confine ourselves to the simplest case in which a number of agents have access to shared data. The data are changed via contractions, which may in principle be proposed by any one of the agents. We will explore some of the options and problems that present themselves. A central question of this paper is: assuming that multiple agents, each guided by a familiar set of rationality postulates, propose or perform (single agent) contractions for a shared theory, — what are the laws governing the global contractions?

The rest of the paper is organized as follows. In Section 2 we briefly outline the general set-up. Section 3 contains an informal discussion of concurrent contractions, and Section 4 recalls some facts from the standard Gärdenfors framework. Then, in Sections 5, 6 and 7 we present our formal approach to concurrent contractions, based on the idea of interleaving. We conclude the paper with comments and suggestions for further work in Section 8.

2 General Set-up

There have been many proposals to alter or extend the basic Alchourrón, Gärdenfors, Makinson (AGM) framework of theory change (see [5] for an overview), but most of the literature in the AGM tradition focuses on a single agent changing a theory as she receives new information. The actions of this solitary agent are usually specified in terms of functional input/output behavior:

$$(T, \phi) \mapsto T', \tag{1}$$

where the input consists of a collection of sentences T (the material to be changed) and a sentence ϕ (the newly received information), and the output is a collection of sentences T' (the result of the cognitive action). Traditionally, three forms of theory change are considered: *expansions*, where we add ϕ to T and close under logical consequence; *contractions*, where we remove ϕ from T while preserving as much of T as possible; and *revisions*, where we add ϕ to T while maintaining or restoring consistency. In this paper we change the format given in (1), and consider concurrent contractions that are specified by expressions of the form

$$T \sim \begin{pmatrix} \phi_1 \\ \vdots \\ \phi_n \end{pmatrix}, \tag{2}$$

or $T \sim \vec{\phi}$, where T is as before, and $\vec{\phi}$ is a vector of formulas to be contracted from T; $\sim$ is the concurrent contraction action whose principles we want to understand. The basic assumption here is that there are n agents $A_1, \ldots, A_n$, each of whom proposes or performs a contraction of T in accordance with her own contraction operation. That is, A_1 proposes or performs a contraction of T by $\phi_1, \ldots, A_n$ proposes or performs a contraction of T by ϕ_n, where each agent A_i has her own contraction operation $-_i$. The expression in (2) denotes the result(s) of an operation on T that is somehow composed of contractions of T by $\phi_1, \ldots, \phi_n$ performed by, respectively, $A_1, \ldots, A_n$ using their respective contraction operations $-_1, \ldots, -_n$. The key questions we address are:

- How can we model concurrent contractions?
- Which laws govern the concurrent contraction operation $\sim$?
- How can $\sim$ be understood in terms of the single agent operations $-_1, \ldots, -_n$?

Below we will explore concurrent contractions. We leave the much more complicated (and realistic) case of *heterogeneous* concurrent theory change in which multi-agent contractions, revisions and expansions may take place concurrently to later publications.

3 Why Contract Concurrently?

Before proceeding we give an informal discussion of concurrent contractions. As outlined above, our basic picture is one where n agents $A_1, \ldots, A_n$ simultaneously want to remove information from a given background theory T, that is: each agent proposes or performs a contraction, using her private contraction operation.

To give an example of concurrent theory change at work, one can think of a patient's record in a medical database. Various agents contribute to the theory contained in the database: a family doctor's report, various laboratories with their test results, specialists with further information.... Clearly, it is important that consistency be preserved. One may conceptualize this is by personifying consistency checking in terms of a checker that performs consistency checks at certain discrete intervals. If the checker detects an inconsistency in the shared theory, she rings the alarm bell, asking the agents to suggest contractions that will help remove the inconsistency. The agents then perform or suggest a contraction. Having different areas of expertise, the agents are likely to base their suggested contractions on different notions of which information is more reliable (or 'epistemically entrenched') than other. In other words, when agents suggest a contraction for the shared theory they suggest both *which* information should be given up, and *how* this should be done in their opinion. Therefore, the global change that is to be made to the theory is in general composed out of a finite number of 'private' contractions being performed concurrently.

In the special case where all agents employ the same contraction function, there is a clear connection with the *multiple contractions* proposed by Fuhrmann and Hansson [3], and with forms of *iterated belief change* that have recently been described by Lehmann and others (see [9]).

Ideas related to concurrent contraction also appear in non-epistemic settings. For example, co-authoring and joint research are processes in which concurrent contractions occur frequently. They seem especially appropriate when bugs or inconsistencies are discovered in cases where agents have sole responsibilities for certain parts of the work, and each author can perform contractions on the parts for which she holds responsibility. And of course, in concurrent databases concurrent transactions occur all the time. It is difficult, however, to find pure cases of concurrent contractions that are substantially different from the above ones.

4 Laws and Models for Single Agent Contractions

In this section we describe the laws governing the contraction operations of individual agents taking part in a concurrent contraction; as explained above, we assume that each agent comes equipped with her own contraction function. We start with some technical preliminaries.

Our background language is simply propositional logic, equipped with a classical consequence operator Cn that satisfies all the usual properties (see [4]). A *theory* is a set of formulas T that is closed under Cn; a *belief base* K is a set of formulas that needs not be a theory. In the AGM tradition there are two ways of reasoning about contraction functions, a *syntactic* way which specifies postulates that reasonable contraction functions should satisfy, and a *semantic* way that defines contractions functions obeying those laws. Here's a list of the standard AGM postulates for contraction.

$T - \phi$ is a theory (logically closed) whenever T is (Closure)
$T - \phi \subseteq T$ (Inclusion)
If $\phi \notin T$, then $T - \phi = T$ (Vacuity)
If $\not\vdash \phi$, then $\phi \notin T - \phi$ (Success)
If $\phi \in T$, then $T \subseteq \mathrm{Cn}((T - \phi) \cup \phi)$ (Recovery)
If $\vdash \phi \leftrightarrow \psi$ then $T - \phi = T - \psi$ (Extensionality)

We refer the reader to [4, 5] for a discussion. The above laws constrain how contraction functions $-$ should operate on a single, fixed theory T. But when n agents each come up with a formula ϕ_i to be contracted from a theory T, they should not only provide the system with a contraction function $-_i$, but, since the actual implementation of $T \sim \vec{\phi}$ may deal with several 'intermediate' results T' from which some of the ϕ_i's still have to be contracted, their contraction functions should indicate how to remove ϕ_i from arbitrary theories.

Hansson [6] gives a formal account of contraction functions able to deal with arbitrary theories. His approach is formulated in terms of belief bases K rather than theories T, and he moreover allows for contractions with *sets* of formulas rather than single formulas. We reformulate Hansson's original postulates for the 'base/set' case for the 'theory/formula' case.

Definition 1 (Postulates for single agent contraction) We propose the following postulates for a single agent contraction function $-$ that is defined for any theory T and formula ϕ:

$T - \phi$ is a theory (logically closed) whenever T is (Closure)
$T - \phi \subseteq T$ (Inclusion)
If $\psi \in T \setminus (T - \phi)$ then there exists T' with $T - \phi \subseteq T' \subseteq T$ (Relevance)
and $T' \not\vdash \phi$, but $T', \psi \vdash \phi$
If $T' \vdash \phi \leftrightarrow \psi$ for all subtheories $T' \subseteq T$, then $T - \phi = T - \psi$ (Uniformity)
If $\not\vdash \phi$, then $\phi \notin T - \phi$ (Success)

Relevance ensures that if a formula ψ is excluded from T when ϕ is rejected, then ψ plays a role in the fact that T implies ϕ. Whereas Success ensures that formulas that should be given up are in fact given up, Relevance blocks the deletion of formulas that need not be deleted. Uniformity ensures that the result of contracting T with ϕ depends only on the subsets of T that imply ϕ; if all subsets derive a given formula ϕ iff they derive ψ, then contracting with either ϕ or ψ produces the same result. Observe that Vacuity is derivable from Inclusion and Relevance.

In the setting of concurrent contractions it may well be that some agents want to refrain from action. The next proposition shows how we can mimic this situation.

Proposition 2 *If a contraction function $-$ satisfies the postulates of Definition 1, then, for any theory T and tautology $\top$, we have $T - \top = T$.*

The best known model of a contraction function in the AGM theory is partial meet contraction. It is defined as follows. Let $T \perp \phi$ denote the set of maximal subsets of T that fail to imply ϕ. A *one-place selection function* for T is a function s such that for all formulas ϕ, if $T \perp \phi$ is non-empty, then $s(T \perp \phi)$ is a non-empty subset of $T \perp \phi$. When $T \perp \phi$ is empty, $s(T \perp \phi) = \{T\}$. Then, an operation $-$ on a theory T is a *partial meet contraction* if $T - \phi$ is the intersection of the selected maximal subsets of T that fail to imply ϕ: $T - \phi = \bigcap s(T \perp \phi)$.

One-place selection functions are specific for a particular theory; if s is a one-place selection function for T, and $T \neq T'$, then s is not a one-place selection function for T' (see Hansson [6]). Selection functions that work for arbitrary theories are obtained by extending them with an additional argument; thus we will assume that each agent i is equipped with a two-placed selection function s, where, for each theory T and set of theories $(S \perp \psi)$, we have $s(T, (S \perp \psi)) \subseteq (S \perp \psi)$.

The following result links up the postulates for $-$ with two-placed contraction functions; a proof is given in the Appendix.

Theorem 3 *A single agent contraction function $-$ satisfies the postulates of Definition 1 iff there exists a two-placed selection function s with $T - \phi = \bigcap s(T, (T \perp \phi))$, for any theory T and formula ϕ.*

Now that we have shown how an agent's contractions can be modeled using two-placed contraction functions s, we pause a moment and reflect upon the desired effects of the first argument of s. Recall that $S \perp \psi$ denotes all maximal sub-theories of S that do not entail ψ (if $\not\vdash \psi$). When contracting ψ from S, the function s should make a selection from these sub-theories. This selection should principally reflect the agent's preferences among the theories in $(S \perp \psi)$. Thus, if we have

$$(S \perp \psi) = (U \perp \chi) \neq \emptyset,$$

it seems natural to require that

$$\bigcap s(S, (S \perp \psi)) = \bigcap s(U, (U \perp \chi)).$$

In other words, the common parts of the selections agrees whenever possible.[1] Hansson calls a selection function with this property *unified*. When working with belief bases this property doesn't come for free. Hansson comes up with a condition on contraction functions called *redundancy* to characterize unified partial meet contractions. In our set up this redundancy principle reads as follows:

Redundancy reformulated. Suppose T is a theory, and $\nvdash \phi$. Suppose furthermore that Z is a set of formulas, satisfying: (i) $T \cup Z$ is a theory, and (ii) for all $\zeta \in Z$: $\vdash \zeta \rightarrow \phi$. Then we have: $T - \phi = (T \cup Z) - \phi$.

Theorem 4 *If a contraction function $-$ satisfies the postulates of Definition 1, then it also satisfies redundancy.*

Theorem 4 (the proof of which is to be found in the Appendix) guarantees that we do not have to add redundancy as a separate postulate, so that we can now formulate the main result of this Section; its proof is given in the Appendix.

Theorem 5 *A single agent contraction function $-$ satisfies the postulates of Definition 1 iff there exists a two-placed* unified *selection function s with $T - \phi = \bigcap s(T, (T \perp \phi))$, for any theory T and formula ϕ.*

In the sequel, we will assume that selection functions are unified, and we will often suppress their first argument.

5 From Sequential to Interleaved Contractions

In many models of situations in which multiple agents need to access shared resources, one finds a reduction to a sequential, non-deterministic scheme. Our model of concurrent contractions will be based on the same idea. To see how we arrive at our model, consider the following diagram in which a contraction by a singel agent i is depicted by a line segment labeled with i. It pictures how multiple agents might — in principle — act on a single theory T to perform their individual contractions as time progresses: their actions might or might not overlap in arbitrary ways. But what does it mean for an agent i to start a contraction *while another agent j is still performing her contraction*? To what should i apply her selection function?

[1] Note that the first argument of s is still relevant: when modeling a contraction $T-\phi$, we calculate $s(T, (T \perp \phi))$.

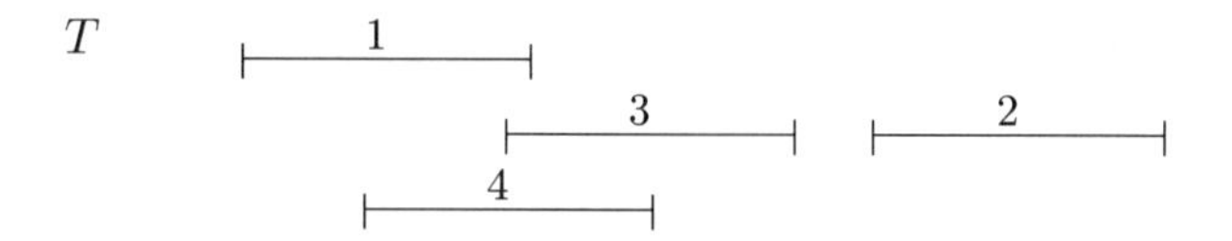

Figure 1: Overlapping contractions

What should she act on, if not on the outcome of j's actions? To perform a concurrent contraction one should execute the individual single agent contractions, one at a time. Thus, instead of Figure 1, Figure 2 seems to offer a much more realistic picture.

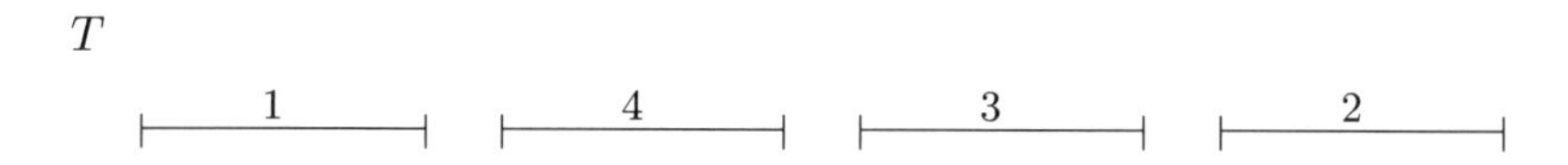

Figure 2: Interleaved contractions

To understand this situation, it may help to observe that there is a clear analogy with some forms of concurrent computation. Specifically, the situation is reminiscent of the concurrent execution of several independent programs on a single processor (see e.g. [2]). In a popular formal model *concurrency* is represented by *interleaving*. This means that parallel processes are never executed at precisely the same instant, but take turns in executing atomic transitions. When one of the participating processes executes an atomic transition, the others are inactive. Thus, rather than input/output pairs, execution *sequences* of the atomic instructions of sequential processes are at the focus of attention. And rather than talk about input/output pairs, one describes properties of concurrent programs that hold under some or all interleavings of the instructions. Let us briefly expand on this issue.

As parallel execution of sequential processes is modeled by the non-deterministic interleaving of atomic steps of the individual processes in interleaved models of concurrent programs, a program starting in a given state may follow any one of a number of computation paths corresponding to the different non-deterministic choices the program might make. The different computation paths thus represent alternative possible 'futures': at each moment, time may split into alternative courses and thus has a 'branching' tree-like structure. A semantic theory of computations provides a formal basis for describing or deducing properties of programs under *all* possible interleavings (see [2] for further details).

A similar concern is found in concurrent database theory, where one studies mechanisms for controlling the execution of several transactions at the same time. Here, one of the main interests lies in describing all possible executions of transactions and in identifying serializable transactions, that is: transactions that are equivalent in some sense to serial (consistency preserving) database transactions (see [8, Chapter 10] for an introductory overview).

In our setting of multi-agent contractions, we take a similar interleaved approach. Concurrent contractions will be viewed as (collections of) sequences of 'atomic' single agent contractions that don't overlap and that don't interfere. This interleaved approach calls for new ways of thinking about theory change. For a start, if we reduce concurrent contractions to non-deterministic sequential contractions, instead of single one step contractions we should be considering collections of sequences of contractions that are organized in a tree as in Figure 3. But then, we also have to give

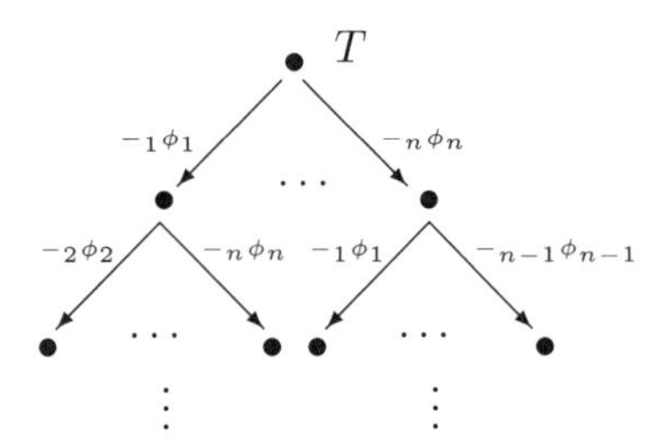

Figure 3: Interleaved contractions

up the idea of concurrent contractions as functions. For, in general, even singleton contractions are not commutative, that is: the identity

$$(T - \phi_1) - \phi_2 = (T - \phi_2) - \phi_1$$

is not universally valid (see [6] for a plausible counterexample). Hence, even in the case where we only have two agents that share the same contraction operation, say $-_1$, the global concurrent contraction $T \sim \binom{\phi_1}{\phi_2}$ may have two interleaving executions leading to different results: $(T -_1 \phi_1) -_1 \phi_2$ and $(T -_1 \phi_2) -_1 \phi_1$. As a consequence, it makes little sense to talk about *the* outcome of interleaved contractions. As a further consequence, we have to abandon the idea that contractions can be specified in terms of preconditions and postconditions. Instead, we need to reason about intermediate stages of a concurrent contraction, as these clearly have internal structure in the interleaved approach.

6 Laws for Interleaved Contractions

Traditionally, theory change in the AGM tradition has had two main concerns: (1) Constraints and axioms that rational operations of theory change should satisfy; and (2) Explicit constructions of operations of theory change that satisfy those constraints or axioms. In our approach to concurrent theory change we follow the same strategy. In particular, in this section we present a list of postulates that any reasonable operation of interleaved contraction should satisfy. Then, in Section 7 below, we present a model for interleaved contractions that satisfies these postulates.

We first need some notation. Let $\vec{\phi}$ and $\vec{\psi}$ be vectors of formulas of length n. We write $\vec{\phi} \equiv \vec{\psi}$ for 'for all $i \leq n$: $\vdash \phi_i \leftrightarrow \psi_i$,' and for a set of formulas Z, $\vec{\phi} \equiv_Z \vec{\psi}$ means that for all $i \leq n$: $Z \vdash \phi_i \leftrightarrow \psi_i$. Special vectors are $\vec{\top}$ and $\vec{\bot}$, consisting only of the formulas $\top$ and $\bot$, respectively; $\vec{\phi}\,[\chi/\phi_i]$ denotes the result of replacing the i-th component ϕ_i by χ. A *concurrent contraction function* is a function that produces a set of theories when given a theory T and a vector $\vec{\phi}$ as input. Finally, we write $(T \sim \vec{\phi}) \propto S$ for 'S is a result of concurrently contracting T with $\vec{\phi}$.'

Definition 6 (Basic postulates for interleaved contractions) Let T be a set of formulas, and let $n > 1$ be a natural number (the number of agents taking part). We assume that each $-_i$ satisfies the postulates for single agent contraction given in Definition 1.

If T is a theory and $(T \sim \vec{\phi}) \propto S$, then S is also a theory (C-closure)

If $(T \sim \vec{\phi}) \propto S$ then $S \subseteq T$ (C-inclusion)

Suppose $(T \sim \vec{\phi}) \propto S$ and $\psi \notin S$, $\psi \in T$. Then there are T' and $i \leq n$ such that $S \subseteq T' \subseteq T$, $T' \nvdash \phi_i$, and $T', \psi \vdash \phi_i$ (C-relevance)

If $\vec{\phi} \equiv_{T'} \vec{\psi}$ for all subtheories $T' \subseteq T$ then $(T \sim \vec{\phi}) \propto S$ iff $(T \sim \vec{\psi}) \propto S$ (C-uniformity)

For all $i \leq n$, if $\nvdash \phi_i$ and $(T \sim \vec{\phi}) \propto S$, then $\phi_i \notin S$ (C-success)

If for all $j \neq i$, we have $\phi_j = \top$, then $(T \sim \vec{\phi}) \propto S$ iff $S = T -_i \phi_i$ (Solo)

C-closure and C-inclusion are straightforward concurrent versions of their single agent counterparts. C-Relevance says that for every formula ψ that is given up in a concurrent contraction from T, there is an agent that is 'responsible' for this removal; according to the individual obedience to Relevance, this agent can determine a part of T from which the formula she proposed for contraction has indeed been removed, and in the process of doing this, ψ had to be given up. C-uniformity says that if no subtheory T' of T can distinguish any component ϕ_i of $\vec{\phi}$ from the corresponding component ψ_i of $\vec{\psi}$, then concurrently contracting $\vec{\phi}$ from T cannot be distinguished from concurrently contracting $\vec{\psi}$ from T. This uniformity postulate implies the following condition of C-extensionality:

$$\text{If } \vec{\phi} \equiv \vec{\psi} \text{ then } (T \sim \vec{\phi}) \propto S \text{ iff } = (T \sim \vec{\psi}) \propto S \qquad \text{(C-extensionality)}$$

It guarantees that only the *content* of the individual's proposal for contraction matters, not the actual *form*. C-success guarantees that, as long as an agent does not propose to contract a tautology, her request for contraction will have been granted in each of the possible results. Thus, whereas C-relevance says that each formula that is given up in a concurrent contraction should be due to one of the agents, C-success guarantees that all of the agents' wishes will be met as far as they are reasonable. Finally, the Solo postulate shows that interleaved contractions really build on the individual contraction strategies: when only one agent comes up with a non-trivial formula to be removed, it will be her strategy that determines the result of the concurrent contraction.

The postulates in Definition 6 provide no means to reason about possible 'intermediate' results of interleaved contractions, and they certainly don't impose the condition that the concurrent contraction process can be unraveled into successive single agent contractions. To make up for this, we consider two further laws: Decomposition and Composition.

$$(T \sim \vec{\phi}) \propto S \Rightarrow \begin{cases} \textit{either } T = S \text{ and } \vec{\phi} \equiv \vec{\top} \\ \textit{or } \text{there exist } S' \text{ and } i \text{ with } i \leq n \text{ and } S \subseteq \\ S' \subseteq T \text{ such that } \phi_i \not\equiv \top, S' = T -_i \phi_i \text{ and} \\ (S' \sim \vec{\psi}) \propto S, \text{ where } \vec{\psi} = \vec{\phi}[\top/\phi_i] \end{cases}$$

(Decomposition)

Decomposition says that concurrently contracting with $\vec{\top}$ is a void action and that a concurrent contraction with $\vec{\phi} \neq \vec{\top}$ can be decomposed in an individual contraction $-_i$ followed by another, yet simpler, concurrent contraction.

$$(T \sim \vec{\phi}) \propto S \Leftarrow \begin{cases} \textit{either } T = S \text{ and } \vec{\phi} \equiv \vec{\top} \\ \textit{or } \text{there exist } S' \text{ and } i \text{ with } i \leq n \text{ and } S \subseteq \\ S' \subseteq T \text{ such that } \phi_i \not\equiv \top, S' = T -_i \phi_i \text{ and} \\ (S' \sim \vec{\psi}) \propto S, \text{ where } \vec{\psi} = \vec{\phi}[\top/\phi_i] \end{cases}$$

(Composition)

Composition states that if one recursively unravels a concurrent contraction $T \sim \vec{\phi}$ into an individual contraction $T -_i \phi_i$ followed by a concurrent contraction of a vector $\vec{\psi}$ (obtained from $\vec{\phi}$) from the theory $(T -_i \phi_i)$, one ends up with a theory S that will be a result of the initial concurrent contraction. Notice that the Solo postulate is a consequence of Decomposition. If we think about interleaving contractions in an algorithmic way, we can view the Composition and Decomposition postulates as halting criteria: to contract $\vec{\phi}$ from T, try to turn all components of $\vec{\phi}$ into the formula $\top$ by successively contracting with one ϕ_i after another until $\vec{\phi}$ equals $\vec{\top}$.

Observe that the conjunction of Decomposition and Composition is equivalent to the following statement; let n be the length of $\vec{\phi}$.

$$(T \sim \vec{\phi}) \propto S \quad \text{iff} \quad \text{there exists a permutation } f \text{ of } \{1, \ldots, n\} \text{ such that } S = ((\cdots(T -_{f(1)} \phi_{f(1)})\cdots) -_{f(n)} \phi_{f(n)}).$$

Theorem 7 *Assume that a set of individual contraction functions* $-_i$ $(1 \leq i \leq n)$ *and a concurrent contraction* $\sim$ *are connected via the Decomposition and Composition laws. If all the* $-_i$*'s satisfy the postulates from Definition 1, then* $\sim$ *satisfies all the Concurrent postulates from Definition 6.*

Proof. Suppose all the $-_i$'s satisfy the postulates from Definition 1. As pointed out above, we have that $(T \sim \vec{\phi}) \propto S$ iff for some permutation f of $\{1, \ldots, n\}$

$$S = ((\cdots(T -_{f(1)} \phi_{f(1)})\cdots) -_{f(n)} \phi_{f(n)}).$$

Let $T_0 = T$ and $T_i = T_{i-1} -_{f(i)} \phi_{f(i)}$, for $i > 0$. Note that, by Inclusion, we have $T_i \subseteq T_{i-1}$ (for $1 \leq i \leq n$). Now, $\sim$ satisfies C-closure trivially: if T is a theory then, by n applications of Closure, $T_0, T_1 \ldots T_n = S$ are all theories. C-inclusion follows similarly.

For C-relevance, suppose all $-_i$'s satisfy Relevance, and suppose that $(T \sim \vec{\phi}) \propto S, \phi \notin S, \psi \in T$. Since each $-_i$ satisfies Inclusion, there must be some j such that $\psi \in T_j$, $\psi \notin T_{j+1}$. Since $T_{j+1} = T_j -_{f(j)} \phi_{f(j)}$ and $-_{f(j)}$ satisfies relevance, we find a T' with $T_j -_{f(j)} \phi_{f(j)} \subseteq T' \subseteq T_{j+1}$, $T' \nvdash \phi_{f(j)}$ and $T', \psi \vdash \phi_{f(j)}$. Using Inclusion, we see that $S = T_n \subseteq T' \subseteq T_0 = T$. From this we can conclude that $\sim$ satisfies C-relevance.

For C-uniformity, suppose that $\vec{\phi} \equiv_{T'} \vec{\psi}$ for all subtheories $T' \subseteq T$. Thus, for all $i \leq n$, $T' \vdash \phi_i \leftrightarrow \psi_i$. Suppose furthermore that $(T \sim \vec{\phi}) \propto S$: we have to demonstrate that $(T \sim \vec{\psi}) \propto S$. But, since all the $-_i$'s satisfy Relevance, we immediately see that

$$S = (T -_{f(1)} \phi_{f(1)})\cdots) -_{f(n)} \phi_{f(n)}) = (T -_{f(1)} \psi_{f(1)})\cdots) -_{f(n)} \psi_{f(n)}),$$

which proves that $(T \sim \vec{\psi}) \propto S$.

For C-success, suppose $\nvdash \phi_i$ and $(T \sim \vec{\phi}) \propto S$. Let $i = f(k)$, then, by Success, $\phi_i \notin T_k = T_{k-1} -_i \phi_i$ and, by Inclusion, $\phi \notin T_n = S$.

Finally, we prove that $\sim$ satisfies Solo: suppose that for all $j \neq i$, we have $\phi_j = \top$. Let k be such that $i = f(k)$. Then, by Proposition 2, we have for any $m \neq k$, that $T_m = T_{m_1}, m > 0$. Thus, we have

$$T = T_0 = T_1 = \ldots = T_{k_1}, T_k = T_{k_1} -_i \phi_i = T_{k+1} = \ldots = T_n = S.$$

Thus, $S = T -_i \phi_i$. $\dashv$

Theorem 7 expresses a *transfer* property: if we define a concurrent contraction $\sim$ via Composition and Decomposition using individual contractions $-_i$, we get the rationality postulates for $\sim$ if we impose rationality postulates on all the $-_i$'s. Theorem 8 expresses a *projection* principle going in the converse direction.

Theorem 8 *Assume that a set of individual contraction functions $-_i$ ($1 \leq i \leq n$) and a concurrent contraction $\sim$ are connected via the Decomposition and Composition laws. If $\sim$ satisfies the Concurrent postulates from Definition 6, then all the $-_i$'s satisfy the postulates from Definition 1.*

Proof. We note the following. For any formula ϕ, let $\vec{v}(i,\phi)$ be the vector with ϕ at index i, and with $\top$ at all other places: $\vec{v}(i,\phi)_i = \phi$ and $\vec{v}(i,\phi)_j = \top, i \neq j$. Using the interleaved contraction postulate Solo, we immediately obtain:

$$(T \sim \vec{v}(i,\phi)) \propto S - \ S = T -_i \phi \tag{3}$$

Equation (3) expresses that a single agent contraction can be modeled by the multiple-agent contraction, provided that all agents but one refrain from acting. Now, let $\sim$ satisfy the properties of Definition 6. Then, using (3), one easily reads off the $-_i$ properties Closure, Inclusion and Success from C-closure, C-inclusion and C-success for $\sim$, respectively. For Relevance, suppose that $\psi \in T \setminus (T -_i \phi)$. Using C-relevance, we find a $j \leq n$ and a T' with $(T -_i \vec{v}(i,\phi)) = S \subseteq T' \subseteq T$ such that $T' \not\vdash \vec{v}(i,\phi)_j$ and $T', \psi \vdash \vec{v}(i,\phi)_j$. Since for all $k \neq i$, $\vec{v}(i,\phi)_k = \top$, we must have $j = i$. Since $\vec{v}(i,\phi)_i = \phi$, we have $T' \not\vdash \phi$ and $T', \psi \vdash \phi$ for some T' with $(T -_i \phi) \subseteq T' \subseteq T$, expressing that $-_i$ satisfies Relevance. To check Uniformity for $-_i$, suppose that $T' \vdash \phi \leftrightarrow \psi$ for all subtheories $T' \subseteq T$. By definition of $\vec{v}(i,\phi)$, we immediately see that $\vec{v}(i,\phi) \equiv_{T'} \vec{v}(i,\psi)$ so that C-uniformity yields

$$(T \sim \vec{v}(i,\phi)) \propto S \text{ iff } (T \sim \vec{v}(i,\psi)) \propto S.$$

Using (3) we conclude that $(T -_i \phi) = S = (T -_i \psi)$, which proves Uniformity. $\dashv$

Combining Theorems 7 and 8 we see that the Composition and Decomposition postulates properly link individual and concurrent contractions together: postulates for individual operations are guaranteed by imposing postulates on the concurrent one, and vice versa.

Requiring that single agent contractions $-_i$ and a concurrent contraction $\sim$ are related through Composition and Decomposition is a non-trivial requirement, even if the single agent contractions satisfy the postulates of Definition 1, and the concurrent contraction satisfies the postulates of Definition 6. The main reason is that the postulates for contraction don't, in general, uniquely pin down its actual implementation. One can have different single contractions $-_1$ and $-_{1'}$ satisfying the postulates of Definition 1 (for example, $-_1$ can be a full meet contraction $-_1$, and $-_{1'}$ a partial meet contraction). Now, assume that $-_1$ and $-_{1'}$ are composed with $-_2$, $\ldots$, $-_n$ (all satisfying the postulates of Definition 1) into $\sim$ and $\sim'$, respectively, using Composition. Then, the single agent contractions $-_1, \ldots,$ $-_n$ and the concurrent contraction $\sim'$ are not related via Composition and Decomposition.

7 A Model for Interleaved Contractions

Let $n > 1$ be the number of agents. We assume that for each i, agent i's contraction function is defined using a selection function s_i, as outlined in Section 4. The models we are about to define are called selection systems; they are based on the selection functions contributed by the individual agents. Roughly, a selection system is a collection of compositions of single agent selection functions that satisfies certain constraints.

More precisely, a *selection system* $(\mathcal{S}, \mathcal{T}, \mathsf{s}_0)$, intended to represent interleaved contractions, is given by the following components:

- $\mathcal{S}$, a set of states. Each state s is labeled with a theory $\mathrm{Th}(\mathsf{s})$. These are the theories that the theory of the *initial state* $\mathrm{Th}(\mathsf{s}_0)$ can evolve into by applying sequences of single agent contractions. Two states may be labeled with the same theory.
- $\mathcal{T}$, a set of possible transitions built up from the individual agents' single contraction: $\mathcal{T} = \{(\mathsf{s}, \mathsf{s}') \mid \exists i \exists \phi \, (\bigcap_i s_i(\mathrm{Th}(\mathsf{s}), (\mathrm{Th}(\mathsf{s}) \perp \phi)) = \mathrm{Th}(\mathsf{s}'))\}$. Here s' is called a *successor* (or $-_i\phi$-*successor*) of s; notation: $\mathsf{s} \xrightarrow{i,\phi} \mathsf{s}'$. For technical reasons we will assume that all successor steps are irreflexive: if $\mathsf{s} \xrightarrow{i,\phi} \mathsf{s}'$ then $\mathsf{s} \neq \mathsf{s}'$.

A state $\mathsf{s} \in \mathcal{S}$ is *terminal* if it has no successors. A *choice sequence* of a selection system $(\mathcal{S}, \mathcal{T}, \mathsf{s}_0)$ is a finite sequence $\sigma : \mathsf{s}_1, \ldots, \mathsf{s}_m$ satisfying the following requirements. First, the *Initiation* requirement says the state s_1 is the initial state of the selection system, that is: $\mathsf{s}_1 = \mathsf{s}_0$. Second, the *Consecution* requirement says that for each pair of consecutive states

s_j, $\mathsf{s}_{j+1} \in \sigma$ there is a selection function s_i and a formula ϕ such that $\mathsf{s}_j \xrightarrow{i,\phi} \mathsf{s}_{j+1}$. (Observe that two states may be connected by multiple transitions.) Finally, the *Termination* requirement says that the final state s_m is a terminal state.

A *prefix* is a sequence $\mathsf{s}_1, \ldots, \mathsf{s}_k$ satisfying the requirements of initiation and consecution, but not necessarily of termination. The *length* of a prefix is its number of states.

Let T be a theory, n the number of agents, and $\vec{\phi} = (\phi_1, \ldots, \phi_n)$ a sequence of formulas. Our next aim is to determine what it means for a selection system $\mathbf{T} = (\mathcal{S}, \mathcal{T}, \mathsf{s}_0)$ to model or represent the interleaved contraction $T \sim \vec{\phi}$. We will impose three constraints. First, the *Start* constraint says that $\mathrm{Th}(\mathsf{s}_0)$, the theory of the initial state, should equal T. Second, the *Tightness* constraint requires that for every choice sequence σ in $\mathbf{T}$ and every $i \leq n$ there exists at most one pair of consecutive states s_j, s_{j+1} in σ such that $\mathsf{s}_j \xrightarrow{i,\phi} \mathsf{s}_{j+1}$. Intuitively, the tightness property says that no attempt is made to carry out a single agent contraction in $T \sim \vec{\phi}$ twice. Third, the *Fairness* constraint says that for every choice sequence σ in $\mathbf{T}$ and every $i \leq n$ there is a consecutive pair s_j, s_{j+1} in σ such that $\mathsf{s}_j \xrightarrow{i,\phi} \mathsf{s}_{j+1}$. holds. The fairness property expresses that every single agent contraction in $T \sim \vec{\phi}$ will eventually be carried out.

Let $\mathbf{S} = (\mathcal{S}, \mathcal{T}, \mathsf{s}_0)$ be a selection system. $\mathbf{S}$ is called a *model* for $T \sim \vec{\phi}$ if it satisfies the starting, tightness and fairness conditions for $T \sim \vec{\phi}$. Given a model $\mathbf{S} = (\mathcal{S}, \mathcal{T}, \mathsf{s}_0)$ for $T \sim \vec{\phi}$, a *proper choice sequence* of $T \sim \vec{\phi}$ is simply a choice sequence in $\mathbf{S}$. What this definition boils down to is that we view interleaved contractions as generators of proper choice sequences.

To be able to express the connection between concurrent contraction functions and selection systems, we say that a contraction function $\sim$ *generates a full selection system for T and $\vec{\phi}$* if there are single agent selection functions $s_1, \ldots, s_n$ such that $-_j$ is defined in terms of s_j $(1 \leq j \leq n)$, and for all S such that $(T \sim \vec{\phi}) \propto S$ there exists a sequence $S_0, \ldots, S_n$ such that $S_0 = T$, $S_{i+1} = \bigcap s_{f(i+1)}(S_i, (S_i \perp \phi_{f(i)}))$, where f is a permutation of $\{1, \ldots, n\}$, and $S_n = S$.

Theorem 9 *Let $\sim$ be a concurrent contraction function, and $-_i$ a set of single agent contractions. Then $\sim$ satisfies the C-postulates from Definition 6, and $\sim$ and $-_i$ are related via the Composition and Decomposition laws from Section 6 iff, for every theory T and vector of formulas $\vec{\phi}$, $\sim$ generates a full selection system for T and $\vec{\phi}$.*

Proof. First, suppose that $\sim$ satisfies the C-postulates of Definition 6, and suppose also that $\sim$ and $-_i$ are related via Composition and Decomposition. Let S be such that $(T \sim \vec{\phi}) \propto S$. Just as in the proof of Theorem 7 we find a permutation f of $\{1, \ldots, n\}$ such that

$$S = ((\cdots(T -_{f(1)} \phi_{f(1)}) \cdots) -_{f(n)} \phi_{f(n)}).$$

Now, define $S_0 = T$ and $S_{i+1} = (S_i -_{f(i+1)} \phi_{f(i+1)})$. Theorem 8 guarantees that each individual contraction $-_{f(i+1)}$ satisfies the postulates of Definition 1 and hence, we may use one direction of Theorem 3 to conclude that each contraction $S_i -_{f(i+1)} \phi_{f(i+1)}$ corresponds to taking the intersection of the selection that agent $f(i)$ generates, using S_i and $S_i \perp \phi_{f(i)}$, so that we have $S_{i+1} = \bigcap s_{f(i+1)}(S_i, (S_i \perp \phi_{f(i)}))$. This proves that every T and $\vec{\phi}$ generate a full selection system.

For the converse, suppose that for every T and $\vec{\phi}$, the operator $\sim$ generates a full selection function. Let S, T and $\vec{\phi}$ be such that $(T \sim \vec{\phi}) \propto S$. We know that, semantically, this gives rise to a sequence $S_0 = T$ and $S_{i+1} = \bigcap s_{f(i+1)}(S_i, (S_i \perp \phi_{f(i)}))$, where each $s_{f(i+1)}$ is a selection function. Now, we use the other direction of Theorem 3 to lift this semantic result to a syntactic level: we can associate a single agent contraction $-_{f(i+1)}$ satisfying the postulates of Definition 1 with each selection function $s_{f(i+1)}$, and we may write $S_{i+1} = (S_i -_{f(i+1)} \phi_{f(i+1)})$. Hence, by an application of Theorem 7 we conclude that $\sim$ satisfies the C-postulates of Definition 6. Finally, by observing that each sequence $S_0, \ldots, S_n$ in a full selection system determines a permutation f of $\{1, \ldots, n\}$ such that $S_{i+1} = S_i -_{f(i+1)} \phi_{f(i+1)}$, it follows that $\sim$ and $-_i$ are related via the Composition and Decomposition postulates. $\dashv$

With the above result we can 'complete the square' in the diagram shown in Figure 4. By walking around the diagram we see that full selection systems are a model for our postulates for interleaved contraction, and any full selection system for T and $\vec{\phi}$ is given by the postulates for $\sim$.

8 Concluding Remarks

We have shown that concurrent contractions are well-behaved in that they satisfy a set of fairly transparent rationality postulates, on the assumption that all the underlying single agents contract in a rational way, and that concurrency is modeled in an interleaving manner.

In the course of the paper we have had to make explicit and alter some of the assumptions underlying the AGM approach to theory change as they seem no longer appropriate in our setting:

- In our interleaved setting theory change operations need not be functional; they are always defined but they need not have a unique

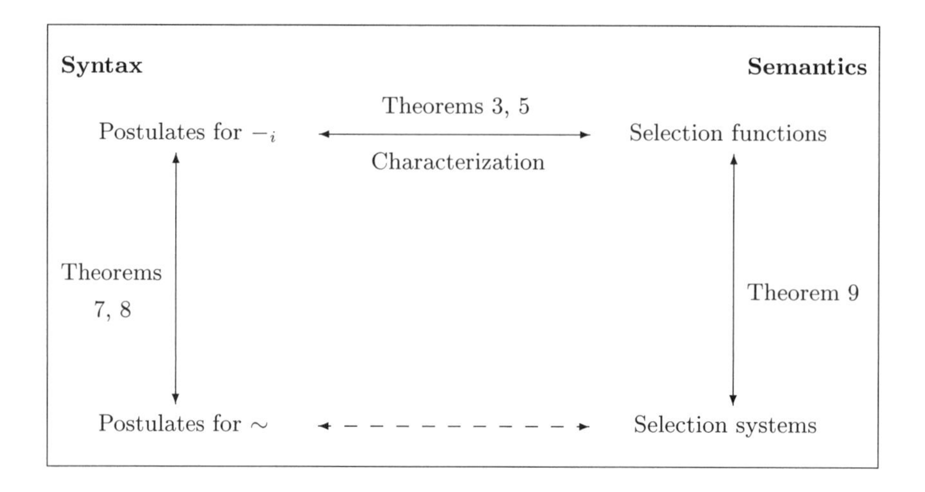

Figure 4: Linking the approaches

outcome. (A similar deviation from the AGM assumptions in the context of single agent theory change is explored by Lindström and Rabinowitz [10].)

- In our interleaved setting theory change no longer is a one step operation. Although theory change typically occurs in dynamic environments in which agents may learn new information in a continuous process, the traditional AGM framework consistently avoids mentioning iterations of its operations. Recently a number of authors have abandoned this assumption, and considered forms of iterated theory change; see for example Lehmann [9] and Kfir-Dahav and Tennenholtz [7].

- In our interleaved setting theory change operations have internal structure, and they are no longer fully characterized by their preconditions and postconditions. In contrast, the traditional AGM postulates have nothing to say about the internal mechanisms by which operations of theory change achieve their goals.

To formulate it in a single sentence, the reason that the above assumptions are no longer valid is that we have been considering collections of sequences of (single agent, one-step) contractions that are organized in a tree as in Figure 3 above.

In our ongoing work we consider alternative models for interleaved contractions called *entrenchment systems* that are based on compositions of single agent entrenchment relations. One can prove a representation result to the effect that every selection system can be represented as an entrenchment system, and vice versa.

Our future work revolves around the idea of using other models of concurrency than interleaving, and determining the effects this has on the postulates describing multi-agent theory change.

Acknowledgments. This work was supported by a grant of the Netherlands Organization for Scientific Research (NWO) and the British Council under the UK-Dutch Joint Scientific Research Programme JRP 333. Wiebe van der Hoek was partially supported by ESPRIT III BRWG project No. 8319 'ModelAge'. Part of the research was carried out while Maarten de Rijke was with CWI, Amsterdam; during this period he was supported by the Netherlands Organization for Scientific Research (NWO), project NF 102/62-356.

A Proofs

Below we give proofs for results that were stated without proofs in earlier sections.

Theorem 3. *A contraction function $-$ satisfies the postulates of Definition 1 iff there exists a two-placed selection function s such that, for any theory T and formula ϕ,*

$$T - \phi = \bigcap s(T, (T \perp \phi)).$$

Proof. Let us first assume that $-$ is defined using a selection function s: we show that $-$ satisfies the required postulates. The Closure and Inclusion conditions follow immediately from the definition of s and the fact that theories are closed under intersection. Uniformity follows because if $T' \vdash \phi \leftrightarrow \psi$ for all subtheories $T' \subseteq T$, then $(T \perp \phi) = (T \perp \psi)$ and hence

$$s(T, (T \perp \phi)) = s(T, (T \perp \psi)).$$

Success is also clear: if $\not\vdash \phi$, then $\phi \notin X$ for any $X \in (T \perp \phi)$, so $\phi \notin \bigcap s(T, T \perp \phi))$. As to the Relevance postulate, suppose $\psi \in T \backslash \bigcap s(T, (T \perp \phi))$. Then, there must be a $T' \in s(T, (T \perp \phi))$ with $\psi \notin T'$. By the definition of s, $T' \subseteq T$ and, by the definition of $(T \perp \phi)$ we must have $T', \psi \vdash \phi$.

Conversely, let $-$ satisfy the postulates of Definition 1. For any theories T and T' and formula ϕ such that $T - \phi = T'$, we have to guarantee that

$\bigcap s(T,(T \perp \phi)) = T'$. We define $s(T,\Theta)$, with $\Theta \in 2^{\mathcal{T}}$ as follows.

$$s(T,\Theta) := \begin{cases} \{T\}, & \text{if } \Theta = \emptyset \\ \{S \in \Theta \mid T - \theta \subseteq S\}, & \text{if } \Theta = (T \perp \theta) \neq \emptyset \text{ for some } \theta \\ \Theta, & \text{otherwise.} \end{cases}$$

To see that s is a selection function, we first observe that $s(T,\emptyset) = \{T\}$ and if $\Theta \neq \emptyset$, then $s(T,\Theta)$ is a non-empty subset of Θ. It is also a function: suppose $T_1 = T_2$ and $\Theta_1 = \Theta_2$. If (T_1,Θ_1) is not a matching pair, then neither is (T_2,Θ_2) and we have

$$s(T_1,\Theta_1) = \Theta_1 = \Theta_2 = s(T_2,\Theta_2).$$

Otherwise, we may assume that $\Theta_1 = T_1 \perp \theta_1$ and $\Theta_2 = T_2 \perp \theta_2$ for some formulas θ_1 and θ_2. Thus $(T_1 \perp \theta_1) = (T_2 \perp \theta_2)$. Let T' be an arbitrary subtheory of $T_1 = T_2$. If $T' \vdash \theta_1$, we have $T' \notin (T_1 \perp \theta_1)$ and hence $T' \notin (T_2 \perp \theta_2)$. Thus, we have either that $T' \vdash \theta_2$ and then also $T' \vdash \theta_1 \leftrightarrow \theta_2$, or some $S \supseteq T'$ with $S \in (T_2 \perp \theta_2)$. The latter is impossible, since it would yield $S \in (T_1 \perp \theta_1)$ and $S \vdash \theta_1$. This proves that for any subtheory T' of T_1 we have $T' \vdash \theta_1 \leftrightarrow \theta_2$. By Uniformity and $T_1 = T_2$, we then have $T_1 - \theta_1 = T_2 - \theta_2$. From this we immediately get

$$s(T_1,\Theta_1) = \{S \in \Theta_1 \mid T_1 - \theta_1 \subseteq S\} = \{S \in \Theta_2 \mid T_2 - \theta_2 \subseteq S\} = s(T_2,\Theta_2).$$

Finally, we have to show that $\bigcap s(T,(T \perp \phi)) = T'$, whenever $T - \phi = T'$. We immediately have $T' \subseteq \bigcap s(T,(T \perp \phi))$. For the other inclusion, we first assume $\phi \in T$. Suppose we have some $\psi \notin T'$. By Relevance, we find an S' with $T' \subseteq S' \subseteq T$, for which $S',\psi \vdash \phi$ and $S' \not\vdash \phi$. This S' can be expanded to an $S \supseteq S'$ such that $S \in (T \perp \phi)$ and $\psi \notin S$. We thus have

$$\psi \notin \bigcap s(T,(T - \phi)).$$

Finally, if $\phi \notin T$, then by Vacuity (which follows from Inclusion and Relevance), we have $T' = T$ and thus $\{T\} = s(T,(T \perp \phi))$, so that $\bigcap s(T,(T \perp \phi)) \subseteq T$. $\dashv$

Theorem 4. *If a contraction function $-$ satisfies the postulates of Definition 1 it also satisfies redundancy.*

Proof. Let T be a theory and suppose $\not\vdash \phi$. Suppose furthermore that Z is a set of formulas, satisfying: (i) $T \cup Z$ is a theory, and (ii) for all $\zeta \in Z$: $\vdash \zeta \to \phi$. We have to prove: $T - \phi = (T \cup Z) - \phi$. If $Z \subseteq T$ the equation holds trivially, so let us assume the existence of a $\zeta \in Z \setminus T$. We now first show that $\phi \notin T$: if we would have $\phi \in T$, we reason as follows. Since

$\zeta \in Z$, we have $(\zeta \vee \neg\phi) \in Z \cup T$. But $(\zeta \vee \neg\phi) \notin T$, since otherwise we would have, by $\phi, \phi \to \zeta \in T$ that $\zeta \in T$. Thus, $(\zeta \vee \neg\phi) \in Z$. By definition of Z, we have $\vdash (\zeta \vee \neg\phi) \to \phi$. Since $\vdash ((\zeta \vee \neg\phi) \to \phi) \to \phi)$, we would have $\vdash \phi$, contradicting one of the premisses. Thus, $\phi \notin T$.

Now we can prove that $(T \cup Z) - \phi = T$. For $\subseteq$, suppose $\psi \in (T \cup Z) - \phi$. By Inclusion, we have $\psi \in T \cup Z$. If ψ would be in Z, we would have $\vdash \psi \to \phi$ and hence $\phi \in (T \cup Z) - \phi$, contradicting Success. Thus, $\psi \in T$. To see that also $T \subseteq (T \cup Z) - \phi$, let $\psi \in T$. Then $\psi \in T \cup Z$. If $\psi \notin (T \cup Z) - \phi$, by Relevance, we find a U with

$$(T \cup Z) - \phi \subseteq U \subseteq T \cup Z$$

and such that $U, \psi \vdash \phi$ and $U \nvdash \phi$. By the assumptions on Z, the latter implies that $U \subseteq T$. Since $\psi \in T$ and $U, \psi \vdash \phi$ we have $T \vdash \phi$ — a possibility we already excluded. Thus, $\psi \in (T \cup Z) - \phi$. $\dashv$

Theorem 5. *A contraction function $-$ satisfies the postulates of Definition 1 iff there exists a two-placed unified selection function s such that, for any theory T and formula ϕ,*

$$T - \phi = \bigcap s(T, (T \perp \phi)).$$

Proof. If s is a selection function, by Theorem 3 we find a contraction function $-$ satisfying the postulates of Definition 1. Conversely, suppose $-$ satisfies the postulates of Definition 1. By Proposition 4 we know that it also satisfies redundancy. We will show that the selection function s whose existence is guaranteed by Theorem 3, is unified. To do so, suppose

$$(U \perp \phi) = (V \perp \psi). \tag{4}$$

We have to show that $\bigcap s(U, (U \perp \phi)) = \bigcap s(V, (V \perp \psi))$. If $(U \perp \phi) = \emptyset$, the conclusion follows from the definition of s. So suppose $(U \perp \phi) \neq \emptyset$. We will first argue that

$$(U \perp \phi) = ((U \cap V) \perp \phi). \tag{5}$$

Suppose that $\chi \in U \backslash V$. Then $\chi \notin V$ and hence $\chi \notin Y$ for any $Y \in (V \perp \psi)$ and, by (4), $\chi \notin X$ for any $X \in (U \perp \phi)$. Since χ has been removed from all maximal subsets of U that do not entail ϕ, we must have $\vdash \chi \to \phi$. Thus

$$\alpha \in U \setminus V \Rightarrow \ \vdash \alpha \to \phi. \tag{6}$$

To prove the $\subseteq$-direction of (5), suppose $X \in (U \perp \phi)$. Then $X \nvdash \phi$ and by (6) we must also have $X \subseteq V$, and so $X \in ((U \cap V) \perp \phi)$. Conversely, suppose $X \in ((U \cap V) \perp \phi)$. Then $X \not\vdash \phi$. Let χ be any formula in $U \setminus X$. If we can show that $X, \chi \vdash \phi$, we may conclude $X \in (U \perp \phi)$. Firstly, if $\chi \in V$, then, since $X \in ((U \cap V) \perp \phi)$, we have $X, \chi \vdash \phi$. If $\mu \notin V$ we have $\mu \in U \setminus V$, and by (6), $X, \mu \vdash \phi$. This proves (5), and, by a similar argument, we of course have $(V \perp \psi) = ((U \cap V) \perp \psi)$. Combining this with (4), we get

$$((U \cap V) \perp \phi) = ((U \cap V) \perp \psi). \qquad (7)$$

Taking $T = U \cap V$ and $Z = U \setminus V$, Redundancy guarantees that $(U \cap V) - \phi = U - \phi$. Since $-$ is modelled by a selection function s, we have

$$\bigcap s\,(U \cap V, ((U \cap V) \perp \phi)) = \bigcap s\,(U, (U \perp \phi)).$$

Similarly, $\bigcap s(U \cap V, (U \cap V) \perp \psi) = \bigcap s(V, (V \perp \psi))$. From (7) we infer

$$\bigcap s(U \cap V, (U \cap V) \perp \phi) = \bigcap s(U \cap V, (U \cap V) \perp \psi),$$

so that we can finally conclude that $\bigcap s(U, (U \perp \phi)) = \bigcap s(V, (V \perp \psi))$, as required. $\dashv$

References

[1] C.E. Alchourrón, P. Gärdenfors, and D. Makinson. On the logic of theory change: partial meet contraction and revision functions. *Journal of Symbolic Logic*, 50:510–530, 1985.

[2] M. Ben-Ari. *Principles of Concurrent and Distributed Programming.* Prentice Hall, 1990.

[3] A. Fuhrmann and S.-O. Hansson. A survey of multiple contractions. *Journal of Logic, Language and Information*, 3:39–75, 1994.

[4] P. Gärdenfors. *Knowledge in Flux.* The MIT Press, 1988.

[5] P. Gärdenfors and H. Rott. Belief revision. In D.M. Gabbay, C.J. Hogger, and J.A. Robinson, eds, *Handbook of Logic in AI and Logic Programming, Vol IV*, Oxford University Press, 1992.

[6] S.O. Hansson. Reversing the Levi identity. *Journal of Philosophical Logic*, 22:637–669, 1993.

[7] N.E. Kfir-Dahav and M. Tennenholtz. Multi-agent belief revision. In *Proceedings TARK 96*, 1996, pages 175–194.

[8] H.F. Korth and A. Silberschwartz. *Database System Concepts.* McGraw-Hill, second edition, 1991.

[9] D. Lehmann. Belief revision, revised. In *Proceedings IJCAI 95*, 1995.

[10] S. Lindström and W. Rabinowicz. Epistemic entrenchment with incomparabilities and relational belief revision. In A. Fuhrmann and M. Morreau, eds, *The Logic of Theory Change*, Springer, 1991.

[11] D. Makinson. How to give it up: A survey of some formal aspects of the logic of theory change. *Synthese*, 62:347–363, 1985.

Proving Through Commutative Diagrams

Yoshiki Kinoshita and Koichi Takahashi

Electrotechnical Laboratory
Tsukuba
305 Japan

1 Introduction

In category theory, commutative diagrams are extensively used to help understand (informal) proofs written in natural language and logical formulae. So, it seems natural to require the use of commutative diagrams in *formal* theorem proving in category theory too. In fact, the difficulty in understanding formal proof is one of the major obstacles in formal mathematics, so assistance in obtaining an intuitive idea of what is written would be of great value. Therefore, we are developing a user interface for proof assistant systems, based on commutative diagrams, and we report its functional specification.

In the course of developing our interface, we encountered a general user interface framework based on relations, rather than by simpler functions. This seems a new framework and we believe it is a contribution to visual logical systems, and we describe it in considerable depth.

What we wish is to represent formulae such as

$$\begin{array}{l} (\forall X)\mathrm{Obj}(X) \supset \\ \quad (\forall f)(\mathrm{source}(f) = X \wedge \mathrm{target}(f) = A) \supset \\ \quad\quad (\forall g)(\mathrm{source}(g) = X \wedge \mathrm{target}(g) = B) \supset \\ \quad\quad\quad (\exists h)(\mathrm{source}(h) = X \wedge \mathrm{target}(h) = A \times B) \wedge \\ \quad\quad\quad\quad (f = h\,; p_1) \wedge (g = h\,; p_2) \end{array} \tag{1}$$

by a diagram such as that in Figure 1. The problem of how to deal with very large formulae is now a major topic in formal proof development. Our aim

Logic, Language, and Computation, Vol. II, edited by Lawrence S. Moss, Jonathan Ginzburg, and Maarten de Rijke.

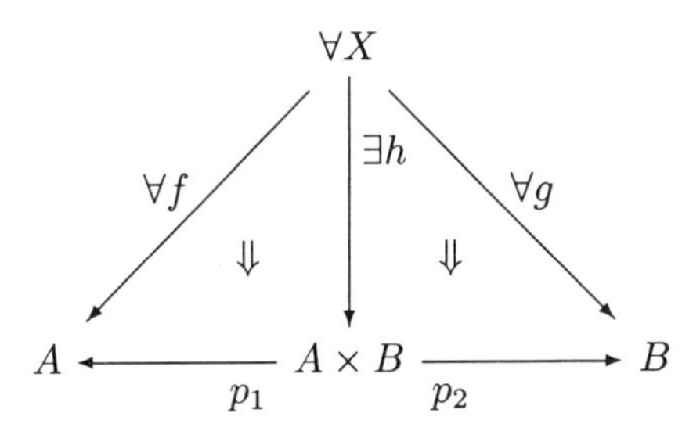

Figure 1: An example of diagram.

is to propose a solution of this problem by providing a good user interface based on commutative diagrams to proof assistants.

While trying to model our commutative diagram interface, we encountered a difficulty which seems to occur when modeling many other user interfaces, too. Our model of user interface have two attributes; items which are the object of computation and items used to communicate with the user. We call the former "computed objects" and the latter "communication objects." The user gives a communication object to the system which then translates it to a *corresponding* computed object and manipulates (or computes) it. Or alternatively, the system has a computed object as a result of some computation, translates it to a *corresponding* communication object, and then shows it to the user. In the case of our system, computed objects are logical formulae and communication objects are commutative diagrams. We have two "correspondences" here, and it would seem natural to search for *functions* to describe them. However, there are no such functions which are satisfactory. Instead, we have to live with *relations*, and the model must be more complicated. So, we propose what we call a *relational model*, and we describe it in Section 4.

This paper is organised as follows. First, we give a formulation of commutative diagrams in Section 2, in terms of directed graphs with cells. Next, we describe the commands of the system. The commands are classified into draw commands and proof commands. We list up all draw commands, and then show how they work by example in Section 3. We then discuss the general framework of the user interface, and propose our own based on relations in Section 4. In Section 5, we explain the proof commands, which are also classified into "pasting" commands, to prove equations, and "logical" commands for each logical inference rule, to prove logical formulae. Finally, related works are reviewed in Section 6.

2 Commutative diagrams

Figure 1 shows a commutative diagram. But what exactly are commutative diagrams? In order to treat them by machines, we need their formal definition. We define commutative diagrams to be directed acyclic planar graphs (i.e., planar DAG) with labels attached to some nodes and edges, together with specified *commutative cells*. A directed graph consists of nodes and edges with the assignment of the source and target node to each edge. By "acyclic," we mean that there are no cycles. So Figure 2(a), for example, is not a commutative diagram. "Planar" means the graph can be embedded

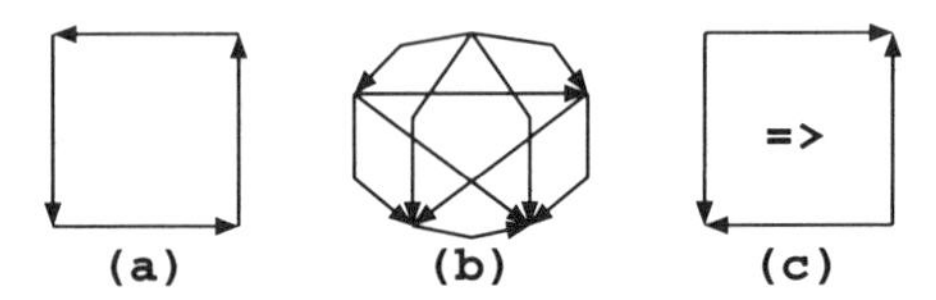

Figure 2: Non-examples of diagrams.

in the plane. Figure 2(b) has no such embeddings, so it is not a commutative diagram. Labels are attached to nodes and edges. Typical labels are variable names, possibly quantified by $\forall$ or $\exists$.

To define commutative cells (Figure 2), we need some auxiliary terminology. We define *cells* to be the minimal region surrounded by edges. Also,

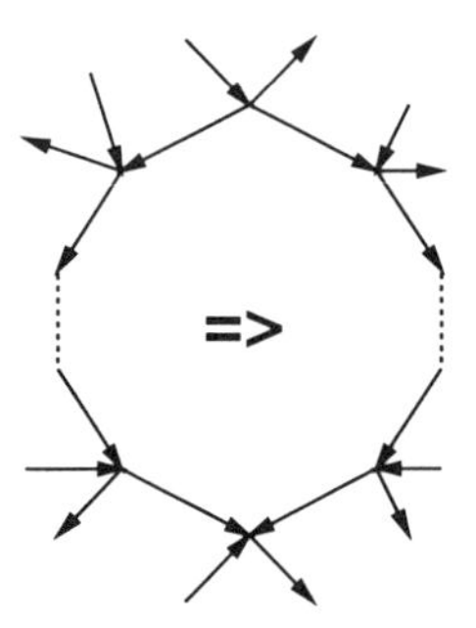

paths of a graph are defined as usual. We construct a *commutative cell* by placing a double arrow in a cell surrounded by just two paths. Figure 2(c) is not a commutative cell, because it is surrounded by four paths.

For example, Figure 1 is a diagram in this sense; it has two commutative cells.

command	description
$\cdot$	put a node
$\rightarrow$	draw a edge
$A, \forall X, \exists h$	labeling to a node or edge
$\Rightarrow$	draw a double arrow

Figure 3: Draw commands.

3 Draw commands

Figure 3 shows the list of draw commands. These commands are used to draw commutative diagrams. The commands are "to put a node," "to draw an edge from a node to another," "to put a label on a node or edge," and "to draw a double arrow in a cell, to make a commutative cell." Clearly, an arbitrary diagram can be drawn using a sequence of these commands.

Figure 4 shows how our example diagram from Figure 1 can be drawn. In (2), we put nodes "A," "$A \times B$," and "B," and draw edges "p_1" and "p_2." We then put another node and attach a label "$\forall X$" to it in (3). We then draw an edge from $\forall X$ to A, and attach a label "$\forall f$" to that edge in (4). Similarly we draw two edges "$\forall g$" in (5) and "$\exists h$" in (6). Since the left cell is surrounded by two paths, we put a double arrow in the cell, to make it a commutative cell in (7). The right cell is similarly made commutative in (8). Thus, we get the diagram as in Figure 1.

Before proceeding further, we give an intuitive meaning to the commutative diagrams. Suppose a node in the diagrams represents a set; so, X, A, B, and $A \times B$ are sets in our example. Also, suppose an edge represents a function from its source to its target, so, f is a function from set X to set A in Figure 1. A path is a composition of functions represented by the component edges. A commutative cell asserts the equality of functions represented by the surrounding paths. The diagram in Figure 1 says the composition of h and p_1 is equal to the function f. Meaning behind the diagram in Figure 1 thus is "for any set X, for any function f from X to A, for any function g from X to B, there exists a function h from X to $A \times B$ such that f is equal to the composition of h and p_1 and g is equal to the composition of h and p_2." The formula (1) is equivalent to this.

The draw commands create not only diagrams but also incrementally create formulae with a specified position, as depicted by the cursor $\triangle$ in Figure 4. We explain with the last drawing example how a formula in each step is generated in Figure 4.

(2): the diagram contains no predicate, so the corresponding formula is the truth $\top$ with the specified position left to itself. (3): "$(\forall X)$" is first

$$A \xleftarrow{p_1} A \times B \xrightarrow{p_2} B \qquad \vartriangle \top \tag{2}$$

$$\begin{array}{c} \forall X \\ \\ A \xleftarrow{p_1} A \times B \xrightarrow{p_2} B \end{array} \qquad (\forall X)\mathrm{Obj}(X) \supset \ \vartriangle \top \tag{3}$$

$$\begin{array}{ccccc} & & \forall X & & \\ & \swarrow^{\forall f} & & & \\ A & \xleftarrow{p_1} & A \times B & \xrightarrow{p_2} & B \end{array} \qquad \begin{array}{l} (\forall X)\mathrm{Obj}(X) \supset \\ \quad (\forall f)(\mathrm{source}(f) = X \wedge \mathrm{target}(f) = A) \supset \ \vartriangle \top \end{array} \tag{4}$$

$$\begin{array}{ccccc} & & \forall X & & \\ & \swarrow^{\forall f} & & \searrow^{\forall g} & \\ A & \xleftarrow{p_1} & A \times B & \xrightarrow{p_2} & B \end{array} \qquad \begin{array}{l} (\forall X)\mathrm{Obj}(X) \supset \\ \quad (\forall f)(\mathrm{source}(f) = X \wedge \mathrm{target}(f) = A) \supset \\ \quad\quad (\forall g)(\mathrm{source}(g) = X \wedge \mathrm{target}(g) = B) \supset \ \vartriangle \top \end{array} \tag{5}$$

$$\begin{array}{ccccc} & & \forall X & & \\ & \swarrow^{\forall f} & \downarrow{\exists h} & \searrow^{\forall g} & \\ A & \xleftarrow{p_1} & A \times B & \xrightarrow{p_2} & B \end{array} \qquad \begin{array}{l} (\forall X)\mathrm{Obj}(X) \supset \\ \quad (\forall f)(\mathrm{source}(f) = X \wedge \mathrm{target}(f) = A) \supset \\ \quad\quad (\forall g)(\mathrm{source}(g) = X \wedge \mathrm{target}(g) = B) \supset \\ \quad\quad\quad (\exists h)(\mathrm{source}(h) = X \wedge \mathrm{target}(h) = A \times B) \wedge \ \vartriangle \top \end{array} \tag{6}$$

$$\begin{array}{ccccc} & & \forall X & & \\ & \swarrow^{\forall f} \ \Downarrow & \downarrow{\exists h} & \searrow^{\forall g} & \\ A & \xleftarrow{p_1} & A \times B & \xrightarrow{p_2} & B \end{array} \qquad \begin{array}{l} (\forall X)\mathrm{Obj}(X) \supset \\ \quad (\forall f)(\mathrm{source}(f) = X \wedge \mathrm{target}(f) = A) \supset \\ \quad\quad (\forall g)(\mathrm{source}(g) = X \wedge \mathrm{target}(g) = B) \supset \\ \quad\quad\quad (\exists h)(\mathrm{source}(h) = X \wedge \mathrm{target}(h) = A \times B) \wedge \\ \quad\quad\quad\quad (f = h\,; p_1) \wedge \ \vartriangle \top \end{array} \tag{7}$$

$$\begin{array}{ccccc} & & \forall X & & \\ & \swarrow^{\forall f} \ \Downarrow & \downarrow{\exists h} & \Downarrow \ \searrow^{\forall g} & \\ A & \xleftarrow{p_1} & A \times B & \xrightarrow{p_2} & B \end{array} \qquad \begin{array}{l} (\forall X)\mathrm{Obj}(X) \supset \\ \quad (\forall f)(\mathrm{source}(f) = X \wedge \mathrm{target}(f) = A) \supset \\ \quad\quad (\forall g)(\mathrm{source}(g) = X \wedge \mathrm{target}(g) = B) \supset \\ \quad\quad\quad (\exists h)(\mathrm{source}(h) = X \wedge \mathrm{target}(h) = A \times B) \wedge \\ \quad\quad\quad\quad (f = h\,; p_1) \wedge (g = h\,; p_2) \wedge \ \vartriangle \top \end{array} \tag{8}$$

Figure 4: An example of drawing.

inserted to the left of the cursor position, resulting the formula $(\forall X) \vartriangle \top$, and then "$\mathsf{Obj}(X) \supset$" is inserted to the left of the cursor position. (4): likewise, "$(\forall f)$" is first inserted to the left of the cursor, then "$(\mathrm{source}(f) = X \wedge \mathrm{target}(f) = A) \supset$" is inserted. We put the precedence of the quantifier to be the weakest, so this formula is parsed as

$$(\forall X)\Big(\mathsf{Obj}(X) \supset (\forall f)(((\mathrm{source}(f) = X) \wedge (\mathrm{target}(f) = X)) \supset \top)\Big).$$

The creation of formulae goes on in a similar fashion. Note that h is quantified by an existential quantifier, so the logical connective used is the conjunction, instead of the implication. The draw command for the double arrow creates equations such as $f = h; p_1$ ($h; p_1$ denoting the composition of h and p_1), and we finally get the requisite formula.

4 Relational model of user interface

Our system displays a logical formula as a character sequence in a traditional way, or as a commutative diagram, as explained earlier. Now, given a commutative diagram typically displayed by the system, it is natural to expect that there is a unique formula represented, so that the user can determine which formula the diagram is showing. In other words, we can determine a function from diagrams to formulae. In terms of implementation, we can also reasonably expect a function from formulae to diagrams, a function which guides the implementation of the display algorithm.

This leads to what we call a *functional model of user interface*, as shown in Figure 5. We not only require the existence of functions f and g back and forth between diagrams and formulae, but also the canonical equivalence relations R between diagrams and S between formulae, as well as the following properties for all formulae ϕ and diagrams δ:

$$f(g(\delta))\ R\ \delta, \qquad \text{and} \qquad g(f(\phi))\ S\ \phi.$$

Here, g followed by f maps a diagram to an equivalent diagram, and f followed by g maps a formula to an equivalent formula. Also, we request a function p from draw command sequences to diagrams.

However, there is no successful definition of such functions! They can be very well defined, but the result (by our definition, at least) has been always unsatisfactory. There are clearly no reasonable functions from commutative diagrams to formulae (Figure 8), but there are functions from formulae to diagrams. However, all the functions that we could define map some formulae to a very cumbersome diagrams. But such over-complicated diagrams are pointless; diagrams are introduced in order to help, not hinder, understanding (Figure 7).

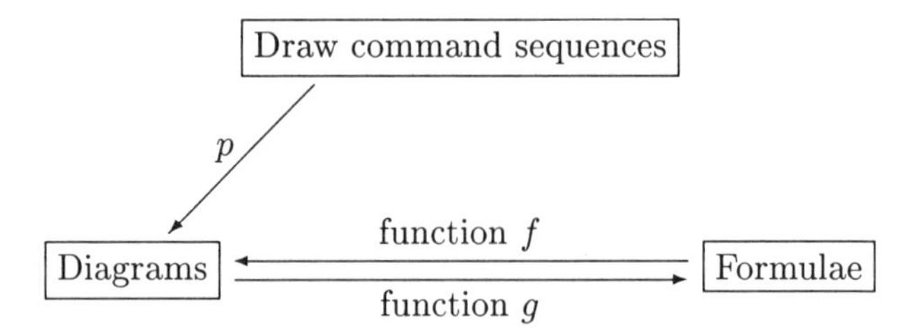

Figure 5: A functional model of user interface.

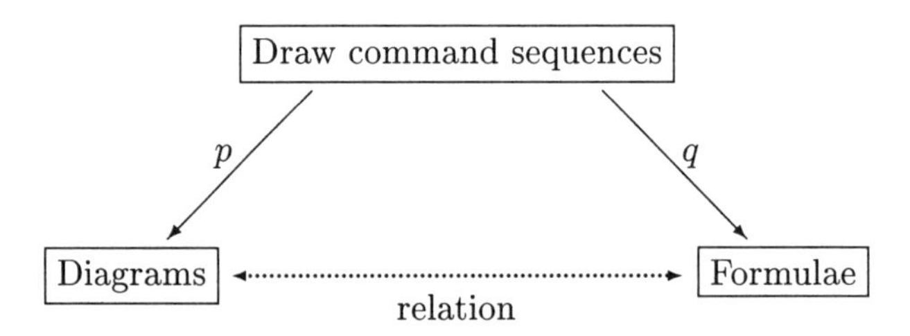

Figure 6: A relational model of user interface.

This shows that the functional model does not work for our case. Instead, we propose a *relational model of user interface*, shown in Figure 6. Here, there is a relation, rather than two functions, between formulae and diagrams. In addition to function p, we also have a function q from draw command sequences to formulae. In the functional model, q was available as the composition of p followed by g, but in the relational model, we need to give q explicitly. We also need canonical equivalences R between diagrams and S between formulae, as in the functional model. For all draw command sequences σ and σ', we postlate the following property:

$$q(\sigma)\ S\ q(\sigma') \supset p(\sigma)\ R\ p(\sigma'). \tag{9}$$

There are only finitely many command sequences mapped to a given commutative diagram. Also, given a logical formula, there are only finitely many command sequences mapped to it. Moreover, those formulae are equivalent in many cases. For example, by changing $(\forall f)$ and $(\forall g)$ draw commands, the resulting formula is equivalent. So, we conjecture that in fact there is a limited, tractable number of logical formulae corresponding to a given commutative diagram.

Since a similar difficulty in finding functions arises in the design of other user interface [8], we believe our relational model is worth generalizing to some class of user interfaces. Interestingly, property (9) recalls us logical relations, the central notion in the theory of parametricity of programming languages, which might support the use of our relational model.

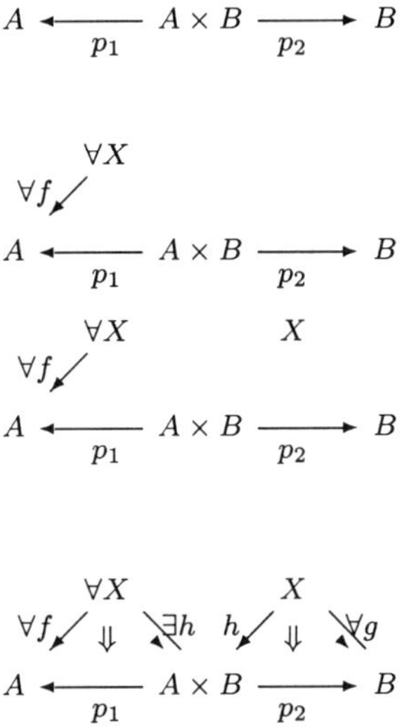

$$A \xleftarrow{p_1} A \times B \xrightarrow{p_2} B \qquad \vartriangle \top \tag{10}$$

$\vdots$

$$\begin{array}{c} \forall X \\ \forall f \swarrow \\ A \xleftarrow{p_1} A \times B \xrightarrow{p_2} B \end{array} \qquad \begin{array}{l} (\forall X)\mathrm{Obj}(X) \supset \\ \quad (\forall f)(\mathrm{source}(f) = X \wedge \mathrm{target}(f) = A) \supset \ \vartriangle \top \end{array} \tag{11}$$

$$\begin{array}{c} \forall X \qquad\qquad X \\ \forall f \swarrow \\ A \xleftarrow{p_1} A \times B \xrightarrow{p_2} B \end{array} \qquad \begin{array}{l} (\forall X)\mathrm{Obj}(X) \supset \\ \quad (\forall f)(\mathrm{source}(f) = X \wedge \mathrm{target}(f) = A) \supset \ \vartriangle \top \end{array} \tag{12}$$

$\vdots$

$$\begin{array}{c} \forall X \qquad\qquad X \\ \forall f \swarrow \Downarrow \nwarrow \exists h \quad h \swarrow \Downarrow \nwarrow \forall g \\ A \xleftarrow{p_1} A \times B \xrightarrow{p_2} B \end{array} \qquad \begin{array}{l} (\forall X)\mathrm{Obj}(X) \supset \\ \quad (\forall f)(\mathrm{source}(f) = X \wedge \mathrm{target}(f) = A) \supset \\ \quad\quad (\forall g)(\mathrm{source}(g) = X \wedge \mathrm{target}(g) = B) \supset \\ \quad\quad\quad (\exists h)(\mathrm{source}(h) = X \wedge \mathrm{target}(h) = A \times B) \wedge \\ \quad\quad\quad\quad (f = h\,;p_1) \wedge (g = h\,;p_2) \wedge \ \vartriangle \top \end{array} \tag{13}$$

Figure 7: A formula does not determine diagram.

5 Proof commands

There are two types of proof commands: pasting commands for proving equations and logical inference commands for proving compound formulae.

5.1 Pasting command—proving equations

There are commands to prove equations between paths of diagrams by *pasting* the cells. As an example, assume we wish to prove $f = g$ from four assumptions $f = (\mathrm{id}\,; f)$, $\mathrm{id} = (g\,; h)$, $(h\,; f) = \mathrm{id}$ and $(g\,; \mathrm{id}) = g$. An informal proof would look like the following.

Assumptions: $f = \mathrm{id}\,; f$, $\mathrm{id} = g\,; h$, $h\,; f = \mathrm{id}$, $g\,; \mathrm{id} = g$.
Goal: $f = g$.
Proof: $f = \mathrm{id}\,; f = g\,; h\,; f = g\,; \mathrm{id} = g$.

$$A \xleftarrow[p_1]{} A \times B \xrightarrow[p_2]{} B \qquad \vartriangle \top \tag{14}$$

$\vdots$

$$\begin{array}{c} \forall X \\ \downarrow \exists h \\ A \xleftarrow[p_1]{} A \times B \xrightarrow[p_2]{} B \end{array} \qquad \begin{array}{l} (\forall X)\mathrm{Obj}(X) \supset \\ \quad (\exists h)(\mathrm{source}(h) = X \wedge \mathrm{target}(h) = A \times B) \wedge \ \vartriangle \top \end{array} \tag{15}$$

$\vdots$

$$\begin{array}{c} \forall X \\ \swarrow \forall f \quad \Rightarrow \quad \downarrow \exists h \quad \Leftarrow \quad \forall g \searrow \\ A \xleftarrow[p_1]{} A \times B \xrightarrow[p_2]{} B \end{array} \qquad \begin{array}{l} (\forall X)\mathrm{Obj}(X) \supset \\ \quad (\exists h)(\mathrm{source}(h) = X \wedge \mathrm{target}(h) = A \times B) \wedge \\ \quad\quad (\forall f)(\mathrm{source}(f) = X \wedge \mathrm{target}(f) = A) \supset \\ \quad\quad\quad (\forall g)(\mathrm{source}(g) = X \wedge \mathrm{target}(g) = B) \supset \\ \quad\quad\quad\quad (f = h\,; p_1) \wedge (g = h\,; p_2) \wedge \ \vartriangle \top \end{array} \tag{16}$$

Figure 8: A diagram does not determine formula.

We show how to prove this using pasting commands. The proof goes on in a "goal-oriented" manner. There is always a goal formula (usually only one) and some assumptions. Our assumptions and goals are shown in Figure 9. The goal diagrams are pasted using some assumptions. In Figure 10, we

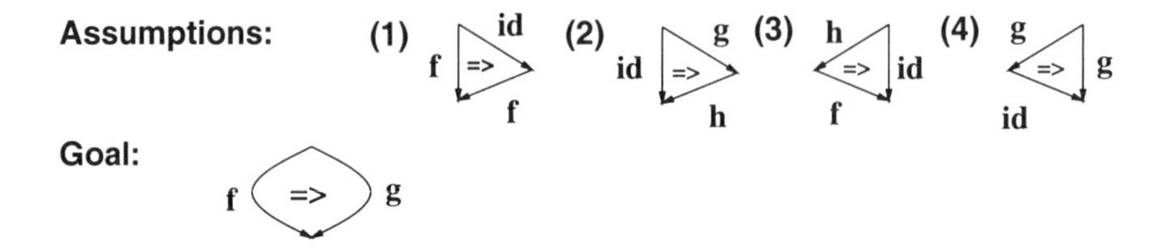

Figure 9: A representation using diagrams.

explain how the pasting command works by example. We paste the diagram (1) in the assumption to the goal diagram. Edge f matches, so we paste it along edge f. Pasted cells are coloured; uncoloured cells become the new goal. This means that we can now prove $id; f = g$ instead of $f = g$. Similarly, the diagrams (2) and (3) are pasted. Finally, the goal cell is the same as in diagram (4) in the assumption. In this case, we can paste the whole diagram. The goal diagram is now completely coloured, so it has been proved. The pasting command is used to prove the diagram by pasting.

We now describe the exact condition under which the pasting commands can be applied. If the assumption diagram exactly matches the goal diagram, the whole diagram is then pasted. In general, when the whole left path of the assumption diagram matches a part of the left path of the goal diagram, we can paste. In this case, the assumption diagram is pasted in the goal diagram like in Figure 11. The pasted part is coloured, and the uncoloured part becomes a new goal commutative cell which is surrounded by two paths.

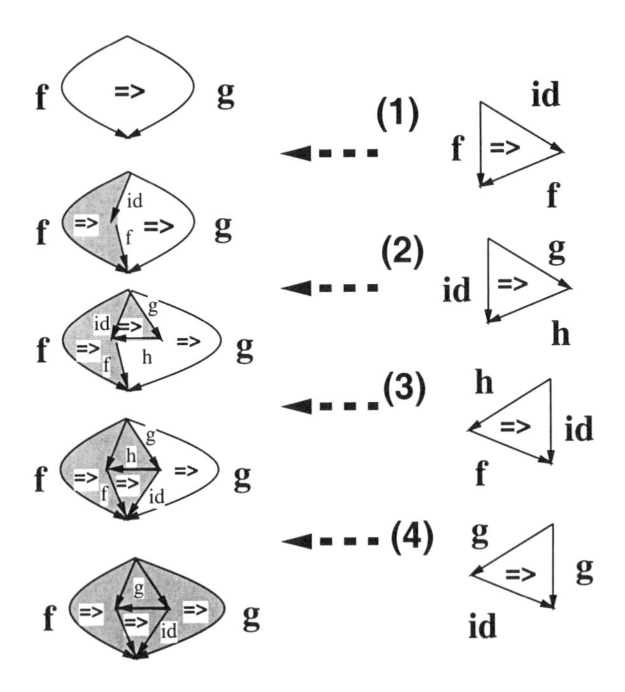

Figure 10: Non-examples of diagrams.

As in Figure 12, if a part of left path of diagram matches the goal, we can not paste it. Our example proof is written within the realm of proof trees shown in Figure 13. It is clear that there are many steps and that it is more cumbersome than the proof by pasting!

5.2 Proof commands for the compound formulae

So far, we dealt only with equations, but we also treat compound formulae as appears in the system LK introduced by Gentzen[5]. We make a usual simplification to LK: we consider the hypotheses of sequents to be a *multiset*

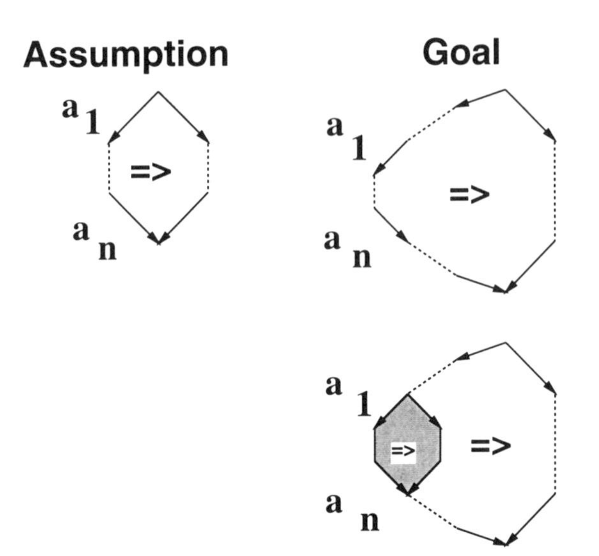

Figure 11: Condition of pasting.

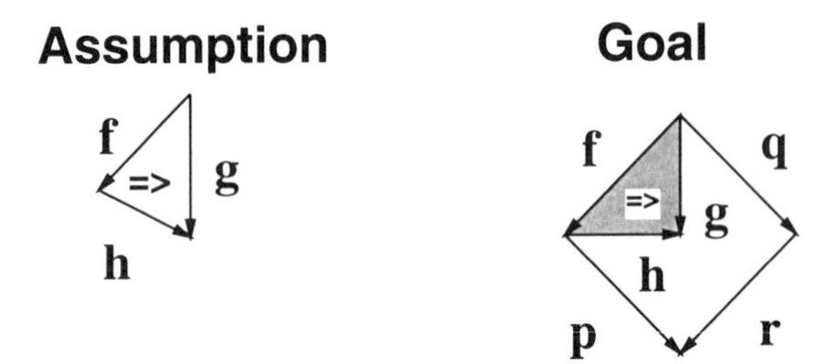

Figure 12: A non-example of pasting.

$$\frac{f = \mathrm{id}\,; f \qquad \dfrac{\dfrac{\mathrm{id} = g\,; h}{\mathrm{id}\,; f = g\,; h\,; f} \qquad \dfrac{\dfrac{h\,; f = \mathrm{id}}{g\,; h\,; f = g\,; \mathrm{id}} \quad g\,; \mathrm{id} = g}{g\,; h\,; f = g}}{\mathrm{id}\,; f = g}}{f = g}$$

Figure 13: Proof tree.

rather than a sequence of formulae.

As we showed in the previous section, our pasting command supports a forward reasoning of equations. In contrast, our proof command supports backword reasoning in the level of predicate calculus. In a proof session, the system maintains a multiset of goal sequents, which are displayed as windows called *tasks.* A task window has typically three subwindows: the *Hypo* and *Consequence* windows containing the multiset of formulae in the hypothesis and consequence part, respectively, of the sequent, and the *Data* window containing the set of variables, which should be made into eigenvariables of the proof tree of the current goal (Fig. 14). So, task windows are displayed during the proof session, but the system attaches a sequent to each task window; that sequent is called the *underlying sequent* of the task window. Such seemingly redundant book keeping is in fact not redundant because of the argument we gave in Section 4.

A proof session proceeds by applying a logical rule (backwards) to the current goal. For instance, if $\Gamma \vdash \Delta, \mathcal{A} \wedge \mathcal{B}$ is the current goal, one may apply the Right-$\wedge$ rule to split the current goal to be $\Gamma \vdash \Delta, \mathcal{A}$ and $\Gamma \vdash \Delta, \mathcal{B}$. If the current goal is in the form of $\mathcal{A} \vdash \mathcal{A}$, it can be resolved by applying the Axiom rule. One repeats such backward inferences until all goal is resolved.

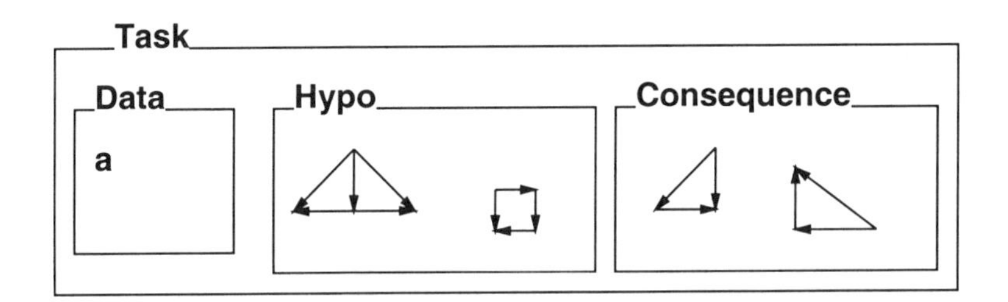

Figure 14: An example of Task window.

Now let the current task T be as follows: its Data set and Hypo multiset are empty, its Consequence is

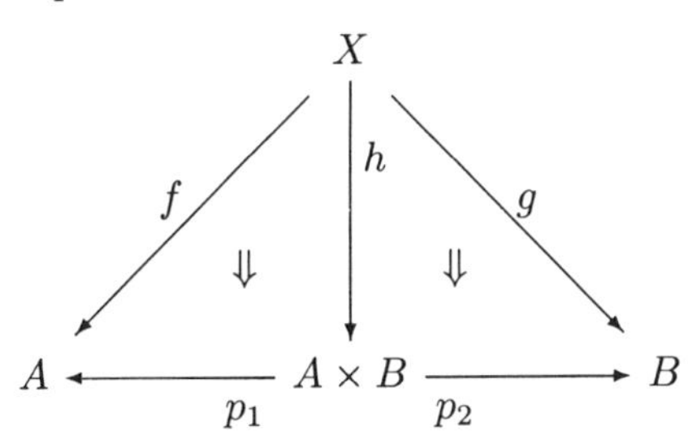

and the underlying sequent is $\vdash (f = h\,; p_1) \wedge (g = h\,; p_2)$. An application of Right-$\wedge$ command with respect to the unique conjunction formula of the underlying sequent splits T into two tasks T_1 and T_2 which are described as follows. The Data and Hypo of T_1 and T_2 are the same as those of T, but the Consequence of T_1 is

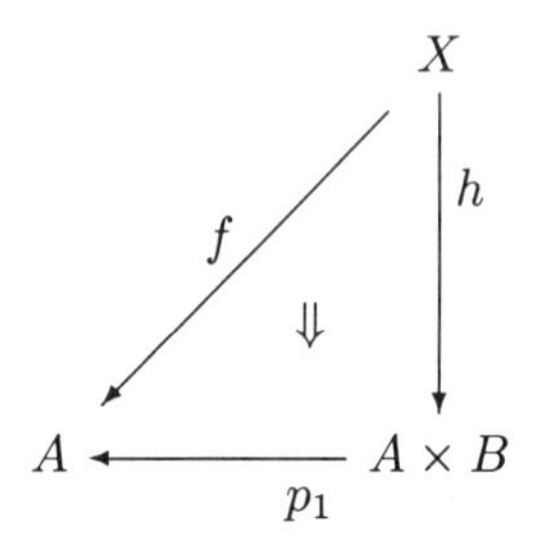

and its underlying sequent is $\vdash f = h\,;p_1$, while the Consequence of T_2 looks like

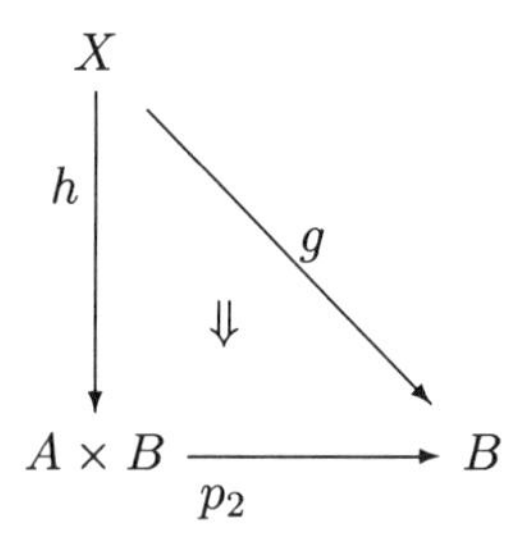

and its underlying sequent is $\vdash g = h\,;p_2$.

In this way, the execution of proof command changes not only the underlying sequent of the task, but also the diagram being displayed.

6 Related works

Freyd and Scedrov defined diagramatic representation for categorical property [4]. In his representation, a sequence of diagrams represents a unique formula. Our system only use one diagram for formula, so one diagram don't decide one formula. The Freyd-Scedrov diagrams are bigger than ours, but the meaning is clearer. Freyd-Scedrov diagrams have an explicit order with which the quantifiers should appear in the logical formula, but the order is ambiguous in our diagrams. We think the merits of our small representaion is more important than the demerits of the ambiguities.

Shin gave a formal system for Venn diagrams, VENN, whose primitive objects are diagrammatic, not linguistic [7]. He also gave a semantic analysis for VENN. Cooper argued for a particular treatment of generalised quantifiers [3]. Barwise and Etchemendy developed a visual logical system Tarski's World which treats first-order logic [2, 1]. Yamamoto et al. showed the formalisation of planar graphs which seems to be applicable to our system [9]. Kinoshita and Takahashi gave a theoretical background of pasting commands [6]. They used an algebraic structure over groupoids arising from the pasting style proofs.

7 Conclusion

We introduced the functional specification of our interface using a commutative diagram for proving predicates in category theory. In the design process, we encountered the notion of relational models for user interfaces, making this worth of further investigation.

Acknowledgments

We very thank Peter Aczel and Steven Vickers for the discussion after our presentation in the ITALLC 96. We thank Masami Hagiya, Mitsuhiro Yamamoto, and Shin-ya Nishizaki for their comments, especially letting us know of their work on formalisation of planar graphs. Hidemoto Nakada implemented an experimental system according to the functional specification given here. We also thank Satoru Tomura and Yutaka Sato for their support in various forms.

References

[1] J. Barwise and J. Etchemendy. *The Language of First-Order Logic (including the Macintosh program, Tarski's World).* Number 23 in CSLI Lecture Notes. CSLI, 1990.

[2] J. Barwise and J. Etchemendy. Hyperproof: Logical reasoning with diagrams. In *AAAI Spring Sympo. Series, Reasoning with Diagrammatic Representations*, pages 80–84, 1992.

[3] R. Cooper. Generalized quantifiers and resource situations. In *Situation Theory and its Applications, Vol. 3*, number 37 in CSLI Lecture Notes, pages 191–211, 1993.

[4] P. J. Freyd and A. Scedrov. *Categories, Allegories.* North-Holland, 1990.

[5] Gerhard Gentzen. Untersuchungen über das logische schliessen. *Mathematische Zeitschrift*, 39:176–210, 405–431, 1934–5.

[6] Yoshiki Kinoshita and Koichi Takahashi. Gropoid of equational proofs. *Bulletin of the Electrotechnical Laboratory*, 60(11):711–718, 1996.

[7] S. Shin. A situation-theoretic account of valid reasoning with venn diagrams. In *Situation Theory and its Applications, Vol. 2*, number 26 in CSLI Lecture Notes, pages 581–605, 1991.

[8] J. Tanaka, December 1995. Personal communication.

[9] M. Yamamoto, S. Nishizaki, M. Hagiya, and Y. Toda. Formalization of planar graphs. In *8th International Workshop on Higher-Order Logic Theorem Proving System and Its Applications*, volume 971, pages 369–384, 1995.

96304
165-

Putting Channels on the Map: A Channel-Theoretic Semantics of Maps?

Oliver Lemon and Ian Pratt

Department of Computer Science
University of Manchester

Abstract We develop a theory of graphical representations (GRs), respecting their spatial nature and their properties of approximation to truth, or verisimilitude. A formal semantics for simple maps, measuring their verisimilitude with respect to regions of the world, is given in terms of "spatial representation systems" formalized using channel theory (CT). The semantic value of a map is relativized to the perspective taken by an agent interpreting it. Several problems are then raised for the channel-theoretic account of representation, which are highlighted by the consideration of more complex maps.
Keywords: graphical representation, formal semantics, channel theory, versimilitude, spatial logics, Geographical Information Systems (GIS).

1 Outline

Various authors have sought a formal account of the functioning of graphical representations (eg: [Pra93, Ham95]). Rather than investigating the semantics of diagrams, we wish to develop a theory of richer graphical information systems such as maps, which describe complex parts of the world. We show that channel theory [SB93, BS94, BS97], with its fine-grained account of representation systems and a notion of spatial "local logics", enables us to handle many (but, as the theory stands, not all) of the special non-standard properties of maps. We concentrate on the graphical phenomena of approximation to truth (verisimilitude) and exploitation of the spatial structure of the graphical medium. The semantics is also parameterised (definition 8) so that the semantic value of a map depends on the interpretational stance of the agent reading it. In addition, the consideration of maps raises some new issues for the channel-theoretic account of representation systems (section 8).

A formal semantics of graphical representations requires an understanding of representational systems in general – a theory of representation.

Logic, Language, and Computation, Vol. II, edited by Lawrence S. Moss, Jonathan Ginzburg, and Maarten de Rijke.

Maps, in particular, exhibit complex phenomena for which a theory of representation should account. Our analysis of graphical representation, in terms of spatial constraints on representations, turns out to have some similarities with the *constraint hypothesis* recently proposed by Shimojima [BS95, Shi96]. However, we also identify the important property of *verisimilitude* of GRs, which we relate to their capacities for representational error.

Channel theory is argued ([BS97] lecture 20) to be a promising formal framework in which to discuss such complex representational phenomena – we wish to explore this claim in relation to maps, in all their glory.

2 Desiderata on a Theory of Representation

Before we introduce the formalism, it is as well to understand the job we intend it to do. We seek a theory of representation general enough to cover graphical representation. Such a theory should:

1. Describe the relation between a representation, interpretational conventions, and the structure which is being represented.

2. Account for the restrictions in expressive power of representational systems which are due to their particular representational media.

3. Be fine-grained enough to describe the variety of errors that are possible in a given class of representation.

4. Account for the robustness of representational systems under error – the fact that they continue to represent when flawed.

5. Systematically describe the differing semantic properties of different representations and interpretations.

6. Be able to predict the semantic effect of processing representations in various ways.

7. Be sufficiently formal to admit of logical and mathematical analysis.

Thus we shall not be restricted in the ways that traditional purely truth-conditional approaches to representation are, although the truth of a representation will remain an important limiting concept. Moreover, a map carries spatial information in virtue of spatial relations between its symbols, and it carries information about the geographical types of objects so represented, in virtue of conventional interpretations of those symbol types (often made explicit by its key).

3 Graphical representation, versimilitude, and imperfect information flow

Representation systems are a focus of study in the search for an understanding of the "logic" of imperfect information flow. The notion of imperfect information, at least as it pertains to representation, has previously been studied by philosophers (see eg: [Odd86]) in the guise of verisimilitude measures. Systems exhibit imperfections in information flow when the connections which render them informative also admit of exceptions, whence mis-information. Real representations, in particular maps, are often (in fact nearly always) imperfect, but they are more often than not good *enough* for most purposes. That a map puts one tree out of place, for example, does not render it useless. So to dismiss all maps as misrepresentations would be rash. Maps are specifically mentioned (along with blueprints, photographs, x-rays, and radar screens) as imperfect information systems, in [SB93, BS94, BS97].

To illustrate further,

> "Consider the radar screen with one blip caused by a sunspot. The positions of the other blips correctly indicate the position of planes in the sky, so there is a sense in which the radar screen is [a] reasonably good representation of the sky, despite the errant blip. Likewise, the photo of Claire's yard represents the area as being hilly and wooded, and it is right; but it shows a tree which has been removed since the photo was taken. It would be quite wrong to say that the photo no longer represents the yard, or even that it is a misrepresentation." (page 22, [BS94])

As stated above, a suitable theory of representation will be able to describe such imperfections and exceptions, and account for the superiority of some representations over others.

Graphical representation systems differ from other representation systems (for example, Natural Languages) in that they are non-ambiguous (there is no ambiguity about the location of their symbols, see eg: [SO95] on *specificity*) and non-sequential (their constituents can be processed in almost any order, as observed by [LS87]). We make an additional claim though; that GRs have these special characteristics as consequences of a more general property – their exploitation of spatial structure. Thus, we claim that –

- GRs have spatial structure[1].

[1] Whereas spoken sentences, for example, are "acoustic objects", which exploit only temporal/sequential structure.

- GRs employ the mereological, topological, and geometrical structure of the plane in representation.
- GRs exhibit "imperfect information flow"; they operate effectively under various degrees of imprecision.

The above phenomena raise a challenge for any candidate theory of representation.

4 Spatial representation systems

In previous research we have established a working definition of diagrammatic representations and some basic concepts in the semantic analysis of languages such as Euler's Circles [Pra93, LP97c, LP97a, SL98]. We can be more precise about the type of representation system in which we are interested here. A "Spatial Representation System" (SRS) is one in which representing tokens are objects whose mutual spatial relations represent spatial relations in the target structure (that structure which is to be represented). Thus many diagrams (eg: physics and engineering diagrams, architect's blueprints) are SRSs, but not all of them are (eg: Venn diagrams). Maps are a type of SRS too, and are particularly interesting in virtue of the types of information which they may *not* explicitly carry: universal, conditional, negative, and disjunctive information are all either implicit in a map (ie: an agent has to reason about the map to derive such statements), or are completely absent. This fact is due to the specificity property of graphical representations, which is due in turn to their exploitation of the spatial medium. One is forced, in conveying any information at all, to make specific commitments about the locations of particular objects. Thus, maps are limited to expressing positive, conjunctive, and existential information, under spatial constraints on relations between their tokens. Furthermore, they may exhibit representational errors.

4.1 The structure of space

A general problem with the logical analysis of GRs (see [Ham95, Lem97]) is that their symbolic representations (being linear structures) do not exhibit (or in general preserve) the appropriate spatial structure of the graphical domain. Indeed, in applications involving visual information it seems that we require formalisms which allow us to encode and process the spatial relationships between representational tokens. In this vein, we have investigated spatial logics, and developed some complete spatial description languages [LP97b, PS97][2]. At issue, then, are the types of constraints that spatial structure imposes on a representation system. For example, topology and geometry impose the following constraints (this list is not exhaustive).

- equidistance (there are at most n+1 mutually equidistant points in an n-dimensional space)
- proximity (in the plane there are at most 6 points which are not near to each other but which are near to points near to a single point[3])
- planarity (certain patterns of connection or overlap between objects cannot be realized in the plane - i.e. the non-planar graphs, see [Kur30])

Thus, depending on which spatial relations are of representational import, different constraints operate on different types of representation. For instance, certain patterns of connection simply cannot be represented in the plane, so a two-dimensional representation which uses connection between its symbols to represent certain properties of a target domain will be restricted in expressive power (see [LP97a, LP97c]). Such properties also lead to efficacy phenomena such as "free rides", as well as problems such as "over-specificity" (see [BS95, BS97]).

5 Verisimilitude and error

As noted, GRs are seldom completely faithful to the structures which they represent. Nevertheless, they approximate those structures closely enough to be useful for a wide range of practical purposes. One way of giving a grade of their versimilitude is to quantify over the various types of *errors* in a representation. As we shall show, the capturing of error phenomena is a major advantage of channel theory over traditional formalisms. Moreover, we show that a verisimilitude measure based on representational error satisfies proposed properties of a verisimilitude ordering. The basic intuition is that those representations exhibiting fewer and less serious errors relative to the structure to be represented, exhibit greater verisimilitude.

5.1 Postulates on verisimilitude

We list the following minimal requirements on any rational notion of versimilitude of representations. Let $S_W PQ$ stand for "P is more similar to the

[2] Note that such logics are rare eg:[AV95].

[3] Where "near" is understood to mean "$\leq$ distance d away".

world (W) than Q is". Where P, Q, R are representations (sets of claims in some logical vocabulary), $v(R)$ stands for the verisimilitude of R, the real world W is the maximally consistent representation, and the ordering $v(P) \leq v(R)$ stands for "R is closer to the truth than P is". Obviously then, we can define $S_W PQ$ iff $v(P) > v(Q)$. The operation $P \cap Q$ is to be understood as producing a representation which consists of all and only those claims upon which P and Q agree. We list the following requirements, for all representations R;

1. $v(R) \leq v(W)$ (top element) "you can't get better than the real thing". i.e. $S_W W R$

2. S_W is transitive, irreflexive, and almost connected (see eg:[vB88] p.26, [Odd86])

3. $v(R) \leq v(R \cap W)$ ("errors are bad"). i.e. $S_W (R \cap W) R$

Thus we claim that, intuitively, verisimilitude is a partial ordering on representations with a top element (the world considered as a representation) and where verisimilitude is the inverse of error. Note that failure to convey certain information is construed here as a type of error. One could opt for a more partial approach, where lack of information is not such a serious defect as mis-information. See [Odd86] for more alternatives here.

5.2 Seven types of error

Given that our approach to verismilitude is based upon the errors in a representation, we now consider the imperfections possible in graphical representations. Maps commonly exhibit the following kinds of errors[4]:

1. Depiction of non-existent objects; eg: a tree is depicted where there is not one in reality.

2. Ommission: non-depiction of (existing) objects; eg: the non-appearance of secret government installations on maps[5].

3. "Thematic" error (incorrect taxonomy); eg: a castle depicted as a church.

4. Spatial error; eg: an object depicted as being in a different place to that which it actually inhabits, or in a different spatial relation with other objects. (This is a specific type of thematic error – one where there is a misclassification with respect to spatial taxonomy).

[4]Sources of error are temporal change, scaling, digitising, initial measurement error, projection. For literature on map error, see [Sch91, Chr91].

[5]The secret air force base "Area 51", for example.

5. Plan error (incorrect depiction of shape); eg: depiction of a square house as circular. This is a particular sort of spatial error.

6. Over-generalisation (depiction of a group of objects as one object); eg: depiction of a group of trees as a forest region. This phenomenon might also be termed "multiple representation", since a single token represents many objects in the target domain[6].

7. Under-generalisation (depiction of one object as a collection of objects); eg: depiction of a city as a group of buildings.

This classification of error is, as yet, non-formal. Indeed, a satisfactory taxonomy of error in GRs can only be given on the basis of a formal theory of representation. Thus, we now turn to the main task at hand; the use of channel theory to formalize the above account of graphical representation.

6 A channel-theoretic semantics of simple maps

A formal semantics of maps and map-like representations is the subject of the pilot study [Pra93] and of ongoing research[7]. Such a semantics describes the relationships between symbols on a map and some region of the world. In the following development, both of these structures (map and region) are construed as *classifications* (in the sense of channel theory); the map is construed as a classification of certain symbols as being of certain types, and the region is a classification of certain objects in a space by way of their geographical properties. An interpretation of a map as being a map *of* a certain region depends on two things; (1) how each symbol on the map is linked to objects in the region, and (2) how the type of symbol corresponds to the properties of objects in the region. Employing the language of channel theory, these two interpretational devices shall be called the *signalling* and *indicating relations* of an "information channel" which exists between the two structures. Indicating relations – that is relations between symbol types and object properties – are either purely conventional (often explicitly defined by a map's key), or in the case that they operate over spatial types, they are objective (since spatial relations within the map correspond to spatial relations in the world[8]). Such relations allow spatial information

[6]Note that this type of error, and the next, only make sense with respect to some predetermined notion of what the correct representation is.

[7]See *http://www.cs.man.ac.uk/ai/oliver/mapsem.html* for developments.

[8]See the "isomorphism thesis" of [Ham95], eg: if symbol A is inside symbol B, then the map claims that the referent of A really is inside the space enclosed by the referent of B.

about a region to be carried by a map. Signalling relations, those relations stating which symbols (tokens of the representation) refer to which objects (tokens of the "world" classification) are less simple; there may be a variety of ways to connect map symbols with objects in the region (for example, if a map user is not sure how to orient a map).

Thus, our semantics of maps operates at the point where each token object in the map has been associated with both spatial and graphical types, and each geographical object (in the world/region) has been identified and classified geographically.

6.1 Representation systems in channel theory

We employ the channel-theoretic account of representation systems, with some additional devices in the graphical context. A "map channel" is taken to be an example of a representational channel (see eg: [BS94], page 23.)

There are two classifications linked by a map channel; (1) an image interpretation, consisting of typed symbols (eg: symbol at map-location l is a castle symbol), and (2) a classification of objects in a region by geographical type (eg: the object at world-location L is a church). Note that each token supports some types which carry spatial information about the token. These structured types are to be governed via spatial "local logics" (which we discuss shortly). Thus each token in each classification must have two and only two[9] types: a "thematic type" (the type of symbol or geographical object that it is) and a location type.

Definition 1 *(Spatial Representation System in CT)*
a) A representation system is a structure $\langle C_1, c, C_2 \rangle$, consisting of two "classifications" (C_1, C_2) and a "channel" (c).
b) A classification $C = \langle S, T, \models \rangle$ is a set T of types, S of tokens, and a classification relation $\models$ between their elements.
c) A channel consists of indicating relations $T_1 \Rightarrow T_2$ between types T_1, T_2 of different classifications, and signalling relations $s_1 \mapsto_c s_2$ between tokens s_1, s_2 of different classifications.
d) A spatial representation system is a representation system in which every token is classified as being of some spatial type.

Thus $s_1 \models T_1$ states that token s_1 is of type T_1; for example "symbol (token) s_1 is a castle-symbol", or "object (token) s_1 is a castle". Channels are typed by indicating relations, so that $c \models T_1 \Rightarrow T_2$ states that channel c is of the type which supports that constraint. Signalling relations are

[9] We assume that thematic types are exclusive, as are location types. For instance – a church is not a castle, a campsite icon is not a carpark icon.

relativised to channels, so that $s_1 \mapsto_c s_2$ states that token s_1 is a signal for s_2 (the "target") relative to channel c.

It is natural to think of these classifications as "representing" and "represented" worlds respectively, and of channels as sets of interpretational conventions (indicating relations) and reference relations (signalling relations). Thus, keeping conventions (the indicating relations) static, different referential interpretations of the representation can be given by varying the way its tokens are associated with those of the represented world (varying the signalling relations). So, keeping the signalling relations and the represented and representing worlds static, we may evaluate different sets of interpretational conventions (constraints). Later (in section 8), we shall look at graphical representation slightly differently, and consider the "distributed" and non-referential character of many graphical tokens in maps.

6.2 Spatial "local logics"

As mentioned, we use the notion of a special structure on types in a classification, called a "local logic"[BS97], in order to capture the spatial nature of the medium. Unfortunately, there is not room here to investigate specific spatial logics (but see [Lem96, LP98a, LP98b]), so an indication of our proposal will have to suffice. The idea of "spatial representation systems" in CT is that all tokens are classified as being of types which encode their position and spatial relations to other tokens. For example, a rail station symbol may lie on a rail track symbol, so the information conveyed by such a map is that the station connects with the track. As discussed, the connection type must obey certain constraints (especially in the 2D case) and these are to be captured by axioms which structure the spatial types in the appropriate way (again, in the connection case, non-planar connection structures must be ruled out in 2D). Some appropriate "local logics" have been identified in terms of modal and first-order axiomatizations, see [LP97b, PS97], although there is much scope for further investigation here.

6.3 Varieties of graphical error

Here the account of error provided by standard channel theory is extended in order to deal with the range of errors occurring in real maps. Thinking about the types of errors which crop up in maps brings to light new issues for the taxonomy of exceptions in channel theory. The following definitions are routine from the literature on channel theory (see eg: [Bar92]).

Definition 2 *(Error of type e_1) Channel $c \models T_1 \Rightarrow T_2$ has a "pseudo-signal" s_1 iff $s_1 \models T_1$ but there is no s_2 such that $s_1 \mapsto_c s_2$.*
(Error of type e_2) Channel $c \models T_1 \Rightarrow T_2$ has a "multi-signal" s_1 iff $s_1 \models T_1$

and there is more than one s_i *such that* $s_1 \mapsto_c s_i$.
Channel $c \models T_1 \Rightarrow T_2$ *has a "clear signal"* s_1 *iff* $s_1 \models T_1$ *and there is a unique* s_i *such that* $s_1 \mapsto_c s_i$.

These definitions correspond to (1) a map which depicts non-existent objects, (2) a map exhibiting a symbol which refers to a variety of different objects (over-generalization), and (3) a map with a symbol which refers to one and only one object in the world. In other words, these definitions cover the cases of representation of non-existent objects, over-generalization, and an "ideal" symbol respectively.

However, consideration of map errors brings to light the following new definitions, covering errors of taxonomy, object ommission, and under-generalization (or "multiple representation").

Definition 3 *(Error of type* e_3*) Channel* $c \models T_1 \Rightarrow T_2$ *has a "null-signal"* s_1 *iff* $s_2 \models T_2$ *but there is no* s_1 *such that* $s_1 \mapsto_c s_2$

Thus, a null signal is associated with a classification when it fails to contain a token which "ought to be there" with respect to the channel and target classification. For example, a map of all the pubs in Edinburgh exhibits a null-signal (with respect to a channel which links icons to pubs) if it fails to display an icon signalling the "Pear Tree" public house.

Definition 4 *(Error of type* e_4*) Channel* $c \models T_1 \Rightarrow T_2$ *has a "classification error"* s_1 *iff* $s_1 \models T_1$ *when* $s_1 \mapsto_c s_2$ *and* $s_2 \not\models T_2$

Thus a token is a classification error if it is incorrectly typed with respect to the token to which it is linked. For example, a pub-type token which signals an object which is not a pub, is misclassified.

Definition 5 *(Error of type* e_5*) Channel* $c \models T_1 \Rightarrow T_2$ *has a "multi-targets"* s_i *iff* $s_2 \models T_2$ *and there is more than one* s_i *such that* $s_i \mapsto_c s_2$

This type of error correponds to the phenomenon of under-generalization or 'multiple representation' - where an object (represented token) is represented by more than one token.

One might think that all the errors accounted for so far here are *non-spatial.* However, note that spatial errors are just a subclass of the classification errors. That is, it is a mistake of taxonomy if an object is misrepresented as touching another one, for example. The spatial errors so captured will depend on the spatial predicates of local logics employed in the classifications. So far we have considered only qualitative spatial relations, but metric errors may also be captured, as we shall show.

The table below summarises:

Map errors	Channel exceptions
non-existent objects	pseudo-signals
object ommission	null-signals
over-generalization	multi-signals
under-generalization	multi-targets
thematic error	classification error
spatial error	classification error

Note that all errors are exhibited by the representing classification – they are all associated with tokens (or missing tokens[10]) in the representation.

Shortly, we shall need to quantify errors in the representations, so we introduce some simple notation; $es_i(M, c, R)$, the error set of type i of a representing classification M with respect to a channel c and a target classification R is the set of errors of type i exhibited by that classification, relative to c and R. The error size of type i (written e_i) in a representation M, is simply the size of this set:

Definition 6 *(Error sets and sizes)*
(Where S_M is the set of tokens of representing classification M)
$es_i(M, c, R) = \{s \in S_M \mid s \text{ is an error of type } i \text{ w.r.t. } c \text{ and } R\}$
$e_i(M, c, R) = \mid es_i(M, c, R) \mid$

6.4 Metric spatial errors in channel theory

We noted that the spatial errors captured so far encode mistakes in qualitative spatial information. However, we have the intuition that an object depicted as being 1 mile from its actual location is a less serious error than its being represented 10

miles adrift. Areal or 'plan' errors exhibit the same behaviour; we want to be able to measure the verisimilitude of ill-shaped symbols. In order to record locative error (and its close cousin, plan error) in CT, we use spatial representation systems as described above. Recall that each token of each classification of an SRS is typed as being at a certain location.

Denote the set of location types in the "representing world" or map by $l_1 \ldots l_n$. Denote the set of location types in the geographical classification by $L_1 \ldots L_N$. Then $s_1 \models l_1$ states that token s_1 is classfied as being at location l_1. Now, an adequate channel between a map and a region will contain indicating relations which systematically connect location types in maps to location types in the geographical classification (systematic in the sense that they preserve the topology of, and scale the metrics on,

[10] Of course, missing tokens only make sense with respect to a represented classification, and a channel.

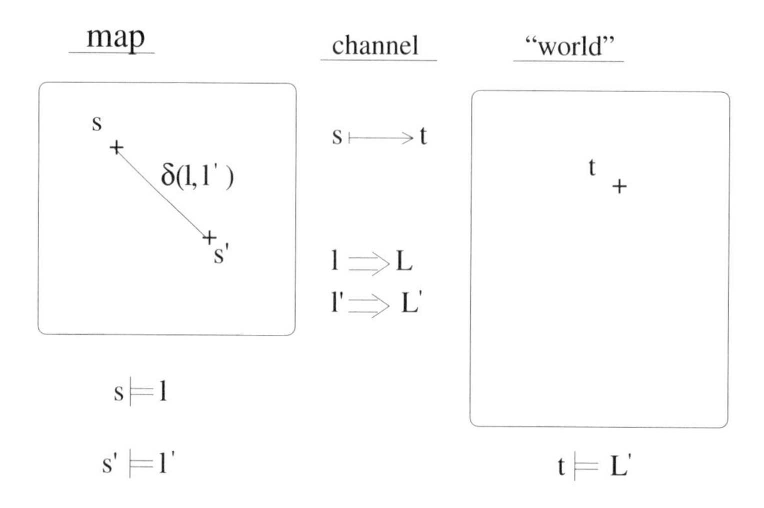

Figure 1: Locative error in channel theory

location types). In the simplest case, assume that $n = N$ (the map and "world" classification contain the same number of locations), and that the constraints of the channel connect them in the standard way - so that there is a 1-1 mapping between locations which preserves the structure of the spaces (eg: if L_1 is right of L_2 in the 'world' and $c : l_1 \Rightarrow L_1$ and $c : l_2 \Rightarrow L_2$, then l_1 is right of l_2 in the map, and so on)[11]. We now measure locative error in the following way: impose a metric $\delta(L_1, L_2)$ between location types internal to classifications. This function can be defined in respect of the structure on location types (ie: assuming a structure on locations, quantify the locations in the shortest path between L_1 and L_2). Then define the locative error of a signal s for target t to be the distance $\delta(l, l')$ between its actual location l and the "correct location" l' (ie: in the representation). The "correct location" l' is determined by the location type L' of the target $t \models L'$ and the constraint $c : l' \Rightarrow L'$ between locations. In order to keep things simple, we rule out multi-signals here (multiply427427 representing token s may have a locative error for each token which they signal). Thus, keeping in mind figure 1;

Definition 7 *(Metric Locative Errors, e_6)*
Where token $s \models l$ (for l a location type), s is not a multi-signal, and for some token t of the target classification R and location type L such that

[11]One could imagine some strange variations here, possibly corresponding to certain projections.

$t \models L$, $s \mapsto_c t$, $c : l \Rightarrow L$, and $c : l' \Rightarrow L'$. Where δ is a metric over location types, the locative error of token s, with respect to c and R, is

$$le_c^R(s) = \delta(l, l').$$

Then for a map M with a set of tokens S,

$$e_6(M, c, R) = \Sigma_{s \in S} le_c^R(s).$$

Locative error is thus a quantitative kind of classification error, where the degree of error is measured with respect to the (spatial) structure on types. We leave open the modification of this definition so as to accomodate plan/areal error.

6.5 Verisimilitude in channel theory

The quality of a representation M, its semantic value $\mathrm{Val}(M, c, R)$, with respect to a channel c and the classification R of which it is a representation, can be given as below.

Definition 8 *(Total Error of a map M)*
for classifications M and R, channel c, errors e_i as defined above[12]*, and parameters α_i such that $\Sigma\alpha_i = 1$.*
The error of M with respect to R under c is given by:

$$\epsilon(M, c, R) = \Sigma_{i=1}^{6} \alpha_i e_i(M, c, R).$$

The numbers $\alpha_1 \ldots \alpha_6$ reflect the context in which the representation is being evaluated (i.e. the "interpretational stance" or "perspective" of an agent). The parameters allow the measure to reflect practical importance of certain types of error relative to certain tasks. For instance, an agent using a map of a minefield will be concerned about errors of ommission and location, thus the values of α_2 and α_6 will be high in comparison to the other α_i (and some parameters may even be zero).

We define versimilitude of representations in channel theory in the following way,

Definition 9 *(Verisimilitude of map M)*

$$\mathrm{Val}(M, c, R) = \frac{1}{1 + \epsilon(M, c, R)}.$$

[12]That is, where $e_1(M)$ gives the number of pseudo-signals in M (with respect to c and R), $e_2(M)$ gives the number of null-signals in M, and so on for classification errors, multi-signals, and multi-targets in M. Finally, $e_6(M)$ gives the sum of all locative errors in M.

Note that $0 < \mathrm{Val}(M, c, R) \leq 1$, and $Val(R, id, R) = 1$ where id is the identity channel. We shall write $V_c(M)$ for $\mathrm{Val}(M, c, W)$, where W is some region of the world. Obviously, this is not the only such measure which can be given here – a multiplicative rather than additive error definition might be employed for example. However, this definition satisfies our requirements on a verisimilitude measure. Thus, with respect to the given desiderata:

Theorem 1 *Val is a rational verisimilitude measure.*

Proof: recall the definition given above (section 5.1). Firstly, $V_{id}(W) = 1$ and $V_c(M) \leq 1$. Next, irreflexivity: $\neg S_W xx$ since $V_c(x) \not> V_c(x)$. Transitivity: $S_W xy \wedge S_W yz \Rightarrow V_c(x) > V_c(y) > V_c(z) \Rightarrow S_W xz$. Almost connectedness: $S_W xy \Rightarrow V_c(x) > V_c(y) \Rightarrow V_c(x) > V_c(z)$ or $V_c(z) > V_c(y) \Rightarrow S_W xz \vee S_W zy$. (In fact, our definition satisfies connectness; cf. [Odd86], p. 12.) Finally, $V_c(M \cap W) \geq V_c(M)$ since $M \cap W$ may only contain errors of ommission, whereas M may also contain other errors – so $S_W(M \cap W)M$.

Representations (classifications) can now be ordered with respect to how well they represent a given classification under a certain metric. For example, an *optimal representation* (non-unique) from a set of competing representations is given by the minimization of all errors over representations with respect to one channel. A *best interpretation* of a given representation is given by the minimization over errors relative to different channels (where both signalling and indicating relations are changeable.)

Having completed this basic exploration of the potential of CT to provide an appropriate theory of representation in the present context, we turn our attention to possible applications of the semantics.

7 Applications

Maps figure among the wide variety of representations processed by computer systems. In particular, map-like representations play a central role in geographical information systems (GISs), which are now used for a wide range of commercial and scientific purposes. Centrally, a GIS is a system for storing and processing spatially related data, of which cartographic information (in the form of electronic maps) is an important part. The computer processing of map-like representations is also important within AI and robotics (eg: [Dav86, DFH93, LN98]). As attempts are made to integrate different spatial reasoning algorithms (e.g. algorithms for map-acquisition, map-updating and route-planning) into such autonomous systems, methods for assessing the individual effectiveness and reliability of appropriate representations and algorithms will become crucial.

The graphical error types captured by the channel theoretic formalism do crop up in the representations of Geographical Information Systems. The table below relates errors in the representations of the GIS Arc/Info (see eg: [Zei94]) and mapping system for a mobile robot employed by [DFH93][13], to the channel theoretic taxonomy of errors.

Arc/Info error type	**Channel Error**	**Robot mapping error type**
dangles	pseudo signals	mis-correspondence
ommissions	null signals	new-looks-old
	multi-targets	old-looks-new
thematic error	classification error	

ARC/INFO processes information, computed from digitised maps, in the form of topological relations between arcs, points and regions (polygons consisting of arcs) representing geographical regions. "Dangles" are small error arcs and nodes which are produced by the process of digitising paper maps. The system performs map-processing algorithms which manipulate information contained in collections of such structures. A good semantics of maps should allow us to consider whether certain of these processes preserve or improve the accuracy/quality of the representations involved. We leave this issue for future work.

8 Further graphical phenomena

We have shown that channel theory is relatively successful in capturing some special properties of graphical representations - in particular their capacity for error. However, there are other, more elusive, graphical properties. Indeed, a detailed consideration of maps throws up issues for the wider account of representation and information presented in channel theory. Considering more complex maps, we query the notion of information sites (tokens) as unstructured objects, and the referential function of tokens as ascribed by signalling relations. Consideration of particular types of map symbol, such as bridges, embankments, and rock sketches, also throws up peculiar problems for a formal semantics of maps based on channel theory. Below we mention six such problems.

When one considers how it is that graphical systems represent, it appears incorrect to construe all representation as occurring via unstructured

[13]Similar representational errors crop up in the "Mercator" system of [Dav86]. Although not officially dubbed a "GIS", Mercator and the system of [DFH93] are systems for the representation, retrieval, and assimilation of geographic knowledge, designed so that mobile robots may build up "cognitive maps" of their environment.

tokens (in CT tokens may be typed in a structured way, but they are not internally structured). For example, a blue-shaded region of a map may not function only as a single representing object (a single token of the map classification) but as a different sort of "distributed" representational device. One may choose points inside the region, or on its boundaries, and consider their properties. The intuition here (very informally) is that rather than being discrete objects, some map tokens may be seen as "substance" or "stuff" representing some geographical type, distributed over certain spatial regions. When one considers only such representational devices as icons (as implicit in the simple system presented above), this intuition is obscured, and the notion of "representing token" seems natural. However, tokens of graphical representations generally exhibit much richer structure than this, so it seems we need some other device (perhaps more complex structured tokens) in order to capture the sorts of "distributed" representational capacity inherent in graphical representations. Tokens of GRs can be thought of as 'geometrical entities', with particular spatial (mereological and geometrical) properties.

Another issue arising for the channel-theoretic account of representation is that of the referentiality of tokens, encoded by signalling relations. Throughout the development of the above system we have used discrete representational tokens which refer to real objects in virtue of signalling relations. However, when one considers the issue of referentiality of map symbols, this stance sometimes seems unnatural. For instance, the presence of a church icon on a map (without additional labelling text[14]) seems not to *refer* to any church, but rather to make an existential claim about the conditions obtaining at a certain location. The semantic value of a map depends on its descriptive, as well as referential, capabilities. So, rather than referring, some map icons may be seen as making descriptive claims. Indeed, it may well be that there are varieties of reference available in maps which have not been considered before. Our question, then is of the ability of CT to deal with elements of representation systems which are descriptive in this way, rather than referential. That is, tokens which do not signal any particular object in the world, but rather, describe properties of locations[15].

But things get more complex still. Consider bridge icons for example. As well as describing the presence of 'bridge-stuff' at certain locations they describe locally non-planar regions of the world, where, for example, a rail line may pass over a road. In such cases the road symbol is often

[14]Our intuitions have it that when there is additional text labelling an icon (eg: "St. Mary's") then reference is secured. This property of 'heterogeneous representation' (see [Bar93, Ham95]) may be captured via parallel channel composition.

[15]Again, this points to taking locations as tokens (rather than types), with a spatial structure over them; i.e. a notion of structured information sites.

occluded by the bridge symbol, but that does not mean that there is no road under the bridge – thus bridge symbols raise more complex issues for a satisfactory formal semantics. There may be parts of linear tokens occluded by bridge symbols (when the bridge goes over them) but these parts ought not to be treated as pseudo-targets (missing tokens). Futhermore, modelling the change in local topology induced by a bridge symbol seems rather problematic in CT.

We need to address the fact that spatial relations between map tokens often carry extra (non-spatial) information about the world. A good example is the convention that a station icon placed *on* a rail line carries the information that the station *serves* the rail line – the relation of 'serving' amounts to more than just the co-location of the station and the line. Thus spatial relations between tokens can encode non-spatial information. This can be captured by allowing some spatial types (e.g. "connects") between certain symbol type (say, train track and train station) to indicate non-spatial types (e.g. "serves"). So some indicating relations have to be restricted to certain classes of tokens.

Particular problems arise when one encounters pictorial tokens on maps – for example, rock sketches and relief representations are drawn so as to create a particular visual impression – and such representational capacities are not captured by the simple notion of representing tokens. Indeed, inspecting many graphical symbols draws out that visual information often has an important phenomenal aspect (see [Lem98a, Lem98b]). For some symbols it is a particular colour, shape, or other visual impression that allows certain information to be carried (and not just their classification as being a symbol of a certain type). In some communication, phenomenal properties of tokens (over and above those that determine syntactic type) are semantically important. Thus, considering phenomenal aspects of the graphical medium raises new challenges for a theory of information flow.

Tokens such as arrows (denoting steepness, flow direction, or a recommended route), slope icons, or embankment regions, present various problems. Some do not describe any physical object, but rather modify the meaning of tokens to which they attach (e.g. a steepness arrow on a road symbol) – so signalling relations again seem inappropriate. All of them crucially involve direction (e.g. of a slope or embankment, or a one-way system), so that directed line segments or regions are the representing (and represented) tokens. So again, more complex geometrical entities need to be invoked as tokens of classifications. While we have reservations about the ability of CT to handle these richer representational phenomena, we do not know of any current formal framework in which they can be accommodated. A model-theoretic semantics for (complex) maps, addressing the above problems, is currently being explored.

9 Conclusion

We began by discussing a formal theory of GRs as *spatial* information systems whose representations exhibit verisimilitude, and found that maps have specific characteristics which make them an interesting test-bed for any candidate theory of representation. A version of channel theory was developed as a candidate theory of representation in this context and a formal semantics for simple maps developed within that framework. It was noted that complete spatial logics are required for the project, in order that 'local logics' may encode the spatial structure of graphical representations. We suggested further possibilities for formal accounts of representation systems.

To summarize, the main issues raised here are:

- Requirements on a formal theory of representation.
- The nature of graphical representation - "imperfect information flow" and spatial constraints due to the planarity of the representing medium.
- Desiderata on rational verisimilitude measures (section 5.1).
- Developments of channel theory.
- Several problems for a channel-theoretic semantics of complex maps.

The "approximation semantics" developed here may have implications for formal semantics more generally; specifically those systems employed in NL semantics (for it seems clear that linguistic representations also exhibit verisimilitude and error phenomena). In addition, work in the semantics of spatial description languages [Lem96, LP97b, PL97] is also of wider application. Future work will involve further formal exploration of the constraints imposed by the structure of the graphical medium, and resulting efficacy phenomena [LP97c, LP98c]. Further non-standard representational properties of GRs, and the problem of capturing them formally, are also under investigation.

Acknowledgements

This research is supported by the Leverhulme Trust (grant no. F/120/AQ) and The British Council's British-German academic research collaboration programme, project number 720. With thanks to the Ordnance Survey, Carl Vogel, and Rodger Kibble.

References

[AV95] Nicholas Asher and Laure Vieu. Toward a Geometry of Common Sense: a semantics and a complete axiomatization of mereotopology. In *International Joint Conference on Artificial Intelligence (IJCAI '95)*, pages 846 – 852. AAAI Press, 1995.

[Bar92] Jon Barwise. Constraints, channels, and the flow of information (second draft). unpublished manuscript, 1992.

[Bar93] Jon Barwise. Heterogeneous reasoning. In Guy Mineau, Bernard Moulin, and John Sowa, editors, *ICCS '93: Conceptual Graphs for Knowledge Representation*, volume 699 of *Lecture Notes in Artificial Intelligence*, pages 64 – 74. Springer Verlag, Berlin, 1993.

[BS94] Jon Barwise and Jerry Seligman. The Rights and Wrongs of Natural Regularity. In James Tomberlin, editor, *Philosophical Perspectives, vol 8* . Ridgeview, California, 1994.

[BS95] Jon Barwise and Atsushi Shimojima. Surrogate Reasoning. *Cognitive Studies: Bulletin of Japanese Cognitive Science Society*, 4(2):7 – 27, 1995.

[BS97] Jon Barwise and Jerry Seligman. *Information flow: the logic of distributed systems.* Cambridge Tracts in Theoretical Computer Science 44. Cambridge University Press, 1997.

[Chr91] N. R. Chrisman. The Error Component in Spatial Data. In Maguire, Goodchild, and Rhind, editors, *Geographical Information Systems, volume 1: Principles.* Longman, 1991.

[Dav86] Ernest Davis. *Representing and Acquiring Geographic Knowledge.* Research Notes in Artificial Intelligence. Morgan Kaufmann, Los Altos, CA, 1986.

[DFH93] Gregory Dudek, Paul Freedman, and Souad Hadjres. Using local information in a non-local way for mapping graph-like worlds. In *13th International Joint Conference on Artificial Intelligence (IJCAI)*, pages 1639 – 1645. Morgan Kaufmann, 1993.

[Dre81] Fred Dretske. *Knowledge and the Flow of Information.* Bradford/MIT Press, Cambridge MA., 1981.

[Gor96] Valentin Goranko. Hierarchies of Modal and Temporal Logics with Reference Pointers. *Journal of Logic, Language, and Information*, volume 5, number 1, 1996, pages 1–24.

[Ham95] Eric M. Hammer. *Logic and Visual Information.* Studies in Logic, Language, and Computation. CSLI Publications and FoLLI, Stanford, 1995.

[Kur30] G. Kuratowski. Sur le probleme des courbes gauches en topologie. *Fund. Math.*, 15:271–283, 1930.

[Lem96] Oliver Lemon. Semantical Foundations of Spatial Logics. In L. C. Aiello, J. Doyle, and S. C. Shapiro, editors, *Principles of Knowledge Representation and Reasoning: Proceedings of the Fifth International Conference (KR '96)*, pages 212 – 219, San Francisco, CA., 1996. Morgan Kaufmann Publishers.

[Lem97] Oliver Lemon. Review of "Logic and Visual Information" by E. M. Hammer (CSLI Publications). *Journal of Logic, Language, and Information*, 6(2):213–216, 1997.

[Lem98a] Oliver Lemon. Review of "Explaining Consciousness: the Hard Problem" (Bradford/MIT press), ed. J. Shear. *Philosophy in Review*, 1998. (in press).

[Lem98b] Oliver Lemon. Review of "Information Flow. The Logic of Distributed Systems" by Jon Barwise and Jerry Seligman (Cambridge University Press). *Erkenntnis*, 1998. (in press).

[LN98] Oliver Lemon and Ulrich Nehmzow. The scientific status of mobile robotics; multi-resolution mapping as case study. *Journal of Robotics and Autonomous Systems*, 1998 (in press). Special issue on scientific methods in mobile robotics.

[LP97a] Oliver Lemon and Ian Pratt. Logical and Diagrammatic Reasoning: the complexity of conceptual space. In Michael Shafto and Pat Langley, editors, *19th Annual Conference of the Cognitive Science Society*, pages 430–435, New Jersey, 1997. Lawrence Erlbaum Associates.

[LP97b] Oliver Lemon and Ian Pratt. On the incompleteness of modal logics of space: advancing complete modal logics of place. In M. Kracht, M. de Rijke, H. Wansing, and M. Zakharyaschev, editors, *Advances in Modal Logic* volume 1, pages 115 – 132. CSLI Publications, Stanford, 1998.

[LP97c] Oliver Lemon and Ian Pratt. Spatial Logic and the Complexity of Diagrammatic Reasoning. *Machine GRAPHICS and VISION*, 6(1):89 – 108, 1997. (Special Issue on Diagrammatic Representation and Reasoning).

[LP98a] Oliver Lemon and Ian Pratt. Complete Logics for QSR: a guide to plane mereotopology. *Journal of Visual Languages and Computing*, 9:5 – 21, 1998. Special Issue on Qualitative Spatial Reasoning.

[LP98b] Oliver Lemon and Ian Pratt. Logics for Geographic Information. *Journal of Geographical Systems*, 1998. Special Issue on Computational Intelligence techniques in Geography, (in press).

[LP98c] Oliver Lemon and Ian Pratt. On the insufficiency of linear diagrams for syllogisms. *Notre Dame Journal of Formal Logic*, (to appear).

[LS87] J. Larkin and H. Simon. Why a Diagram is (Sometimes) Worth 10,000 Words. *Cognitive Science*, 11:65–99, 1987.

[Odd86] Graham Oddie. *Likeness to Truth*. Reidel, Dordrecht, 1986.

[PL97] Ian Pratt and Oliver Lemon. Ontologies for Plane, Polygonal Mereotopology. *Notre Dame Journal of Formal Logic*, 38(2):225–245, 1997.

[Pra93] Ian Pratt. Map Semantics. In Andrew Frank and Irene Campari, editors, *Spatial Information Theory: a theoretical basis for GIS*, volume 716 of *Lecture Notes in Computer Science*, pages 77 – 91. Springer Verlag, Berlin, 1993.

[PS97] Ian Pratt and Dominik Schoop. A complete axiom system for polygonal mereotopology of the real plane. *Journal of Philosophical Logic*, 1998. (in press).

[SB93] Jerry Seligman and Jon Barwise. Channel Theory: Towards a mathematics of imperfect information flow. unpublished manuscript, 1993.

[Sch91] Hansgeorg Schlichtmann. Plan Information and its Retreival in Map Interpretation: the view from Semiotics. In David Mark and Andrew Frank, editors, *Cognitive and Linguistic Aspects of Geographic Space*, volume 63 of *NATO ASI Series D*. Kluwer, Dordrecht/Boston/London, 1991.

[Shi96] Atsushi Shimojima. *On the Efficacy of Representation*. PhD thesis, Indiana University, 1996.

[SL98] Keith Stenning and Oliver Lemon. Aligning logical and psychological perspectives on Diagrammatic Reasoning. *Artificial Intelligence Review*, 1998. (to appear).

[SO95] Keith Stenning and Jon Oberlander. A Cognitive Theory of Graphical and Linguistic Reasoning: Logic and Implementation. *Cognitive Science*, 19(1):97 – 140, 1995.

[vB88] Johan van Benthem. *A manual of Intensional Logic.* Lecture Notes, number 1. CSLI Publications, Stanford, California, 1988.

[Zei94] Michael Zeiler. *Inside ARC/INFO.* OnWord Press, Santa Fe, 1994.

Disjunctive Information

Edwin D. Mares

Department of Philosophy
Victoria University of Wellington
Wellington, New Zealand

Abstract An unresolved disjunction in a situation s is a disjunction made true by the information contained in s such that neither of its disjuncts is contained in s. I suggest that there are reasons to allow the possibility of unresolved disjunctive information in some frameworks and I give a semantics for disjunction that does this task. This semantics postulates paths of situations originating at situations. A disjunction is true at a situation if and only if at least one of its disjuncts is true at some situation on each path. Postulates are given that force the various connectives to behave in an acceptable manner.

1 Introduction

Semantics for non-classical logics typically fuse classical elements with non-classical elements. In my field – relevant logic – most semantic treatments combine non-classical treatments of negation and implication with classical conditions for conjunction and disjuction.[1] Is this appropriate? Of course we cannot answer this question unless we have in mind a particular use or interpretation of the semantics. One tradition in relvant logic begining with Urquhart's [19] takes the indices of its model to be pieces of information or partial information states. On the face of it, pieces of information or partial information states do not resolve all disjunctions. That is to say, we sometimes learn that p or q without learning that p or learning that q. We call such disjuctions *unresolved*.

The problem of unresolved disjunctions is not merely an epistemic problem. Sometimes a partial information state (which I will call a "situation" here because of the debt that the current theory owes to situation theory)

Logic, Language, and Computation, Vol. II, edited by Lawrence S. Moss, Jonathan Ginzburg, and Maarten de Rijke.

[1]One is and should be always afraid of making sweeping claims like this. Fine's [9], discussed below, is a striking counterexample to this, as is the work of Stephen Read, who treats conjunction and disjunction as intensional operators.

contains the information that a certain disjunction holds, but does not contain the information as to which of its disjuncts hold. As we shall see below, indeterministic causal processes yield examples of this sort. If an event e has two possible effects e' and e'', and a situation s in which e occurs does not determine which of e' and e'' will occur, then the s contains the information that either e' will occur or that e'' will occur, but does not contain the information that e' will occur or that e'' will occur. (We shall also see that indeterminacy is not the only source of unresolved disjunctions.)

My aim in this paper is to provide a semantics in which unresolved disjunctions can hold at situations. The basic idea of this semantics is quite simple. A situation contains information about what its extensions look like. In the example of the event e causing one of either two events, we have an unresolved disjunction about the future. According to the disjunctive information in s, there will be a situation in which e' occurs or a situation in which e'' occurs. Here we have a branching structure that represents the information that s contains about the future. We are used to branching models from temporal and tense logics. Here we adopt these models to account for unresolved disjunctions more generally, not just those that contain information about the future.

2 The Problem: Some Examples

The following are four cases of unresolved disjunctions.

Example 1: Suppose that we have a beam of light entering a polariser. The angle of oscillation of the light beam is 45° to the horizontal, that is, all photons in the beam are polarised at a 45° angle. The polariser is a horizontal-vertical polariser. Thus, all the photons emerging from the polariser oscillate either horizontally or vertically. According to no hidden variables interpretations of quantum physics, when each photon that enters the polariser, it is undetermined whether it will emerge oscillating horizontally or vertically.

Now, let situation s be such that there is a beam of photons heading into a polariser each of which is oscillating at 45°. Let us call one of these photons, φ. s is further constrained such that nothing can (according to the laws of physics) interfere with φ's passing through the polariser and ending up oscillating at either the horizontal or the vertical. But s does not include the information that φ emerges from the polariser oscillating at the vertical or the information that it oscillates at the horizontal. *But s does include all the relevant laws of physics.*[2] s, then, is the sort of situation

[2] Jon Barwise has suggested that I could avoid the problem of unresolved disjunctions by distinguishing between the information *contained* in a situation from the information

that Laplace's demon has before it. All the physical information about the world (or a bit of it) until t and the laws of nature. But, contra Laplace, if no-hidden-variables quantum physics is right, the demon does not have sufficient information to predict the path of the electron. (We should also either stipulate the s contains the information that the photon must travel through the polariser and that the apparatus will not be interfered with during its traversal or add a third disjunct meaning '..or the working of the apparatus is somehow disrupted', but for simplicity we will ignore such provisos.)

Thus, it would seem that s contains the information that either φ emerges oscillating at the horizontal or emerges oscillating at the vertical. Let us call this disjunctive piece of information $H(\varphi) \vee V(\varphi)$. It is undetermined in s whether φ will emerge oscillating horizontally or vertically. Thus we can say that it is neither the case that $s \models H(\varphi)$ nor that $s \models V(\varphi)$. In other words, we have an unresolved disjunction.

Example 2. In example 1, the unresolved disjunction appears because of an indeterminacy in the physical system involved. But indeterminism is not essential for there being unresolved disjunctions. Let us suppose that we have an electric circuit with two switches. If both switches are on, then the light attached to the circuit will light, but if either switch is off, then the light will not light. Perhaps the following picture will help.

In s, there is no information about whether switch 2 is off or on. However, s contains the information that switch 1 is on. Thus, in the very short time between when 1 is turned on and when the light is supposed to turn on (or when the information from the lower branch reaches the part of the circuit in s), it is not clear in s whether the bulb will light. But, the information is there that either the bulb will light or it will not. Hence we have an unresolved disjunction.

The difference between examples 1 and 2 is rather interesting. The unresolved disjunction comes about in example 1 from the indeterminacy of

---

carried by that situation. Given this distinction, we could say that s does not contain the unresolved disjunctive information, but rather carries it by means of the laws of nature. In channel theory, for example, this distinction is made rather nicely by distinguishing between sites and channels. Cites carry information relative to channels. Laws of nature, for instance, are considered channels or bases for channels and are not information contained in situations like s.

I hesitate, however, to adopt a semantics that incorporates a rigid form of this distinction because I am worried that I will not be able to fuse it satisfactorily with the work I have done with André Fuhrmann in [14] on counterfactuals. In our semantics for counterfactuals, a counterfactual is evaluated in part by looking at particular situations in which the antecedent is true. Consider a counterfactual that begins 'If it were the case that there were a polarizer and photon entering it and the laws of nature were such that ...'. On our theory, to evaluate such a counterfactual, we need a situation very much like s – one which contains certain initial conditions and certain laws of nature.

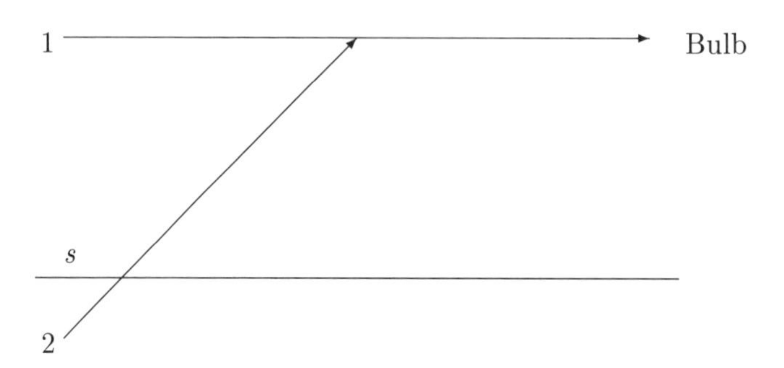

the physical system involved and the partiality of the information involved (i.e. that the information about what occurs after t is not contained in s). In example 2, the unresolved disjunction appears by virtue of the partiality of the information alone. The system in 2 is deterministic.

Example 3. Suppose that the genetic and other biological data available in a situation s is consistent with a given species d being a descendent of species a_1 or of species a_2. The available data rules out that any other species contemporaneous with a_1 and a_2 could be the ancestor of d but does not decide which among a_1 or a_2 is the ancestor. (Again, we probably should add extra provisos to treat other (perhaps remote) contingencies, but the point can more clearly be made with this very simple example.)

Even without formalisation, we can still see quite clearly that a very similar pattern appears to that in examples 1 and 2. We can have unresolved disjunctive information, in this case, about the past.

Example 4. Suppose that when we detect a distant astronomical object it appears quite red. Let us isolate our telescope in a spatially small situation s that does not include the astronomical object. According to s, there is the image of a red object in our telescope. Thus, again simplifying, s contains the information that there is (or was) a red object somewhere in the universe or an object that appears red because it is moving rapidly away from us. In s there is no further information to decide between these two possibilities. Once again, we have an unresolved disjunction.

What is interesting about this fourth example is that it shows that the distance in time between a situation and what its information is about is not the only reason for there being unresolved disjunctions. Spatial distance may also contribute to ambiguity of information available in a situation. Although there is a temporal gap between the state of the astronomical object observed and the observation of it, we can imagine similar cases in which there is ambiguity created by spatial separation in a situation in which the speed of light is infinite (and hence information transfer is instantaneous).

3 The Solution

Before I get to my own formal solution to the problem, let us quickly examine a precursor.

The solution that I propose is a modification of the treatment of disjunction used in forcing methods from set theory. Here I follow the treatment of forcing in [2]. The forcing condition for disjunction is the following:

$$s \models A \vee B \text{ iff } \forall s'(s \trianglelefteq s' \Rightarrow \exists s''(s' \trianglelefteq s'' \;\&\; (s'' \models A \textit{ or } s'' \models B)).$$

It says that a disjunction holds at a point if and only if, for every extension of that point, there is a further extension at which one disjunct or another holds. If we put this into temporal terms (i.e. treating $\trianglelefteq$ as extending information forward temporally), it says that a disjunction is true if and only if, no matter what path the future takes, it is still a possibility that one or other of the disjuncts will come true.

The forcing condition is inadequate in the context of the semantics for relevant logics, at least if one wants to use standard methods to prove completeness. The carrier set for the canonical model for a relevance logic is the set of theories for that logic, including the trivial theory (i.e. the set of wffs). Moreover, $\trianglelefteq$ in canonical models is just the subset relation between theories. In a canonical model so constructed, the trivial theory is greater than or equal to any other theory. In connection with the usual lemma that, for all theories s in the canonical model, $s \models A$ if and only if $A \in s$, we get into a bind. For, given the forcing condition for disjunction, every disjunction would be true at every situation in the canonical model. Clearly, we do not want this.

Another difficulty with the forcing condition (as stated above) is that it looks a bit weak. We want not to say in our analysis of the cases given above that extensions of information in the situation always leave open the possibility that one or other of the disjuncts will hold, but rather that every legitimate path from the situation will eventually contain the information that one or other of the disjuncts holds. This, as we shall see, is exactly the intuition upon which my theory is based. In many contexts, however, we can reformulate a forcing-style condition to look like my sort of condition (choosing as legitimate paths exactly those in which every disjunction true at an origin point is resolved somewhere along the path). So this is not a deep problem, but rather a difficulty about presentation.

As I have said, my solution is couched in terms of a theory of situations. I do not assume much about the constitution of situations, except they carry information by supporting or rejecting states of affairs and by their relations to other situations. As I said, a situation supports or rejects *states of affairs*. A state of affairs is a structure $\ll R, i_1, ...i_n; p \gg$ where R is an n-place relation, the i's are individuals, and p is a polarity (either 0 or 1). If the polarity of the state of affairs is 1, then, according to s, $i_1, ..., i_n$ stand to each other in the relation R. If the polarity is 0, then s says that they do not stand to each other in that relation.

We follow standard practice in situation theory and impose an ordering relation $\trianglelefteq$ on situations. If $s \trianglelefteq s'$, then s is a part of s'. For the purposes of this paper, I assume that all information is persistent. That is, if a piece of information is true at s it is also true at all situations that contain s as a part. Note that this assumption will play a crucial role in the semantics for disjunction. I am unsure how to modify this semantics to treat non-persistent information.

Recall example 1. What seems to be fact about the situation that makes $H(\varphi) \vee V(\varphi)$ obtain in s is that the information is contained in s that there will be another situation s' at $t+1$ such that either $s' \models H(\varphi)$ or $s' \models V(\varphi)$. Let us say that s *determines* that there will be some such situation s', although it does not fully determine which such s' it will be. In other words, s determines what range of futures it will have, but does not determine uniquely what future it will have. To s we assign a set $D(s)$ of sets of situations. Considering the case of the future alone, these sets of situations are futures that s deems admissible. That is, they are the ways in which, according to s, the universe might unfold. In each set of situations $\delta \in D(s)$, there is some s' in which φ has some polarization or other after leaving the polarizer. Thus, in all the futures admissible according to s, φ has a particular polarization (either horizontal or vertical), but this polarization is not the same in all its admissible futures. We generalize from this picture to think of $D(s)$ as the ways in which, according to the information in s, s might be extended (not merely in the temporal sense, but in any sense that we wish).

Formally speaking, my thinking about this topic is a development of ideas found in Fine [9]. Fine's use of "non-prime" theories (i.e. theories that contain unresolved disjunctions) and his techniques for proving completeness for relevant logics was my starting point in developing this theory.

The major difference between my theory and his is that his truth condition for disjunction at a situation s appeals to the cover of prime situations for s. I found this counterintuitive. For, I found no natural way of analysing the examples above in terms of prime extensions.

4 The Formal Theory

In order to give an adequate semantics for disjunction, some constraints must be placed on the function D and its interaction with other elements of our frame theory. The following are some suggested postulates:

1. If $s', s'' \in \delta$ $(\in D(s))$, then either $s \trianglelefteq s'$ or $s'' \trianglelefteq s'$ (i.e. δ is a chain).
2. If $s' \in \bigcup D(s)$, then $s \trianglelefteq s'$.
3. If $\delta \in D(s)$, then $s \in \delta$.
4. If $s' \trianglelefteq s$, then, if $\delta \in D(s)$, there is some $\delta' \in D(s')$ such that
$$\delta \cap \delta' = \{t \in \delta' : \exists t' \in \delta' \; t' \trianglelefteq t\},$$
i.e., $\delta \cap \delta'$ is a final segment of δ'.
5. If $s' \in \cup D(s)$, then
$$D(s') = \{\delta' : \exists \delta \in D(s)(s' \in \delta \;\&\; \delta' = \{s'' \in \delta : \; s' \trianglelefteq s''\})\}.$$
6. If for each $\delta \in D(s)$, there is some $s' \in \delta$ such that s' supports $\ll R, i_1, ...i_n; p \gg$, then s supports $\ll R, i_1, ...i_n; p \gg$.
7. If s supports $\ll R, i_1, ...i_n; p \gg$ and $s \trianglelefteq s'$, then s' supports $\ll R, i_1, ...i_n; p \gg$.

The first postulate states that each way of extending s is itself a series of extensions. Thus, we get a picture of δ as set of concentric "spheres"; of situations that successively contain more and more information. We need this postulate to prove the persistence theorem below. Postulate two states that s is at the centre of these spheres. The second postulate ensures that the inference from $s \models A$ to $s \models A \vee B$ holds throughout the frame. Postulate 3 ensures that, for a situation s with no inconsistent information, disjunctive syllogism holds. That is, if s is consistent, $s \models A \vee B$, $s \models \neg A$, then $s \models B$. Postulate 4 is also needed to prove the persistence lemma and postulates 5 and 6 are needed to prove the absorption theorem stated below.

In addition to frame conditions, we need truth conditions for the various connectives. The only connectives that I will treat are conjunction, negation, and of course disjunction. We begin with some appropriate language. This "language" may be a linguistic entity in the conventional sense used in formal logic, a set of infons, a set of propositions, or what have you. We say that $s \models Ri_1...i_n$ if and only if s supports $\ll R, i_1, ...i_n; 1 \gg$ and $s \models^- Ri_1...i_n$ if and only if s supports $\ll R, i_1, ...i_n; 0 \gg$. Our truth condition for disjunction is

$$s \models A \vee B$$
$$\text{if and only if}$$
$$\text{For all } \delta \in D(s) \text{ there is some } s' \in \delta \text{ such that } s' \models A \text{ or } s' \models B$$

I use a modification of Dunn's four-valued semantics for first-degree entailments (see [7]) that incorporates the possibility of unresolved disjunctions. For each connective in this semantics, there is both a truth condition and a falsity condition. For disjunction, this is the falsity condition:

$$s \models^- A \vee B$$
$$\text{if and only if}$$
$$s \models^- A \text{ and } s \models^- B$$

If we want conjunction to remain the dual of disjunction, we have to amend the usual falsity condition for conjunction to read:

$$s \models^- A \wedge B$$
$$\text{if and only if}$$
$$\text{For all } \delta \in D(s) \text{ there is some } s' \in \delta \text{ such that } s' \models^- A \text{ or } s' \models^- B$$

where the truth condition is as usual, viz.,

$$s \models A \wedge B$$
$$\text{if and only if}$$
$$s \models A \text{ and } s \models B$$

The truth condition for negation itself is made a bit more complicated by the branching semantics. The problem is that with a more standard truth condition, say, $s \models \neg A$ if and only if $s \models^- A$, we cannot prove lemma 2 below. So we adopt the following truth condition.

$$s \models \neg A$$
$$\text{if and only if}$$
$$\text{for all } \delta \in D(s), \text{ there is some } t \in \delta \text{ such that } t \models^- A.$$

This truth condition is not as counter-intuitive as it might seem at first glance. It does, however, force us to provide an alternative to the usual interpretation of $\models^-$. The usual way of reading $\models^-$ is 'supports the negation of'. To understand one possible reinterpretation, let us briefly consider a case of future indeterminacy. Suppose that, the information is currently available that sometime in the future one of two events will occur, α or β. Each of these is incompatable with an event γ. Thus, the information that γ fails to occur is available now. But, which specific event will occur that forces γ not to obtain cannot be known at present. We might read $s' \models^- A$ as saying that some fact incompatable with A obtains at s'.[3]

The falsity condition for negation, on the other hand, is quite simple:

$$s \models^- \neg A$$
if and only if
$$s \models A$$

Now that we have a theory of truth (and falsity) we can prove some simple theorems. These theorems show that disjunction and the other connectives behave normally (at least by the standards of relevant logic).

Theorem 4.1 (Persistence) *If $s \trianglelefteq s'$ and $s \models A$, then $s' \models A$ and if $s \trianglelefteq s'$ and $s \models^- A$, then $s' \models^- A$*

Proof By double induction on the complexity of A.

Case 1. $A = Ri_1...i_n$. The positive and negative cases both follow by semantic postulate 6.

Case 2. $A = B \wedge C$. The positive case follows from the truth condition for conjunction and the inductive hypothesis.

The negative case. Suppose that $s \models^- B \wedge C$. Then, by the falsity condition for conjunction, for each $\delta \in D(s)$, there is some $t \in \gamma$ such that $t \models^- B$ or $t \models^- C$. By semantic postulate 4, there is for each $\gamma \in D(s')$, there is some $\delta \in D(s)$ such that $X \cap Y$ is a non-empty final segment of Y. By the inductive hypothesis, there is some $t \in X \cap Y$ such that $t \models^- B$ or $t \models^- C$. Thus, by the falsity condition for conjunction, $s' \models^- B \wedge C$.

Case 3. $A = B \vee C$. Similar to case 2.

Case 4. $A = \neg B$.

The positive case is similar to cases 2 and 3. For suppose that $s \models \neg A$. Then, for each $\delta \in D(s)$, there is a $t \in \delta$ such that $t \models^- A$. Now, suppose that $\gamma \in D(s')$. By semantic postulate 4, there is some $\delta \in D(s)$ such that $\gamma \cap \delta$ is a non-empty final segment of δ. Thus, by the inductive hypothesis,

[3] A few provisos need to be made here. First, by 'A is incompatable with B' I just mean that it is impossible that A and B obatin together. Now, if we allow impossible situations (as I intend to do), it seems that we should allow that A's being incompatable with B without thereby excluding there being stuations which support both A and B.

Moreover, one might object that

for each $\gamma \in D(s')$, there is some $t' \in \gamma$ such that there is some $t \trianglelefteq t'$ and $t \models^- A$. By the inductive hypothesis, $t' \models^- A$. Generalizing, by the truth condition for negation, $s' \models \neg A$.

The negative case is quite simple. Suppose that $s \models^- \neg B$. Then, by the falsity condition for negation, $s \models B$. By the inductive hypothesis, $s' \models B$. Hence, by the falsity condition for negation, $s' \models^- \neg A$. $\dashv$

Lemma 4.2 *If, for all $\delta \in D(s)$, there is some $t \in \delta$ such that $t \models A$, then $s \models A$.*

Proof By a tedious but, for the most part, easy induction on the complexity of A. The only interesting case is the case in which A is $B \vee C$. So, suppose that, (*) for each $\delta \in D(s)$, there is an $s' \in \delta$ such that $s' \models B \vee C$. Now assume that $\delta' \in D(s)$. By our supposition (*), there is some $s' \in \delta'$ such that, $s' \models B \vee C$. Now, by semantic postulate 5, $(\delta'' =) \{s'' \in \delta' : s' \trianglelefteq s''\} \in D(s')$. By the truth condition for disjunction, there is some $s'' \in \delta''$ such that $s'' \models B$ or $s'' \models C$. Thus, there is some $s'' \in \delta'$ such that $s'' \models B$ or $s'' \models C$. Generalizing, by the truth condition for disjunction, $s \models B \vee C$, ending the proof of the case and the lemma. $\dashv$

Theorem 4.3 (Absorption) *If $s \models A \vee A$, then $s \models A$.*

Proof Follows from lemma 2 and the truth condition for disjunction. $\dashv$

Theorem 4.4 (Double Negation) *$s \models \neg\neg A$ if and only if $s \models A$.*

Proof $\Longrightarrow$ Suppose that $s \models \neg\neg A$. By the truth condition for negation, for each $\delta \in D(s)$, there is some $t \in \delta$ such that $t \models^- \neg A$. By the falsity condition for negation, for each such t, $t \models A$. So, by lemma 2, $s \models A$.

$\Longleftarrow$ Suppose that $s \models A$. Then, by the falsity condition for negation, $s \models^- \neg A$. Let t be an arbitrary situation such that there is some $\delta \in D(s)$ and $t \in \delta$. By the persistence theorem, $t \models^- \neg A$. Generalizing, by the truth condition for negation, $s \models \neg\neg A$. $\dashv$

Theorem 4.5 (Distribution) *$s \models A \wedge (B \vee C)$ if and only if $s \models (A \wedge B) \vee (A \wedge C)$*

The proof of the distribution theorem follows from the persistence theorem and lemma 2.

5 Applications and Extensions

Perhaps the most obvious extension of this theory is to treat the existential quantifier. We obtain unresolved quantified information every day. We see footprints on the beach, and have the information that someone has walked there, but not the information that any particular person has been there. One promising route is to adopt the theory of quantification from Fine [10]. It has many similarities to the present approach. But I will leave the topic of quantification for another occasion. It is too difficult a topic to treat in the same paper as disjunction.

5.1 Probabilistic Causation

To return to example 1, our solution to the problem of that example postulates a branching structure. The situation in which the photon is about to enter the polariser has possible futures of two sorts, i.e., ones in which the photon emerges from the polariser horizontally polarised and futures in which the photon emerges vertically polarised. But, not only are there two possible futures, they have probability values assigned to them. It would be nice if a theory like the present one could be extended to give a semantics for probabilistic causal statements.

There are various ways of extending our theory to include probability. First, we could take the probability value assigned to each branch (at a given time) as primitive. Under some interpretations, this is how the many-worlds interpretation of quantum theory treats probability.[4] Second, we could follow McCall [15] and claim that a proportion of branches (at a given time) carry a certain piece of information and take the probability of that information's obtaining as the proportion itself. The first of these options has all the advantages and disadvantages of allowing unexplained primitives into one's ontology. They tend to do the job, but yield little in the way of understanding. The second option is problematic as well. I will not provide a knock-down argument against it in this paper, but rather make some dogmatic remarks to motivate my favourite solution. McCall's view of probability requires that there be primitive possibilities in the exact number and ratios required by the relevant probabilities. The difficulty for us (and for him) is to provide an identity condition for these possibilities that makes sense of the proportionality thesis. This is difficult if not impossible to do.

Instead, I suggest that we follow Bigelow (see [3] for references) and postulate a metric on situations (in Bigelow's theory, worlds) based on the similarity between situations. Bigelow follows Lewis, et al., in setting a

[4]See, e.g., McCall [15].

similarity relation between worlds. Bigelow, however, takes a further step: he postulates a metric (in the standard measure theoretic sense) between worlds. Using this metric, he develops a probability measurse (in again the standard measure theoretic way) on sets of worlds. In so doing, he treats probability as a measure on UCLA propositions, i.e., sets of worlds, and so treats probability as intimately connected with possibility.

I do not have possible worlds in the present ontology, but we can adopt a similar line to Bigelow's. We can set a similarity metric between situations. We then could look at the set of situations that could obtain at a given time (according to a given situation) and place a probability measure based on this metric on the set of propositions restricted to that set. Then we could not only have a semantics for unresolved disjunctions but one that also attributes a probability to each disjunct of an unresolved disjunction.

5.2 Verificationism

It has always struck me as odd that Dummett and other verificationists adopt the intuitionistic semantics of disjunction. That is, they hold that, in an evidential situation s, $A \vee B$ is verified if and only if A is verified or B is verified.[5] Dummett motivates his view by claiming that a theory of meaning should explain our verbal behaviour, i.e., why we assent to statements when we do (see, e.g., [6] p 216). Dummett holds that a verificationist doctrine does this. We assent to statements when we have the requisite evidence to establish that the statement is correct. The verification conditions for a statement are just the conditions under which we can tell that the statement is assertable. In order, then, to understand the semantics of disjunction, we need first to look at how we use disjunction, i.e., when we are willing to assent to disjunctive statements.

Suppose that I leave my dog, Ramsey, and his sister, Emma, in a room together with a floor lamp. I hear a scuttle of paws, a smash and a yelp. Upon entering the room I see the broken lamp and I know that there has been no earthquake, no wind can get into the room, and so on. I would say that under these circumstances I am entitled to assert:

Emma or Ramsey is responsibe for breaking the lamp.

But I am not entitled to assert

Emma is responsible.

[5] I find some support for this worry in the following passage from Detlefsen's [4]: "It is, however, less clear how or whether the remaining traditional constructivist principles (e.g. the full *disjunction* and existence principles) can be recovered [on Dummett's meaning-as-use motivation for adopting intuitionistic logic]" (p 660 my italics).

or

Ramsey is responsible.

(Once again, if you don't like this disjuction, add to it the disjunct, 'or something else is responsible for breaking the lamp'.) I suggest that we are faced with these sorts of circumstances quite often. Given this, we cannot treat unresolved disjunctions are mere anomalies in any theory that is supposed to explain ordinary language use.

It would seem that we could adopt our semantics to help produce an understanding of assertability conditions for disjunctions. The resulting semantics might just look like Fitting's combining the forcing condition for disjunction with Kripke's semantics for intuitionistic logic (see [11]). But it might not. For it depends on what canons of proof, evidence and inference we integrate into our version of verificationism (we might, for example, follow Tennant [18] and incorporate a version of relevant proof into our theory). Here is a fairly general way, however, to conceptualize our semantics for verificationism: Consider an agent in an evidential situation s. That agent has a knowledge base k. This knowledge base constrains the set of extensions s. That is, according to what the agent knows, s can be extended in various different ways. These ways are the members of $D(s)$. As I have said, exactly what should be our decision about D's relationship to $\trianglelefteq$ and the other elements our semantics cannot be determined outside a more fully worked out verificationism.

References

[1] J. Barwise and J. Perry, *Situations and Attitudes*, Cambridge, MA: MIT Press, 1983

[2] J.L. Bell, *Boolean-Valued Models and Independence Proofs in Set Theory*, Oxford: Oxford University Press, 1985, second edition

[3] J. Bigelow and R. Pargetter, *Science and Necessity*, Cambridge: Cambridge University Press, 1990

[4] M. Detlefsen "Constructivism" in I. Grattan-Guinness (ed.), *Companion Encyclopedia of the History and Philosophy of Science*, London: Routledge, 1994, pp 656-664

[5] K. Devlin, *Logic and Information*, Cambridge: Cambridge University Press, 1991

[6] M.A.E. Dummett "The Philosophical Basis of Intuitionistic Logic" in *Truth and Other Enigmas*, Cambridge, MA: Harvard, 1978, pp 215-247

[7] J.M. Dunn "Intuitive Semantics for First-Degree Entailments and Coupled Trees" *Philosophical Studies* 29 (1976) pp 149-168

[8] H. Everett, ""Relative State" Formulation of Quantum Mechanics" *Reviews of Modern Physics* 29 (1957)

[9] K. Fine, "Models for Entailment" *Journal of Philosophical Logic* 3 (1974) pp 336-346

[10] K. Fine, "Semantics for Quantified Relevance Logic" *Journal of Philosophical Logic* 17 (1988) pp 27-59

[11] M.C. Fitting, *Intuitionistic Logic, Model Theory and Forcing*, Amsterdam: North Holland, 19969

[12] G. Forbes, *The Metaphysics of Modality*, Oxford: Oxford University Press, 1985

[13] T. Langholm, *Partiality, Truth and Persistence*, Stanford: CSLI, 1988

[14] E.D. Mares and A. Fuhrmann "A Relevant Theory of Conditionals" *Journal of Philosophical Logic* 24 (1995) pp 645-665

[15] S. McCall, *A Model of the Universe*, Oxford: Oxford University Press, 1994

[16] A. Rae, *Quantum Physics: Illusion or Reality?* Cambridge: Cambridge University Press, 1994

[17] J.B. Rosser, *Simplified Independence Proofs: Boolean-Valued Models of Set Theory*, New York: Academic Press, 1969

[18] N. Tennant, *Anti-Realism and Logic*, Oxford: Oxford University Press, 1987

[19] A.A. Urquhart "Semantics for Relevant Logics" *The Journal of Symbolic Logic* 37 (1972) pp 159-169

Information, Relevance, and Social Decisionmaking: Some Principles and Results of Decision-Theoretic Semantics

Arthur Merin

Lehrstuhl für Formale Logik und Sprachphilosophie,
Institut für Maschinelle Sprachverarbeitung
Universität Stuttgart

Abstract I propose to treat natural language semantics as a branch of pragmatics, identified in the way of C.S. Peirce, F.P. Ramsey, and R. Carnap as decision-theory. The notion of relevance plays a key role. It is explicated traditionally, distinguished from a recent homophone, and applied in its natural framework of issue-based communication. Empirical emphasis is on implicature and presupposition. Several theorems are stated and made use of. Items analyzed include 'or', 'not', 'but', 'even', and 'also'. I conclude on parts of mind.

This paper submits an approach to meaning, with a focus on broadly non-truth-conditional aspects of natural language. Semantics is treated as a branch of pragmatics, identified as decision-theory in the way of C.S. Peirce, F.P. Ramsey, and of Rudolf Carnap in his later work.

A key theoretical notion, distinguishable from, but intelligibly related to, information is the positive or negative relevance of a proposition or sentence to another. It is explicated in the probabilistic way familiar from Carnap and traditional in the philosophies of science and rational action. This makes it a representation of local epistemic context-change potential that is directional in a precisely specifiable sense and naturally related to utterers' instrumental intentions.

Relevance so defined is proposed as an explicans for Oswald Ducrot's insightful '*valeur argumentative*'. In view of possible confusion among some students of language, it is contrasted with a more recent and idiosyncratic pretender to the appellation, due to Dan Sperber and Deirdre Wilson. The latter proposal turns out, at best, to paraphrase H.P. Grice's non-directional concepts of 'informativeness' and 'perspicuity'. (More informative designations are suggested for it, and for the eponymous linguistic doctrine emanating from parts of CNRS Paris and of UC London.)

Logic, Language, and Computation, Vol. II, edited by Lawrence S. Moss, Jonathan Ginzburg, and Maarten de Rijke.

Another key notion to be introduced as closely related to signed relevance is that of 'issue-based communication'. The paradigmatic communicative situation here, unlike in Grice's work on meaning and in its direct developments, is a bargaining game. Participants with actually or virtually inverse preferences regarding an issue negotiate constraints on joint conduct or epistemic commitments. They make claims and supply incentives. A special case thereof is evidence, the utility of which is given by its signed relevance to issue outcomes.

The first three sections below outline (a) basic assumptions, notably on the use of probability under its judgmental interpretation; (b) some relations to other approaches, notably dynamic semantics and (Neo)-Gricean pragmatics; and (c) reasons for considering the more recent homophone doctrine non-pertinent. The next four sections add some detail, with outlines of a range of results wide enough to justify the term 'approach'. Phenomena addressed with novel accounts include those familiar under the following labels: 'conversational implicature', both particularized and generalized, the latter exemplified with 'or' and the Jespersen/Ducrot/Horn facts on negation in scalar predicate contexts (Section IV); 'conventional implicature', illustrated by 'even' and 'but', an implicational universal of which is shown to engage fairly deep properties of inductive reasoning (Section V); 'presupposition' and 'accommodation' with emphasis on 'also', so as to treat 'but also' (Section VI). Assumptions guaranteeing compositionality of relevance are seen to militate for dichotomous issues (Section VII). I conclude with a thesis on the 'sociomorph' nature of some aspects of language and of parts of mind attaching to it.

I. Introduction: Probability in Meaning and Cognition

'Information' is a partial synonym of 'meaning', and there is little of either without reasoning. The most familiar framework for reasoning is that of deductive theories within a classical logic. Let such a theory, T, represent an epistemic state.[1] Changes of epistemic state will be theory changes. Information brings about changes of epistemic state. Classical theory change is monotonic: what is derivable stays derivable. New things may become derivable; and will so become, if the state transformation is not vacuous. A transformation is vacuous if induced by an uninformative and, in this sense, meaningless addition.

Let $\mathrm{Cn}(\Sigma)$ be the smallest set of consequences of a set Σ of sentences that includes Σ and is closed under conjunction and classical consequence.

[1] Or doxastic state. In strict parlance, the more familiar 'epistemic' relates to knowledge. This is belief (*doxa*) which is, at the very least, also true. I usually say 'epistemic' when meaning 'doxastic'; as many writers do.

Thus, Cn is a Tarskian closure operation. One might thus identify the meaning of a sentence E in epistemic context T with the set-difference $\mathrm{Cn}(T, E) - T$ i.e. with $\mathrm{Cn}(T, E) - \mathrm{Cn}(T)$.

Already for sentential languages and disregarding obstacles to computing inferential closure, this is not a felicitous picture of human beliefs; nor, hence, of meaning.[2] There are two main reasons. (i) Deductively independent sentences A, B may stand in inferential relations—that is: in *relevance* relations—reflected in verbal and other conduct. Indeed, if Hume was right, all empirical knowledge is of this contingent nature. (ii) Epistemic state-change is, in real life, highly *non-monotonic*. Beliefs can be given up or weakened.

The oldest though far from ontologically leanest formalism for inductive, non-monotonic reasoning is the probability calculus. Detailed theories of meaning for natural languages have not, so far, made much use of it. I suspect that some reasons for such neglect have been relatively a priori. Before proceeding to demonstrate the explanatory virtues of probability by worked example I shall, therefore, try to dispel some general misapprehensions about it. Let me start with two positive characteristics.

The axioms of probability properly extend those of classical logic. They constrain belief to be 'coherent' in the following sense: not being disposed, in unwitting practice, to consider favourable or fair a Dutch Book, i.e. a system of bets in which one *must* lose come what may (Ramsey [64]).

Such coherence is not easy to maintain in human practice. But coherence does not require the imputation of epistemic states given as unique, point-valued conditional probability functions $P(\cdot|\cdot)$. What matters is imputation of transcontextual (rationality-embodying) and contextual (local) *constraints* on possible such states. This is much as in a deductive framework. We do not require that someone's beliefs, even when represented by a deductive theory, determine a unique, completely specified possible world. A proposition, i.e. a set of worlds, will do. Analogously there is what R.C. Jeffrey ([40], p. 72) didactically called a 'probasition': a set of probability functions. It is such constraints, not numerical values, that will explain our data. Here, then, are some further praises of probability:

1. Probability is well investigated, motivated, and known. As a proper extension of classical logic it is a natural benchmark for competence

[2] Predicate languages make explicit further difficulties: the anaphoricity and denotational holism of dynamically extending, transclausal variable binding. See Ramsey [66], whose account of discourse meaning I have adapted here, closed sententially and thus trivialized for expository use; and see well-known contemporary versions of 'dynamic semantics' (e.g. Gazdar [22], Kamp [43], Heim [31], Groenendijk and Stokhof [28]) alongside Isard [38]. Some subsentential structure is engaged explicitly in Sections V and VI.

models of defeasible reasoning. At the very least it could play scout to leaner formalisms. Letting it do so instantiates a standard mathematical tactic.[3]

2. Probability trades with cardinal utility, a quantitative representation of preferences. It offers the one principled link, so far, between normative accounts of rational belief and conduct under conditions of uncertainty—one that could be relied upon when real issues are at stake. Other non-monotonic formalisms do not, as yet. A fair bit of non-monotonic inference, i.e. jumping to conclusions, can be recovered through probabilistic updating gated by utility-dependent decision-thresholds. On such a framework for statistical decisionmaking, most of the northwestern world's industrial production of middle-sized dry goods has been running profitably since the 1940s. Under the name 'Signal Detection Theory' the framework has informed much of air traffic control ergonomics and of perceptual psychophysics, the hardest branch of psychology. Probability, with decision-theory built around it, also has nice mathematical structure.

3. With Ramsey [64] we can explicate belief in terms of *dispositions to conduct* (rather than by 'holding true' and concomitant assent or dissent). This is, essentially, the Maxim of C.S. Peirce's Pragmatism. Example: if you believe rain to be more likely than dryness, don't dislike carrying an umbrella on a dry day more than being wet and are equally happy dry with or without umbrella, then take the umbrella along. In doing so you may behave in a way 'extensionally' (i.e. by inspection) indistinguishable from the way you would if you were sure it was going to rain. To the extent that meaning is seen in terms of doxastic state change potential, we can define it by the potential of meaning-bearing entities to change such dispositions.

   Pragmatist construal makes probabilistic belief states plausible. They appear less plausible if we demand introspectible accountability and explicit imputability of belief states, as required by much forensic discourse. But forensic discourse is not the only important language game of humankind. It should not, therefore, circumscribe the class of useful theories of meaning. What it should do, is raise the question of how other language games relate to it.

4. Probability functions can be treated as personal (Ramsey [64]), and thus 'judgmental' (Jeffrey [40]). This suggests an hypothesis about

[3] See Barwise [7] for the return of classical logic in a state-subspace setting. Note also: doxastic $\Box A$ amounts to $P(A) = 1$; $\Diamond A$ to $P(A) > 0$. (See fn. 6, below.)

discourse: Competent speakers behave on the basis of assuming that (sets of, or constraints on) such functions characterize the belief states of one or more of the actual or virtual discourse participants and, more importantly, codetermine their *conduct*. A well-known deductive special case of such a constraint set is the ostensible *common ground* (CG) of things firmly taken for granted by speaker and addressee (Stalnaker [75]). This suggests extending CG to a *common prior* (CP): a probability distribution that also assigns degrees of commitment to less firmly entertained beliefs or ostensible commitments. (I use 'prior' in the everyday sense that will include effects of empirical information accumulated up to some present, here the point of utterance.) CP is an idealizing construct. But so is CG; and so is, in physics, a classical particle mechanics. The suggestion for adopting a construct is, here as there: adopt, if doing so affords predictive, explanatory gains.

5. There is a base in personal probability even for a boolean semantic component or interpretive projection, and its subsentential extension. An example is Hartry Field's [18] 'conceptual role' semantics for classical predicate logic. We need not, thus, refer to extramental *sive* extrasocial truth-conditions to account for native speaker intuitions on acceptability and paraphrase.

6. Problems of sequential belief revision, related to the failure of conditional probability function spaces to be closed under conditionalization, arise analogously in non-probabilistic settings (Gärdenfors [23]).

7. Although the general probabilistic constraint satisfaction problem is in the NP-complete computational complexity class, it has company there. Example: checking a truth-table of 20 atomic constants is a long-term prospect. The general problem class is of the NP-complete type; as is generalized anaphora resolution following Ristad [69].

8. Though it is sometimes maintained that D. Kahneman and A. Tversky [KT] (cf. [42]) showed probability not to be a serious candidate component for competence models of cognition, the putative evidence has not gone unchallenged by well-informed commentators.[4] But suppose one does accept KT's central experimental finding, namely that people, in certain explicit probability assessment tasks, tend to ignore probability 'base-rates'. Then it still provides no evidence for

[4]Birnbaum's [9] re-analysis in terms of statistical decisionmaking of one of KT's evidential assessment tasks did cast some doubt on the conclusion that experimental subjects violated probabilistic reasoning so as to exhibit normative irrationality.

the irrelevance of probability theory even to a psychologistic, descriptive theory of relevance.[5] On the contrary. KT's finding, expressed in standard probabilist terminology, is this: Subjects tend to ignore prior odds ('base rates') to the extent of setting them evens and attend only to probabilistic relevance (of current observation reports); i.e. to relevance as defined in Section III below.

Let me add a few introductory words about the specific application to be made of probability. I shall be attending mostly to phenomena which in recent years have been addressed under the heading of 'pragmatics'. Yet I think there is now nothing to be gained by using 'pragmatics' as a label for meaning-related phenomena either putatively distinct from, secondary to, or parasitic on, 'semantics proper'. For suppose we followed the late 1970's slogan 'Pragmatics = Meaning − Truth-Conditions' (Gazdar [22]). Then the central features distinguishing current Dynamic Semantics from, say, classical Montague Semantics would be part of pragmatics. And if we say: semantics is what is in the mental lexicon, then 'also', 'but' and 'even' are the province of semantics. Point (5), above, concurs; Sections IV, V, and VI below will back this up with evidence.

Accordingly we might as well let *semantics*, as an activity, mean what it presumably meant for M. Bréal, coiner of the name: the description of meaning. And that, most ecumenically, means: the specification of constraints imposed by utterances on possible contexts of use. I propose to use the label *pragmatics* for a particular approach to meaning in natural language and scientific discourse (cf. Skyrms [72]). The approach develops the pragmatism of Alexander Bain, founding editor of the journal, *Mind*, and of C.S. Peirce. Its tools and some still under-appreciated applications are, again, in large measure due to F.P. Ramsey [64].

II. Information and Relevance

Here is a brief review of probability. We start with a Boolean algebra or 'field' $\mathcal{F}$ of propositions. $\mathcal{F}$ may be identified with the set Pow(Ω) of subsets of a set Ω. The elements w of Ω may be thought of as possible worlds, models, or maximal consistent sets of sentences. Probability masses (weights) are assigned to propositions and are normalized to distribute a fixed unit

[5] The chosen province of Sperber and Wilson [SW], who cite KT in apparent support of just such a contention. SW's claim, specified to probability and its import to meaning, is that it is implausible that people should be able to do unconsciously what they cannot even do consciously ([74] p.79). However, the claim is not backed by evidence specific to probability. If it is to be construed as other than unsupported special pleading, the presumption must be that it rests on a global generalization. Such a generalization could be maintained only in disregard of experimental cognitive psychology.

mass. An assignment $P(\cdot) : \mathcal{F} \rightarrow \mathbf{R}$ assigns probability values in the real interval $[0, 1]$ to propositions. Palpably, the empty set must receive zero mass $[P(\phi) = 0]$; the universal set unit mass $[P(\Omega) = 1]$. The mass of the union $A \cup B$ is the sum of the masses of A and of B minus that of $A \cap B$ (not to be counted twice!), i.e. $P(A \cup B) = P(A) + P(B) - P(A \cap B)$. These are the axioms of finitely additive probability, in standard imagery. We can also derive them from definitional constraints on rational action (Ramsey [64]).[6] Conditioning[7] a probability function $P^i(\cdot)$ on a proposition E means to transform it into a function $P^{i'}(\cdot)$ by (i) assigning zero probability mass to $\Omega - E$ (under strict coherence this amounts to transformation of Ω by set intersection with E), (ii) redistributing all probability mass previously assigned to $\Omega - E$ among points/subsets of E so as to preserve proportions of mass among them, and (iii) renormalizing so as to set $P^{i'}(E) = 1$ (i.e. $P^{i'}(\Omega \cap E) = 1$). This kinematic procedure motivates the standard, 'static' definition of conditional probability, $P^i(A|E) =_{df} P^i(AE)/P^i(E)$ [for $P(E) > 0$].[8] (Notation: $AE =_{df} A \cap B$ for sets; $AE =_{df} A \wedge B$ for sentences; $\bar{A}$ reads 'non-A'.)

A. Informativeness, SW-Relevance, and Relevance: Preliminaries

Against this background, Carnap and Bar-Hillel [11] [CBH] consider two explicata of *amount of information*, based on improbability.

Definition 1 $\text{cont}^i(E) =_{df} P^i(\bar{E}) = 1 - P^i(E)$.

When Ω is finite and, moreover, $P^i(\cdot)$ assigns equal probability mass to all $w \in \Omega$, then $\text{cont}^i(E)$ reduces to the relative number of worlds in $\bar{E}$, i.e. to $\text{card}\{w : w \in \bar{E}\}/\text{card}\{w : w \in \Omega\}$. The second explicatum is phenomenologically interpreted by CBH as unexpectedness:

Definition 2 $\text{inf}^i(E) =_{df} \log_2[1/(1 - \text{cont}^i(E)] = -\log_2[P^i(E)]$.

[6] Recall the betting criterion for being 'coherent'. One fails to be 'strictly coherent', if one assigns zero probability to non-empty sets of possibilities. One can then be made to contract a Book by which one cannot win but *may* possibly lose. For simplicity I assume strict coherence; but only, and almost trivially so, relative to given small discourse contexts predicated on background assumptions. For strict coherence on infinite sets of possibilities there is non-standard probability measure, the range of which extends to infinitesimals (cf. e.g. Skyrms [70]).

[7] Also known as conditionalizing. Indices i individuate probability functions. In judgmental interpretation they stand for epistemic contexts, irrespective of issue-specific aspects. I make indices or strongest propositions K such that $P^i(K) = 1$ explicit where doing so makes things clearer. I suppress them where not.

[8] Jeffrey [39] generalized epistemic changes to those occasioned by shifts of probability short of epistemic certainty. I leave discussion of their uses in the present framework for another occasion. Hint: a proposition asserted may acquire greater credibility short of full commitment.

Inf is conveniently additive when A and B are stochastically independent with respect to a context i given by a probability function $P^i(\cdot)$. This means: if $P^i(A|B) = P^i(A)$, equivalently $P^i(B|A) = P^i(B)$, then $\mathrm{inf}^i(AB) = \mathrm{inf}^i(A) + \mathrm{inf}^i(B)$.[9] Entailment-based informativeness, which is prominent in truly formal explications of Gricean conversational implicature (Gazdar [22]; Soames [73]) is related to inf thus: If A semantically entails B ($A \models B$) then $\mathrm{inf}(A) \geq \mathrm{inf}(B)$; and if, in addition, $B \not\models A$, then $\mathrm{inf}(A) > \mathrm{inf}(B)$. The respective converses are invalid.

Inf or cont themselves have not, so far, found consequential use in linguistic pragmatics.[10] What they offer is a license to compare informativeness even of $\models$-incomparable propositions. This recalls one aspect of a notion proposed by Sperber and Wilson [74], [SW], under the name of 'relevance' and for which I propose the appellation 'SW-relevance'. I do so in view of two things: its idiosyncrasy, and its unacknowledged posteriority to a well-articulated notion of relevance which has been in common use under that generic name among philosophers and students of cognition for most of this century.

SW assume that utterances U carry an ostensive presumption of optimal relevance. This practically implies, for one: if U has several alternative interpretations $\{E_j\}$, interpret as that E^* which can be deemed most relevant in the context; if these interpretations go beyond what is minimally (literally?) said or obvious, make requisite assumptions. With Grice [26], [27], I take the heuristic value of such a principle for granted when it is stated with reference to the intuitive explicandum of relevance; and its explanatory value when it is spelt out for a pertinent explication.

However, SW have not, as intended ([74] p. 119), managed to define 'relevance' as a useful theoretical concept; i.e. have not come up with an explication of the intuitive or indeed Gricean notion. SW-relevance of a sentence E uttered in a context is large to the extent that its *contextual effects* are, and small to the extent that E in context is hard to *process*. It remains unclear how the interaction of these two notions, on any of their conceivable interpretations as variables, would determine the total relevance ordering required by the principle. Both notions remain heterogeneous and wholly informal—with one exception.

SW's sole original, formally committed and elaborated kind of effect ([74], pp. 93–117) identifies size of effect of an utterance with the *number*

[9] If Q_j is a generic element of a partition, then $\sum_j P^i(Q_j)\mathrm{inf}^i(Q_j)$ measures the average expected information gain from an answer, in i, to the question: Which element of the partition is the actual world in? With an intuitively different kind of sample space this is Shannon's entropy as applied in coding theory.

[10] Mandelbrot [52], reported in Miller and Chomsky [61], who derived the distributional facts of 'Zipf's Law' from statistical a priori truths unrelated to 'least effort', did not run under that label.

of non-trivial deductive consequences engendered by it, given some deductive system and a background context, and within some *lingua mentalis*. The appeal to cardinality of a set of sentences as a criterion of relevance did not, around 1980, appear well-informed to observers with a background in logic. A number of remedial accretions ensued, notably the proscription of connective introduction rules. This rules out disjunctive weakening, an obvious scourge. It also rules out conjunction introduction, i.e. the inference from, say, $\{E, K\}$ to EK. The resultant logic will thus treat the set $\{E, K, \overline{EK}\}$ as consistent. Remedy, $\wedge$-introduction, would in effect make any logically independent K and E SW-relevant (to some non-nil extent).[11] The 1995 edition still exhibits no intelligible specification of the projected logic or 'cognitive' deductive device, nor of how ampliative assumptions of a Gricean kind would be handled by it.

Suppose, then, that one omits the ill-informed detour through cardinalities. Suppose one takes, instead, the usual route of forming equivalence classes of sentences equivalent modulo classical consequence. Then one ends up with the Lindenbaum algebra of a deductive theory: the proof-theoretic correlate of a semantic Boolean algebra of propositions, where propositions may be conceived of as sets of worlds or of models. On this basis a close to closest counterpart to the SW proposal that yields a total ordering of size of effect would be some (non-negative) measure of conditional information. Call, then, E 'non-degenerately K-informative' iff $\mathrm{Cn}(K, E)$ is consistent and $\mathrm{Cn}(K, E) - K$ non-empty, where Cn is Tarskian. Think of K as a background context made explicit. The stochastic extension of $\mathrm{Cn}(K, E) - K$ is then measured by (cf. CBH [11])

Definition 3 $\mathrm{inf}^i(E|K) =_{df} -\log_2[P^i(E|K)]$ is the *relative amount of information* in i of proposition E *conditional on* proposition K (intuitively: additional to that of K).

This yields as a

[11]What SW would appear to have proposed ([74] p. 117n26) along such lines for the prominently featured monotone incrementation case ($\mathrm{Cns}(K) \subseteq \mathrm{Cns}(K, E)$) is that 'effect' is an increasing function of the size (i.e. cardinality) of a set $(\mathrm{Cns}(K, E) - \mathrm{Cns}(K)) - \mathrm{Cns}(E)$ of sentences of *lingua mentalis*. The interpretation of their consequence operation, here notated 'Cns', remains obscure. (My K stands for their C, my E for their P.) But suppose 'deducible' were to have a meaning more specific than, say, 'coming to mind when the mind is set going appropriately'. Then, for a context given by a (set of) sentence(s) K and a distinct simple contribution sentence E, there will be some effect if Cns demands closure under conjunction; and none if $\wedge$-introduction is proscribed. (SW do not offer a theory of nonmonotonic theory change by anything like the standards set in literature available by the early 1980s, though they assume non-monotonic revision capabilities. For something relating to such revision see fn. 12, below.)

Fact 1 $\text{inf}^i(E|K) = \text{inf}^i(EK) - \text{inf}^i(K)$.[12]

This function is, like SW's cardinalities, always non-negative; and zero, when E already follows classically from K. If K is the background and is already absorbed into i, then the function reduces to $\text{inf}^i(E)$. The function also correlates with the abovementioned schema for monotone information incrementation, familiar from Ramsey, Stalnaker, Gazdar, Kamp, and Heim. Such a function (or the variants of fn. 12) is not, of course, what SW proposed. Rather, what it represents is a closest intelligible counterpart of their apparent notion of 'effect' in a theoretical setting which does harbour a notion of relevance meeting Carnap's [10] general desiderata for the explication of a pre-theoretical notion.[13]

The 'effect' component of SW-relevance turns out, at best, to be a paraphrase or rediscovery of the first part of Gricean or Neo-Gricean Informativeness ('Be as informative as needed!'). At best: its distinctive and original features, which may have disguised the relationship to some, are just its technical infelicities. The 'effort' component of SW-relevance is essentially a conflation of the second part ('Be no more informative than needed!') with Grice's Maxim of Manner ('Be perspicuous!').[14] It has not,

[12]Proof: $\text{inf}^i(E|K) = \log_2[1/P^i(E|K)] = \log_2[P^i(K)/P^i(EK)] = \log_2[P^i(K)] - \log_2[P^i(EK)] = \text{inf}^i(EK) - \text{inf}^i(K)$. Subtracting $\text{inf}^i(E)$ as well would yield a symmetric function, $-\log_2[P^i(E|K)/P^i(E)]$, representing, modulo sign inversion, positive or negative relevance of E and K to one another (see below). Would the absolute value of that be another closest counterpart? If K represents the context of what is fully taken for granted, this function is zero. If not, and if K is still a proposition or deductively closed (set of) sentence(s), then see fn. 21, below. If, taking account of SW's mention ([74] p. 117) of erasure and of assumption strength modification, the context were to be represented by $P^i(\cdot)$ itself rather than by a proposition K in its domain, then the closest counterpart of effect size for E would simply be $\text{inf}^i(E)$; i.e. $\text{inf}^i(E|K)$ when K is taken for granted. This is the most likely of intelligible counterparts, all things considered.

[13]The desiderata are closeness to intuition, formal intelligibility, empirical fruitfulness, and (last and least) simplicity. SW-relevance meets none of them. As editorial comment p. 736 of *Behavioral and Brain Sciences* 10 (1987) put it, presumably regarding the first and third: '[A]re we, like St. Exupéry's star-counting businessman, just indiscriminate implication-collectors, restrained only by our limited resources?' No doubt [74] and publications invoking it discuss language phenomena that have much to do with relevance. But it is one thing to engage phenomenological intuitions that one might associate with the pre-theoretical label 'relevance' and relate them under predictively non-committal use of otherwise familiar terminology ('encoded', 'inference', 'procedure', etc.). It is another thing to have a theory of the main explicandum; or indeed one that is not pre-empted by earlier, if technically more demanding theories. The SW doctrine offers no apparent advance over prior, formally intelligible work on context-dependence of meaning, context-change, and relevance. Nor has it ever been argued to outperform, or just predict as specifically as, Ducrot's informal work on non-truthconditional meaning. The main concept of that, 'argumentative value', admits of a pertinent explication (see below).

[14]Apparently concurring now (^{2}1995) on both Informativeness and Manner: SW [74] p. 268n30.

so far, delivered value added to the plain injunction either. What SW-relevance, however, misses out on altogether is relevance.[15]

In a quantitative or qualitative stochastic epistemic setting, *relevance* is a relation between two propositions and an epistemic state ('epistemic context'). Intuitively put, E is *relevant* to H in a context iff learning its truth would affect the probability assigned to H. It is *positively* relevant to the (quantitative) extent that it raises it, *negatively* to the extent that it lowers it. Thus, relevance is relevance *to* a point, and a properly disjunctive notion: of positive and of negative relevance.[16] I shall reserve for this generic notion the equally generic designation, under which it is familiar, e.g. from Keynes [47] p. 54, Carnap [10], and Jeffrey [39]; and indeed familiar to any well-informed philosopher of science and cognition. A formal exposition is given in Section III, including a comparison with conditional informativeness. However, before proceeding there, it is worth looking at the action-theoretic concomitants of relevance thus explicated.

B. Ducrot on the Role of Argumentation

Relevance comes into its own when the context involves uncertainty about the resolution of an *issue*. An issue is a partition of the space of possibilities into two or more propositions to which *conflicting interests* attach. If our paradigmatic discourse situation is not one of simple 'information transfer', but of explicit or tacit *debate*, then adoption of some such proposition at issue will be the speaker's argumentative *goal*.

This idea, implicitly specialized to dichotomic issues $\{H, \overline{H}\}$, informs the approach to 'conversational' and 'conventional' implicature of Oswald Ducrot [16] and his sometime student Jean-Claude Anscombre [2]. Anscombre and Ducrot [AD] explained implicatural phenomena through rankings of propositions by what they called their *valeur argumentative* with re-

[15] As Grice also noted (see fn. 16). Nevertheless, significant numbers of students in the language sciences that have not benefited from something like rigorous prior training are now being misled by the self-conferred appellation 'Relevance Theory' (see also fn. 21). Hence, a response other than impassive neglect seems called for. I suggest, for one, use of the more specific and non-judgmental designation 'Sperber-Wilson Theory' for purposes of reference to the doctrine of which SW-relevance is the centrepiece, and [74] both the origin and citation invariant.

[16] Is not precisely this what is wanted? Recall: 'To judge whether I have been undersupplied or oversupplied with information seems to require ... the identification of a focus of relevance ...; the force of this consideration seems to be blunted by writers like Wilson and Sperber who seem to be disposed to sever the notion of relevance from the specification of some particular direction of relevance.' (Grice [27], p. 371f.) Grice's 'direction' as stated remains unspecific between two properties: being directed at a particular point, and having potentially positive or negative (besides zero) polarity of direction. The notion he criticised instantiates neither property. The traditional, probabilistic explication of relevance among propositions instantiates both: one cannot be had without the other.

spect to an ostensible ulterior conclusion H. Often enough, though not in general, H would be explicitly given by linguistic context. Thus, in AD's favourite didactic format, some sentence 'E' is uttered in support of some immediately prior utterance expressing H. Here is a pulp fictional example, inspired by a real case and for a task little discussed by AD, disambiguation:

> [H] *You will now give me your wallet.*
>
> [E] *If not, I'll let you have it.*
>
> [There is a bag of trash/garbage in common view. Speaker is also seen to be holding a gun.]

What interpretation, as speakers of British English, are we and Addressee (A) to give 'E'? That Speaker will let A hang on to the wallet if A doesn't hand it over? That she will present him with a disvaluable commodity, the garbage? Or with the gun? Or that she will shoot him? Most likely the last, the strongest argument in favour of acting in accordance with H.

Often enough such an H is not given, or not to be trusted as being the real, perhaps ulterior goal. And then A, or any hearer, must 'abduce' a hypothesis H^* such that the relevance of (let us simplify: unambiguous) E to H^* is maximal among the alternatives H_j in the taken-for-granted context K. This is what 'figuring out the speaker's apparent and real intentions' is all, or mostly, about. On this view, speakers S do not engage in idle gossip, do not proselytize for the sake of it, but speak to a *point*.

C. Issue-based Communication and the Bargaining Situation

The partisan, argumentative approach of AD can be generalized to something which, in principle, will encompass purely imperatival discourse. For note that the point just referred to is the ostensible adoption, by the interlocutor A, of a proposition H as a constraint on A's conduct or, indeed, as a *constraint on S and A's joint future*. I propose to call discourse of this kind *issue-based communication*. In game-theoretical terms, the constraint to be negotiated is a 'jointly randomized strategy' (cf. Harsanyi [30]). This joint strategy may well remain non-degenerately probabilistic, and persuasive efforts will always be attempts to shift probabilities. Interpersonal pragmatics on this view is interpersonal decision theory. Here it is, more particularly, the decision theory of bargaining.

This does not at all rule out the rational reconstruction of the stability of conventions as an equilibrium in a game of pure cooperation (Lewis [50]). Social institutions, including any sort of language game, are, as far as I am concerned, conventions in the sense of Lewis. But so are the rules of chess, which is a game of pure competition; and so are the written or unwritten rules of trading, which, if anything, is a bargaining game of mixed motives. Pure cooperation which rationalizes persistence of a convention does not prescribe it to be, in turn, a convention of pure cooperation. We cannot, thus, in general presume what Grice, Lewis and implicitly SW do presume. This is that the linguistically *paradigmatic* forms of social interaction, even when fully conventional, have the structure of a 'team game' (of pure cooperation) where all the actors have identical or at least compatible prima facie preferences in aiming for the efficient exchange of information.

On the contrary. Whenever there is debate—whether or not you should give me your wallet; which is the most suitable meeting place; the tastiest animal to eat; which of H and $\overline{H}$ is true—proponent and sceptic have locally incompatible preferences regarding H for as long as H is an issue between them. Such agents are autonomous in the vulgar (that is: not necessarily in the technical Kantian) sense of asking

*Why ** should I (do or believe that)?*

And the answer will be: Because there are incentives in the form of sanctions, rewards, and—as a special case for the proximally pure, epistemic case—evidence (Merin [54]).

This is what lends the paradigmatic social situation the structure of a *bargaining game* (Nash [62]), proposed by Harsanyi [30] as a widely applicable model of social interaction.[17] For our purposes 'paradigmatic' means: having a theory that has a model in the kind of linguistic phenomena we want to explain. No more; no less. This empirical issue is, I believe, independent of the issue of whether or not there must be, underneath, a Lewis-construable convention.

Types of bargaining games are manifold, tokens notoriously 'brittle'. So chances are that only very robust features, shared by the whole class of bargaining games, can have predictive import for the structure of linguistic meaning. And I think: they actually do.

Even for the degenerate case, an 'ultimatum' game (example: law-abiding customer vs. fixed-price store) we know: if a proponent, Pro, makes a claim, Pro won't object to the respondent, Con, conceding more, i.e. a windfall to Pro, but will mind getting less. Con, in turn, won't mind giving away less than conceded, but will mind giving away more. Put simply: claims are such as to engender intuitions glossable 'at least'; concessions, dually, 'at most'. We shall see (Section IV) that this very robust feature of all bargaining games already affords a lot of linguistically useful structure.

Speaker's *utility* attaching to an imperative/jussive E is, prima facie, the utility attaching to the prospect of its realization. What, then, is it for

[17] Against the backdrop of which other kinds of interaction—check: 'love', 'urbanity'—gain their distinction.

indicative/declarative E? Prima facie, it is the *relevance* of E to an ulterior epistemic or deontic constraint. We can indeed explicate relevance to a given proposition as an instance of a utility, i.e. a real-valued representation of preferences.[18] And the explicans here is simultaneously a representation of epistemic state change potential (see Section III).

So we are back with Peirce's and Ramsey's notion of belief as a disposition to conduct; and with interpersonal efforts to transform those dispositions. This helps to embed into the best-developed theory of rational conduct available today AD's approach to natural language.[19] AD shied away from stochastic interpretation, eschewed formalization, and viewed argumentation as something *sui generis* rather than as a special case of a bargaining situation.[20] In consequence, they were unable to equal the predictive commitment of the most rigorous 1970's work on, say, implicature (Horn [34], as developed by Gazdar [22], Karttunen and Peters [45], Soames [73]). But their prolific and subtle insights are the principal linguistic inspiration behind what follows. Moreover, their central notion admits of an explication in terms of Carnap's notion of relevance, which does afford such commitment.[21]

III. Decision-Theoretic Semantics (DTS)

What are the implications of basing the analysis of meaning on one rather than the other of two conceivable ideologies or super-objectives of communication: (i) the (Neo-)Gricean view of optimal information transfer, apparently shared by SW; and (ii) the partisan, persuasion-oriented view, adopted for linguistic analysis by AD?

[18] The relevance function $\lambda X[P(X|H)/P(X)]$ of W.E. Johnson and J.M. Keynes [47] is a cardinal utility in the technical sense of Jeffrey [39]. Proof in Merin [58].

[19] All in French. Horn [35] and, more narrowly focussed, König [48] offer very useful and, though naturally partisan, fairminded glimpses.

[20] They also distinguished their argumentative domain from a logico-semantic one in ways that underplayed their argumentative hand, as König [48] p. 194n5 demonstrates.

[21] Among possible 'effects' mentioned in passing by SW (cf. [74] p. 109) there is an explicandum for this notion of relevance, namely strengthening or weakening of assumptions. But it is not developed nor, apparently, treated as pertinent. There is, for instance, no apparent recognition that this is in essence the distinctive concept of Ducrot [16], who is noted ([74] p. 14n7) to have 'developed a programme in some ways comparable to Grice's '. (This, indeed, is all the 1986 book says about Ducrot besides citing references. Yet, by 1978, exchanges in French journals, including one published from offices of the CNRS at Paris, had established how deeply he differed from Grice in matters of principle and prediction; cf. [4], cp. also Horn [35]. Three further 1995 footnote references are no more informative on the relevant point.) Nor is there notice given that relevance is addressed by Carnap. His [10] is mentioned (SW [74], p. 79f.) as a treatise on subjective probability, but thereupon reported solely for its general measurement-theoretic preliminaries on concept types (its pp. 8-15), minus their references to probability and subsequent conclusions. Carnap explicates relevance in [10] Ch. 6, 'Relevance and Irrelevance', on pp. 346-427.

The former has going for it common sense and paradigmatic language games such as telling someone the time. Yet, well-rounded philosophical discourse already casts doubt on the paradigmaticity of such games. For instance, 'figuring out real intentions', inevitable once Gricean implicatures by flouting of convention arise, already presumes scepticism, and cause for it.[22] The objective of optimal information transmission does not. The objectives reflected in, say, Tacitus, or in the distribution of the GNP, do. No Cartesian *malin génie* is needed there: just cunning fellow-humans.

But neither common sense nor philosophical stance need be decisively consequential for explanation of robust linguistic intuitions. One may consider a particular one of super-objectives (i) and (ii) more consonant than the other with one's introspections on what talk is really all about. Yet that other one may turn out to condition poorly introspectible linguistic intuitions. And it is those that are, today, agreed-on data for inferring structures of what might be loosely called semantic 'competence'.

Let us, then, move to empirical commitments, first with a formal definition of relevance. We ask: How does the probability $P(H)$ of proposition H (intuitively: 'hypothesis') change in the light of evidence fully represented by proposition E?

Ramsey-coherent epistemic updating on evidence deemed certain proceeds, by most accounts, by conditioning. Supposing that i' is the epistemic context obtained from i when E, and just E, is newly learned (adopted) for certain, we obtain

$$P^{i'}(H) \;=\; P^{i}(H|E) \;=\; P^{i}(H) \cdot [P^{i}(E|H)/P^{i}(E)].^{23}$$

Put in 'odds' form, where $\text{Odds}(H) =_{df} P(H)/P(\overline{H})$, Bayes' rule yields

$$\begin{aligned} P^{i'}(H)/P^{i'}(\overline{H}) &= P^{i}(H|E)/P^{i}(\overline{H}|E) \\ &= [P^{i}(H)/P^{i}(\overline{H})] \cdot [P^{i}(E|H)/P^{i}(E|\overline{H})]. \end{aligned}$$

Posterior odds (on the left) equal *prior odds* times the *likelihood ratio* of H on E. For cognitive convenience (Peirce [63]), take logarithms to any fixed base; say, 2 for common measure. Then

$$\log[\text{Odds}^{i}(H|E)] = \log[\text{Odds}^{i}(H)] + \log[P^{i}(E|H)/P^{i}(E|\overline{H})]$$

The last summand measures the epistemic context change potential [eccp] of E with respect to H in i. This is what it means to say that the eccp thus represented is *directed*. It is computed with regard to an H that is not already taken for granted, and it can be positive or negative (or nil).

[22] Check Austin [6] on 'real'.

[23] In accordance with Bayes' definitionally valid formula, $P(H|E) =_{df} P(HE)/P(E) = [P(E|H)P(H)]/P(E)$ where $P(E), P(H) > 0$.

Definition 4 The *relevance* $\mathrm{r}^i_H(E)$ of proposition E to proposition H in an epistemic context i represented by conditional probability function $P^i(\cdot|\cdot)$ is given by $\mathrm{r}^i_H(E) =_{df} \log[P^i(E|H)/P(^iE|\overline{H})]$, also known as the log-likelihood-ratio for simple H.

Definition 5 E is *positively relevant to* H in i iff $\mathrm{r}^i_H(E) > 0$, *negatively* iff $\mathrm{r}^i_H(E) < 0$, else *irrelevant to* H. E is *relevant to* H iff not irrelevant to H. We say: E is *relevant* (pure and simple) iff relevant to an H at issue.[24]

Relevance is intelligibly related to conditional informativeness, where the conditioning propositions are now the complementary pair $\{H, \overline{H}\}$ at ulterior issue:

Fact 2 $\mathrm{r}_H(E) = \inf(E|\overline{H}) - \inf(E|H)$.[25]

So $\mathrm{r}^i_H(E)$ is *decreasing* in $\inf^i(E|H)$ and *increasing* in $\inf^i(E|\overline{H})$. If E were already taken for granted at i, i.e. if $P^i(\cdot) = P^i(\cdot|E)$, its assertion, a primitive notion shorn of implicatures for now, would be uninformative [$\inf^i(E|HE) = 0$ for all i and H]. E would thus be irrelevant to any H whatever, its epistemic impact having already been absorbed into the probability function. In this special case, relevance—to at least one proposition, never mind which—and informativeness coincide. (In general, they don't.) Both are 0. Relevance to the point at issue implies informativeness. Informativeness does not imply relevance to the point at issue. The distinction between relevance and informativeness is otiose at best for pointless talk.[26]

Here is a more general view of the approach that relates it to kinematic semantics without attention to its partisan aspects. $P^i(\cdot|\cdot)$ represents a

---

[24] I adopt Carnap's convention that relevance of E be zero when it contradicts ostensibly immutable background belief, given explicitly or absorbed in i. This accords with Kolmogorov's definition of probabilistic independence, $P(AB) = P(A)P(B)$. If $P(H|E) = 1\,(0)$ given $0 < P(H) < 1$, then $\mathrm{r}_H(E) = \infty\,(-\infty)$. $\mathrm{r}_H(E)$ is a continuous monotone increasing function of other measures of relevance, e.g. $P(H|E) - P(H)$, $P(H|E)/P(H)$. Advantages of $\mathrm{r}_{(\cdot)}(\cdot)$ are, e.g., that it is additive under certain conditions (cf. Fact 5, below) and values all cases of entailment among contingent propositions alike. I.J. Good [24], [25], the great explorer along with A.M. Turing of $\mathrm{r}_H(E)$, calls it 'weight of evidence' (cp. Peirce [63]), and notates $W(H:E)$. I follow *inter alia* Keynes, Carnap, Jeffrey, Stalnaker and the stochastic literature. My reasons are generality (consider causal or moral relevance; cf. Merin [56], Ch. 4) and science policy. 'Relevance' should retain an explication that is both useful in language analysis and consistent with unambiguous scientific usage.

[25] Making a background K not presently at issue explicit leads to the more general **Definition**: $\mathrm{r}_{H|K}(E) = \log[P(E|HK)/P(E|\overline{H}K)]$ is the *relevance of E to H, given K*. This yields $\mathrm{r}_{H|K}(E) = \inf(E|\overline{H}K) - \inf(E|HK)$.

[26] Recall that indiscriminate implication-collecting has a closest intelligible counterpart in indiscriminate informativeness-maximization. — See Section VI, below, where the distinction relevant/informative serves to explicate 'accommodation' (Lewis [51]).

(continuous) *commitment state*: a partial function

$$P^i(\cdot|\cdot) : Prop \times Prop \to [0,1]$$

mapping pairs of propositions to a real number. It is what Ramsey [65] called a *probability theory*, as distinct from a theory defined by closure under conjunction and classical consequence.

Changing the state $P^i(\cdot) = P^i(\cdot|\Phi \vee \overline{\Phi})$ to $P^{i'}(\cdot) =_{df} P^i(\cdot|E)$ by conditioning on E ($P^i(E) > 0$) will increase / decrease / leave unaltered the probability of $X \in \text{Domain}(P^i(\cdot))$. E is then positively / negatively / not at all *relevant* to X in i. Thus we can say: A proposition E acts as a partial function

$$f_E : \mathcal{P} \to \mathcal{P}$$

on a space $\mathcal{P}$ of probability functions. Its fixed points are states i such that $P^i(\cdot) = P^i(\cdot|E)$, i.e. where E is *presupposed* (see Section VI). And it is undefined at i such that $P^i(\cdot) = P^i(\cdot|\bar{E})$, i.e. where $\bar{E}$ is presupposed.[27] Finally, we can define $\text{BF}(i) = \{X : P^i(\cdot) = P^i(\cdot|X)\}$ as the 'full belief' set at i, represented by a 'theory' closed under conjunction and consequence, and immutable by conditioning.[28]

Next consider the interpersonal component of linguistic explanation, as outlined in the paradigmatic response 'Why should I?' I shall, quite informally, cast it in the form of a

Thesis (BG): A dichotomic epistemic issue $\{H, \overline{H}\}$ in an epistemic state i is modelled by a *bargaining game* (BG) between two actors (call them Pro and Con). Its negotiation set is generated by a set Σ of interpretive or expression alternatives linearly ordered by signed *relevance* at i to H. Pro aims to establish, say, H; and Con, $\overline{H}$. Ordinally *dual* (i.e. inverse) *preferences* of Pro and Con with respect to H becoming a joint epistemic commitment then induce dual preferences over Σ.[29]

The naturalness of preference duality over Σ becomes apparent on deriving from Fact 2 the

[27] At either of such i, both E and $\bar{E}$ will be *irrelevant* to *all* propositions. This explicates a basic 'admittance' condition of the presuppositional literature. See Section VI.

[28] Consequence will be classical granted strict coherence and, with application restricted to such a properly classical, 'full belief' theory, subject to a convention (cf. Adams [1]) for *ex falso quodlibet*. To keep matters simple, I ignore the possibility of mutable *full* belief that is *not* identified with classical probability 1. The phenomena treated, or at least as far as treated, here do not require it. Others will, and I engage in forthcoming work available techniques, both probabilistic and closely related, for representing what amounts to deductive non-monotonicity.

[29] By analogy, think of two *homines oeconomici* faced with a cake on a shelf too high for one alone to reach. Who will get how much, if between them they are to get more than nothing? – *Given* that nothing is to be wasted, their preferences are strictly dual.

Corollary 3 $\mathrm{r}_{\overline{H}}(E) = -\mathrm{r}_H(E)$.

If available constraints on probabilities do not suffice to determine a linear ordering of Σ, the game is one of 'incomplete information' (cf. Harsanyi [30], p. 91). We meet an example below. I assume that we are only dealing with single issue contexts. It is, I think, a brute empirical question whether or not such pointedness or tunnel vision is to be definitional of a linguistically predictive notion of 'context'. Empirical questions no doubt attend ways of extending it to the nestings of contexts examined by early phenomenologists and contemporary computer scientists. The same goes for evolution or change of issues. I leave all this for another occasion. I also attend here only to dichotomous issues. Thus, for present purposes, I offer a

Partial Definition 6 An ordered pair $\langle i, H\rangle$ is a (simple, bi-partisan) *context* only if i determines a probability function $P^i(\cdot)$ (i.e. a common prior epistemic context, CP) and H determines a partition $\{H, \overline{H}\}$ of the domain $\mathcal{F}$ of $P^i(\cdot)$ such that the respective preferences of the (two) parties over the partition are ordinally inverse. Call $\{H, \overline{H}\}$ the *issue*, and H the *discourse topic* (DT), of $\langle i, H\rangle$. [30]

Figuring out the ostensible and/or intended H amounts to 'figuring out intentions'. Indeed, it amounts to figuring out what is being 'meant', in one of Grice's senses, by what is 'said'.[31] Let me propose an explication.

Definition 7 The *Protentive Speaker Meaning* (PSM) of a sentence S in context $\langle i, H\rangle$ is that element of the issue $\{H, \overline{H}\}$ for which the relevance sign of the proposition expressed at i by the utterer equals the (short-term rational) utterer's preference sign at $\langle i, H\rangle$.[32]

[30] To avoid complicating things in view of Thesis BG, I leave 'context' and 'issue' fairly sparse. DT refers preference-neutrally to a proposition: the descriptively most convenient element of the issue bi-partition. (Contrast Def. 7, below.) Nothing is lost in proceeding thus by way of abbreviation, instead of calling $\{H, \overline{H}\}$ the DT, as would be suggested by the discourse-analytic method of Yes-No-Questions. More generally, let a question $Q?$, as usual in the literature, denote a partition π_Q into propositions of the set of possible models. (For a yes-no-Q pick a bi-partition; for most wh-Qs, a 2^n-partition, where n is the number of possible individual constant instances of the wh-morpheme; for why- and most how-Qs, one of each.) Then define: A proposition E is *relevant to a question* Q? iff it is relevant to at least one element of π_Q. Complete or eliminative partial answers will then be special cases. A *question* Q_1? is *relevant to another question* Q_2? iff at least one element of π_{Q_1} is relevant to at least one element of π_{Q_2}. Relevance of, and to, questions is essentially just dependence of random variables. In our simple, short-sighted scheme, call Q_1? a *relevant question* (pure and simple) iff, in addition, Q_2? represents the issue.

[31] That such figuring will be a non-hierarchical constraint resolution process is strongly suggested by the speech-recognition literature. 'Bayesian' probability networks are a stock-in-trade of such literature.

[32] I propose to call the proposition expressed by S in i its *Matter-of-Fact Meaning* (MFM). The requirement on MFM is that it be a subset of, i.e. entails, the CG proposition

PSM links up the discourse topic, DT, with the perennial question of what proposition is being expressed, and it adds structure to the disambiguation example of Section II.B. Recall, from II.C, that if I claim \$5 from you, I won't mind (i.e. won't be committed to applying sanctions) if you give me $\$n$ where $n \geq 5$. If you concede \$5 to me, you won't mind (won't be committed to applying sanctions) if I take $\$m$ where $m \leq 5$. This principle also informs discourse which is predicated on assertions and admissions.

Definition 8 The *upward (relevance) cone* $^{\geq_S}\Phi$ of an element Φ of a subset $\mathcal{S} \subseteq \mathcal{F}$ of propositions in context $\langle i, H\rangle$ is the union of propositions in $\mathcal{S}$ that are at least as relevant to H in i as Φ is. The *downward (relevance) cone* $^{\leq_S}\Phi$ of Φ in context $\langle i, H\rangle$ is, dually, the union of $\mathcal{S}$-propositions at most as relevant to H in i as Φ is.

With $\mathcal{S}$ selected by lexical and local relevance structure, I shall make use of the following explicatory

Hypothesis 1 The upward cone $^{\geq}\Phi$ represents Pro's *claim*, induced by lexically explicit Φ. The downward cone $^{\leq}\Phi$ represents Con's default expected compatible counterclaim (i.e. *concession*).

IV. 'Conversational Implicature' and its Relatives

In what follows, a Boolean semantics is presupposed. The possibility that it might, in turn, be induced by other semantic substrates in suitable contexts of use will be ignored.[33] The set of persons to whom probability functions are imputed which, in turn, engage the CP function is left implicit. The default set will include the Speaker and the actual or virtual Addressee, alongside a representation of transcontextual 'common sense' or Aristotelian *endoxa*.[34] In most cases, only ordinal relations among probabilities will be directly appealed to; actual number values other than 0 and 1, never. Read probability function symbols as shorthand for sets of probability functions satisfying stated constraints and indexed to a context.

A. 'Scalar Conversational Implicature' [SCI]

at i. I should define two notions of sentence meaning: respectively as the union and as the disjoint union of MFMs taken over a universe of possible i. The construction and label for MFM and the first of the sentence meanings derive from work on conditionals by Stalnaker, Harper, and Jeffrey. Disjoint union amounts to an MFM-valued function.

[33] The substrates I am thinking of are linear algebras over the reals: structures familiar to natural scientists, economists, and statisticians. Cf. Merin [53], [58].

[34] Cf. Merin [56], Ch. 2 for some of the complexities.

Thesis BG motivates Hypothesis 1. The two jointly explain the *defeasibility differential* that distinguishes at the level of observation a speaker's 'lower-bounded' claim/assertion, e.g. 'Some men walk' from the 'upperbounding' implicature, e.g. 'Not all do'; cf. Horn [34]. Here is how.

Identify the content of Pro's *assertion* with the *upward cone*, and the *scalar implicature* with the *downward cone* corresponding to the counter-claim of an actual or virtual disputant, Con; cf. Def. 8. Note now that Pro's claim is Pro's responsibility to back, but that Con's counterclaim isn't: Pro bears no responsibility of evidential or other backing for it. Hence a necessary condition for its defeasibility is met. Nonetheless, rationality of Con makes the Scalar implicature a computational *default*, and indeed by *quasi-juridical* default, if Pro is not committed to backing more than the greatest lower bound of his claim. But again, there is no guarantee that Con will not irrationally concede a windfall, or that backing for a stronger least claim might not be forthcoming. (An explication of Horn's [34] gradience intuitions in terms of probabilities over various sub-cones of these cones is conceivable.) Net meaning is, finally, the rationally expected-to-be conceded claim: the intersection of 'assertion' and 'implicature'.

A.1 'Particularized SCI':

Here is an example due to Hirschberg [33]:

> Q: *Do you speak Portuguese*? A: *My husband does.* [$= F$]
> [Scalar Implicature: Speaker A (= Wife) doesn't. [$= \overline{E}$]]

Hirschberg's explanation is: The context of potential Portuguese-speakers is the set of alternatives given as the set Σ of non-empty subsets (of family members),

$$\Sigma \; = \; \text{Pow}\{husband, wife, child\} - \phi.$$

Σ is partially ordered by the subset relation, i.e. $A \leq B$ iff $A \subseteq B$. By a generalization of Gricean 'Quantity', $\overline{E}'$ is implicated by utterance of F for all E' with subject denotata $s \in \Sigma$ such that $s \not\leq \{husband\}$.

But what is the 'way of being given' of this context set? The subset structure is not pertinent. Nor is the concomitant rule independently motivated. And so it won't explain (where $\surd$ means 'O.K.')

> A: $\surd$\{*My brother/servant/Kim/Kim Jones*\} *does* [F', F'', F''', F'''']
> A': *\{*Kim Basinger/The president of Brazil*\} *does.* [G, G']

Let Q = Boss Inc., $A = W$ (Wife) at her job interview. Ask: (a) Why should Q ask? and (b) Why should W answer as above? Because: (a) Q wants utility-relevant information ('Should we hire W?' [$= \{H, \overline{H}\}$?]) and (b) W wants to be hired [H]. $\{H, \overline{H}\}$ is the *issue*. H is W's PSM (cf.

Def. 7). Suppose W doesn't speak Portuguese, but can call upon someone who does, virtually gratis. Typically such a person will be kin or agnate; else a friend (Kim); or an acquaintance (Kim Jones). But presumably *not* Kim Basinger. Hence the *. (This gives us a relevant part of a set $\mathcal{S}$ of Def. 8, i.e $\{S \in \mathcal{S} : \mathrm{r}^i_H(S) > 0\}$.)

Moreover, $\mathrm{r}^i_H(E) > \mathrm{r}^i_H(F) > 0$, surely. Now, treat E and F as potential claims. Recall, furthermore, that claims (demands) in bargaining are always 'at least' and that the strongest compatible response (meanest concession) amounts to 'at most' that. So $\bar{E}$ is critical, sceptical ('Persuade us!') Boss's counterclaim, made and conceded *by default*.[35] Its status as a counterargument is given by the following

Fact 4 $\mathrm{r}_{\bar{H}}(\bar{E}) > 0$ *iff* $\mathrm{r}_H(\bar{E}) < 0$ *iff* $\mathrm{r}_H(E) > 0$.

That's why smart W gives an indirect, (call it) *'second-best'* answer. Surely also $\mathrm{r}^i_H(F) > \mathrm{r}^i_H(G) = 0$, and likewise on substituting F' etc. for F and G', G'' for G. Thus, W's utterance of 'E' will never implicate that the President of Brazil cannot speak Portuguese. And note again: presumably E, F, G are logically independent.

A.2 'Generalized SCI':

Entailment-based informativeness has done rather well with 'strong' Scalar implicature from *or* construed as inclusive disjunction ($\vee$) to local, defeasible strengthening with intuitions appropriate to exclusive 'or', i.e. to XOR (∇) (cf. Gazdar [22]; Soames [73]). To make relevance predict we need

Definition 9 A and B are *independent conditionally* on H and $\bar{H}$ in i $[(A \perp_i B | \pm H)]$ iff

$$P^i(AB|H) = P^i(A|H)P^i(B|H) \text{ and } P^i(AB|\bar{H}) = P^i(A|\bar{H})P^i(B|\bar{H}).$$

Intuitively, $(A \perp_i B | \pm H)$ holds when knowing which of H or $\bar{H}$ is the case will account fully for any interactions between A and B in i. The condition guarantees additivity, i.e. compositionality of relevance:

[35] Ede Zimmermann of Stuttgart University reminded me that here W will know whether or not E. Where this assumption is not met, $\bar{E}$ will not be implicated by default (cf. Soames [73]). I should say, in view of gradience (cf. Horn [34]): The addressee or hearer could not then *rely* on being able to sustain $\bar{E}$. Note that $\bar{E}$ could not be relied upon either if W could not be presumed to prefer E being sustainable. Try: $E?$ = 'Do you have a criminal record?'. Here a follow-up by Q, 'And you don't?' is advisable.

Fact 5 *If* $(A \perp_i B| \pm H)$, *then* $\mathrm{r}^i_H(AB) = \mathrm{r}^i_H(A) + \mathrm{r}^i_H(B)$.[36]

A presumption of conditional independence is widely adopted in Artificial Intelligence applications for cumulation of evidence. It reduces computational load and the need for appeal to world knowledge. Here I propose a linguistic correlate:

Hypothesis 2 Natural language interpretation is predicated, where nothing suggests or requires otherwise, on a *Conditional Independence Presumption* [CIP]: $(X \perp Y|\pm H)$ where X, Y are propositional denotata of syntactic sister nodes in coordinate construction schemata X CONJ Y.

This lends empirical import to a family of propositions which, for brevity and salience, I shall call

Theorem 6 *If* $(A \perp_i B| \pm H)$ *and* $\infty > \mathrm{r}^i_H(A)$, $\mathrm{r}^i_H(B) > 0$ *then*
(a) $\mathrm{r}^i_H(AB) > \max[\mathrm{r}^i_H(A), \mathrm{r}^i_H(B)] > \mathrm{r}^i_H(A \vee B) > 0$, $\mathrm{r}^i_H(A \nabla B)$; *whereas*
(b) $\not\vdash \mathrm{r}^i_H(A \nabla B) \geq 0$; $\not\vdash \mathrm{r}^i_H(A \nabla B) < 0$; $\not\vdash \mathrm{r}^i_H(A \vee B) < \mathrm{r}^i_H(A)$, $\mathrm{r}^i_H(B)$.

By Theorem 6.a, CIP will rule out so-called 'paradoxes of relevance' such as the negative relevance to H of conjunctions or even disjunctions of propositions A, B, that are each positively relevant to H. Conjoined with Thesis BG, it explains 'strong' Scalar implicature for $\vee$. To see how, ask first:

What would be a default set of expression alternative designata in a context of use involving no atomic sentences besides A, B? A reasonable candidate should be the Lindenbaum or proposition algebra $\mathcal{F}$ over $\{A, B\}$. The atoms of $\mathcal{F}$ are represented by AB, $\bar{A}B$, $A\bar{B}$, $\bar{A}\bar{B}$. They finest-partition the semantic space of possibilities. Suppose A and B positive to H, which is preferred by Pro. Under incomplete information (nothing e.g. is presumed known about the comparative relevance of $A\bar{B}$, $\bar{A}B$) Con's natural default counterclaim to Pro's $A \vee B$ is $\bar{A} \vee \bar{B}$. The meet (intersection) of claim and counterclaim is then $A \nabla B$.

Note that a *reduction* in the expected (positive, evidential, partisan) *relevance* of a claim will in general have increased (non-partisan) semantic *informativeness* as a byproduct. (Echoes of Adam Smith's Invisible Hand ideology there!) This is because $A \nabla B \models A \vee B$, while $A \vee B \not\models A \nabla B$. Various epistemic conventions for the default to bite are conceivable: e.g. that Pro should know that AB, if AB holds ([73]).

It is an empirical question, then, whether a conventional, hypothetical imperative 'Be Relevant!' or 'Be Informative!' is our predictor. The two

[36] Without any such independence assumptions we can state as a **Fact**: $\mathrm{r}^i_H(AB) = \mathrm{r}^i_H(A) + \mathrm{r}^i_{H|A}(B)$. Recall the definition in fn. 25 here. The epistemic kinematics appears to parallel analyses of context-change familiar from the literature on dynamic semantics. But see also fn. 64.

notions, granted our distinctive definitions for them, are not monotone increasing functions of one another. The above example is a case in point. If the criterion of success is partisan maximization of evidential relevance, *it can pay to be underspecified*—quite independently of any considerations of brevity or freedom of discourse continuation. The implicature then acts as a countervailing force, giving notice, as it were, that cheap advantages of underspecificity are not automatically available.

Relevance predicts as well as, or better than, Neo-Gricean informativeness on some hard distributional facts. Recall that the so-called 'quantitative scales', of which (*and, or*) is a putative instance, are linguistically given to us most tangibly by their diagnostics (Horn [34]). Let us adopt the following

Hypothesis 3 Expression schema '*X if not indeed Y*' is acceptable iff, for every context of use $\langle i, H \rangle$ and propositions X, Y satisfying CIP (Hypothesis 2) and finitude of relevance to H, either

$$(\mathbf{i})\, 0 < \mathrm{r}^i_H(X) < \mathrm{r}^i_H(Y) \quad \text{or} \quad (\mathbf{ii})\, 0 < \mathrm{r}^i_{\overline{H}}(X) < \mathrm{r}^i_{\overline{H}}(Y).$$

From this we can derive

Prediction 1 *'*A or B, if not indeed A*'.
Proof: Let X:= '*A or B*', Y:= '*A*'. By Theorem 6.b, Hypothesis 3.i is not generally satisfied. Nor is 3.ii with respect to $\overline{H}$.

Prediction 2 $\surd$'*A (or B), if not indeed A and B*'.
Proof: Let either X:= '*A or B*' or else X:= '*A*', and let Y:= '*A*'. By Theorem 6.a, Hypothesis 3.i is satisfied. So is 3.ii with respect to $\overline{H}$.

With or without a lexicalization constraint (Horn, Gazdar), Neo-Gricean informativeness would, by itself, mispredict for one or the other option. *Without* one, it predicts acceptability of the first, since disjuncts entail disjunctions without being entailed by them.[37] *With* such a constraint,

[37]Larry Horn, I presume, would counter: '*A or B*', by implicating ignorance of whether A, would explain the * by redundancy of '*if not indeed A*', which opens the possibility of A. I should reply: Agreed. But the crucial one of the pertinent pair of ignorance-implicatures, as modal-logically explicated (Gazdar [22]), is $\neg\mathsf{Knows}_s\neg A$. Unlike its sister, $\neg\mathsf{Knows}_s A$, it relies for its Gricean motivation derivation on the Maxim of Manner, i.e. on relative expression length. If, in line with Grice's own late doubts ([27], p. 372; cp. also Mandelbrot [52] and fn. 10), one prefers doing without Manner, one should prefer my preferred explanation. If not, then not on grounds local to the argument. For some of the global view, see the following sections.

$\{A,\ A\, and\, B\}$ does not carry a 'quantitative scale' (there is no item contrasting with *and*). If so, such diagnostics appear to diagnose relevance-based rather than informativeness-based scales, under the above definitions.

B. Negation of Scaleable Attributes

Both Ducrot [16] and Horn [34],[35] investigated pairs of intuitively 'positive' and 'negative' ordered subsets of lexical fields (e.g. {*boiling, hot, warm*} vs. {*cool, cold, freezing*}; etc.). A Decision-Theoretic Semantics (DTS) will recover their observation that negations of strong (weak) positives are, respectively, relatively weak (strong) negatives; and so on. It need not merely observe the phenomenon stipulatively, as Ducrot had to. Nor need it assume, with Horn, who did offer an explanation, that lexical meanings of scaleables are generally of a logical form paraphraseable 'at least' and perhaps, for marked environments (e.g. Jespersen's *'live on'*), 'at most'.

Horn had assumed that an attribute space was generated by two disjoint sets of basic, non-negated predicates. These 'positive' and 'negative' 'scales' were each ordered by predicate inclusion. This goes with the intuition that 'boiling' surely entails 'at least hot', which in turn entails 'at least warm'. The assumption, though occasionally unkind to common sense—a dubious counsellor to science—had a real predictive payoff. In hypothesizing an ordering by entailment (set-inclusion) it yielded, by contraposition (cp. Fauconnier [17]), the observed relations among negated predicates; and, by simple complementation, Jespersen's observation that e.g. *not 5 pounds* means, usually, 'less than 5 pounds'.[38] Cases such as 'It's not hot, it's boiling!' were to be treated, in line with Ducrot, as instances of Ducrot's very general notion of 'metalinguistic negation', also instantiated elsewhere, e.g., in correction of phonetic infelicities (cf. [35], Ch. 6).

I assume instead, more plausibly, that attribute spaces given by a scaleable lexical paradigm are *partitions* into attributes of similar grain, induced pointwise by partitions of 'expression alternative' propositions (cp. the for-

[38]Horn's assumption on lexical representation of scaleables has been criticised (cf. [35]), most prominently by Carston [13], p. 174f. She favoured, with paradigmatic reference to numerals, a single linguistic meaning or sense (neither 'at least' nor 'at most' nor 'exactly') such that the proposition, i.e. truth-conditional content, actually expressed (Carston: 'explicature' *apud* SW [74]) is contextually determined. However, this plausible contention, and others concurring, did not engender explications of senses and contextual mechanisms able to yield the Jespersen/Horn/Ducrot facts. Horn's theory—something his reply [36] refrains from stating that baldly—was the only approach to these very non-peripheral intuitions that did explain them. This fact also must have blunted the edge of the formidable objections raised by AD [4] in response to Fauconnier's francophone critique of their approach.

mal definition in Gazdar [22]).[39] A context of use $\langle i, H\rangle$ induces a partition $\pi := \pi^{\langle i,H;\geq\rangle}$ linearly ordered by relevance to H. Constraints on π across possible $\langle i, H\rangle$ are, I believe, part of what counts as knowledge of language: i.e. the ability to *use* those attributes in cogitation and in communication. Assertoric communication coarsens π to an order-homomorphic bi-partition $\pi' := \pi'^{\langle i,H;\geq'\rangle}$. The two elements of π' are each topologically *connected.* I.e. neither of them has a gap, as boolean negation would yield when applied to elements of π that are non-extreme with respect to $\geq$.

The bi-partition π' is induced by the independently motivated virtual bargaining pragmatics. For recall Def. 8 of 'upward cone' ${}^{\geq}\Phi := {}^{\geq_S}\Phi$. Call now *'unmarked scalar complement'* of Φ the boolean complement ${}^{<}\Phi := {}^{<_S}\Phi$ in $\mathcal{S}$ of ${}^{\geq}\Phi$ and construe as the denial of a claim deemed excessive. Call *'marked scalar complement'* the dual ${}^{>}\Phi := {}^{>_S}\Phi$ of ${}^{<}\Phi$ and construe as the rejection of a concession deemed insufficient; thus inheriting concession's phenomenologically marked status.[40] Ducrot's and Horn's observation is then captured in the following

Theorem 7 *If propositions X and Y in $\mathcal{F}$ are elements of a partition π of the space Ω of possibilities, such that π is linearly ordered by i-relevance to an H in $\mathcal{F}$, and $\mathrm{r}^i_H(Y) > \mathrm{r}^i_H(X) > 0$, then $\mathrm{r}^i_{\overline{H}}({}^{<_\pi}X) > \mathrm{r}^i_{\overline{H}}({}^{<_\pi}Y) > 0$.*

In words: Let X, Y belong to such a partition π, relevance-ordered in context $\langle i, H\rangle$. If X is positive to H, and Y more so, then the scalar complement of Y is positive to non-H (i.e. negative to H), and the scalar complement of X more so. Analogous relations among attributes λzX, λzY etc. are induced by abstraction over z-instances and contexts of use $\langle i, H\rangle$. Try with $X = \lambda zX(a) =$ 'It's warm'; $Y = \lambda zY(a) =$ 'It's hot'; $H =$ 'The temperature is high'; or, if you like, $H =$ 'It wants a rest'. Scalar implicature, i.e. expected counterclaims, will then upperbound claims, much as above.

V. 'Conventional Implicature'

[39] Plausibility is here not so much a matter of introspection on meaning, let alone about the form of entries in the mental lexicon. Rather, if spaces of basic attributes were structured as sets of inclusion chains, they could not be partitions if either chain had $n > 1$ elements. One could not, then, define a probability distribution over them. Basic attributes would thus be ill-suited for computing expected values of actions conditioned by propositions they generate. This pragmatic consideration has linguistics shaking hands with philosophy and the design sciences. See Merin [60]. Partitioning will also make better sense of predicates such as 'tepid', featured in famous example pairs such as *This beer is tepid if not downright warm* and *This coffee is tepid if not downright cold.*

[40] Why marked? 1. 'Don't pick a fight you know you're gonna lose' (Sen. Sam Goldwater). 2. There is double anaphoricity: to a concession responding, in turn, to a putative claim. One need not here invoke Ducrot's and Horn's still formally unexplicated 'metalinguistic negation'.

Traditional accounts of degree particle *even* (e.g. Karttunen and Peters [45]) note that *Kim even talks* [abstractly: $even(\alpha, \beta)$ where α is focus of *even*] indicates highest 'surprise/improbability' of *Kim talks* [$= E := (\alpha, \beta)$]; the comparison class being propositions (α', β) with $\alpha' \neq \alpha$ (e.g. *Kim walks*).

Anscombre [2] proposed superior argumentative value (AV) of E over E' with respect to some H. Let us explicate this by relevance, $\mathrm{r}^i_H(\cdot)$. Then, for given H and a set $\{E_j\} = \{(\alpha_j, \beta)\}$ of expression alternatives, $\mathrm{r}^i_H(E) = \max_j[\mathrm{r}^i_H(E_j)]$ iff $\mathrm{r}^i_H(E) = \max_j[\mathrm{inf}^i(E_j|\overline{H}) - \mathrm{inf}^i(E_j|H)]$. In simplified words: the E_j most (positively) relevant to H in i is that of which the truth would be the most surprising in i if $\overline{H}$ rather than H were to be accepted in i. Of course, there is no point arguing for an H already firmly accepted. And indeed, the intuition for the above example pair is that one is arguing with it for an H still at issue, say, H = 'Kim is doing great'. Check for the other intuitions.

Whether or not one takes the AV properties of *even* to be primary, as I do on diachronic grounds,[41] or else derived: intuitions are squared and an issue between Anscombre [2], Kay [46]–who each plead *contra* surprise—and Francescotti [19]—who pleads *pro*—is accountably resolved.

Note: $\mathrm{r}^i_{H|K}(E) > \mathrm{r}^i_{H|K}(E')$ does not imply $E, K \models E'$ for backgrounds K such that $K \not\models E \rightarrow E'$. Nor, even, will it imply $E, K \models_i E'$ when transcontextual entailment, $\models$, is replaced by a context-relative relation, $\models_i$, defined by: $X \models_i Y$ iff $P^i(Y|X) = 1$. Neither of $\models$ or $\models_i$ would license a 'contextual entailment' from E to E', by which Kay [46] intends to explicate superior 'informativeness', his explanatory ordering relation for the implicatures of *even*. But 'entailment', however weakened, is surely a misconception or misnomer—for relevance, it would appear—in view of

> Author: The hero of my novel sleeps with M******.[42]
> Editor: So what. The hero of our Greek deadwood[43] even sleeps with his own mother.

A DTS yields an account of the conjunction *but* and its interaction with *even* that recovers observations and pioneer explanations of Anscombre and Ducrot [4]; along with the much more occult interaction with *also* (Section VI). Consider, for cases not involving *also*, and simplified for exposition,

---

[41] *Even*, like Fr. *même* or Ge. *selbst*, originates as an emphatic particle, indicating 'importance'; as do the latter even now. Importance of objects means: high relevance of propositions involving them. Cp. Section VI on 'salience'.

[42] A 1990s media personality.

[43] 'Dust-gatherer' might be a less idiomatic, academic creole alternative for things that don't sell too well.

Hypothesis 4 '*A but B*' is felicitous in state i only if there is at issue in i a proposition H such that $\mathrm{r}^i_H(A) > 0$ and $\mathrm{r}^i_H(B), \mathrm{r}^i_H(AB) < 0$.[44]

Example:

Kim is a doctor, but she lives in Mozambique
[H = Kim is rich.]

Consider the hypothesis now in conjunction with a further

Hypothesis 5 '*A* CONJ *even(B)*' [:= (α_1, β) CONJ $even(\alpha_2, \beta)$] with VP-focus of *even* (on α_2) is felicitous in state i only if there is at issue an H such that $H \neq B$ and $\mathrm{r}^i_H(B) > \mathrm{r}^i_H(A) > 0$.

And suppose, as we did, that, for any given reading, H must be unique. This implies

Prediction 3 **Kim splurbs but Kim even glurbs* for any English substitution instances of *splurbs* and *glurbs*; explicating AD's claim and explanation.[45]

Frege's view ([20] § 7) was, it seems, that 'A but B' required B unexpected in view of A. A stock example is:

Kim is poor but (she is) honest.

But, as the 'doctor' example shows, such *A-unexpectedness* of B is not intuitively necessary for felicity of 'A but B'.[46] Nor is the obvious explication

[44] $P^i(\cdot)$ represents the CP, the evolution of which is under negotiation; more intuitably (also) the speaker-ostended epistemic state of the Addressee or of third parties ('common sense'). The hypothesis part-explicates Ducrot ([15] p. 128ff.; [16]), for whom 'A but B' presents A as an argument for some R, and B against, such that B inhibits the conclusion, R, which the hearer would draw in view of A. Cf. Merin [57] and [56] for some 270 pp. on 'but', including occurrences translating Ge. *sondern* or Sp. *sino*, which are ignored here.

[45] A referee noted that the * might be due to violation of a requirement for pronominalization, fielding as a likely counterexample *Mary tried as hard as she could but {she/?Mary} even couldn't lift the newspaper.* Looking at *Kim splurbs but she even glurbs*, I still feel that it sounds better with *but then* in place of *but*: compare e.g. *Kim walks but {then/*ϕ} she even talks.* Contrast *Kim walks and {she / Kim} even talks.* So perhaps phrasal heaviness along with the intensionality of *try as hard as one can* linking with that of *cannot* introduces complications. Cf. Merin [56], Ch. 5 on atemporal *but then*; and Section VI, below, on *also*.

[46] Sharper example: [A] *Wilkins is a man of principle, but* [B] *he is now Chancellor of Crowford.* No aspersions need be cast on Crowford, nor Wilkins. Simply imagine yourself discussing the moral decline of Batbridge and suppose that *A but B* presents *A* as an argument *for* some issue proposition H (e.g. 'Wilkins can stop the rot at Batbridge') and B *against*; with B ostensibly carrying the day.

by $P^i(B|A) < P^i(B)$ or, equivalently, $\mathrm{r}^i_B(A) < 0$ formally necessary for $\mathrm{r}^i_H(A) > 0$, $\mathrm{r}^i_H(B) < 0$, and $\mathrm{r}^i_H(AB) < 0$ jointly to hold. The condition $\mathrm{r}^i_B(A) < 0$ is not even necessary, more generally, for A and B to be H-*contrary* at i, i.e. for $-\mathrm{sgn}(\mathrm{r}^i_H(A)) = \mathrm{sgn}(\mathrm{r}^i_H(B)) \neq 0$. (NB: $\mathrm{sgn}(X) =_{df}$ 1/0/-1 for $X > / = / < 0$.) But note the following

Theorem 8 *If* $(A \perp_i B| \pm H)$, *then* $P^i(B|A) < P^i(B)$ *iff* $\mathrm{sgn}(\mathrm{r}^i_H(A)) = -\mathrm{sgn}(\mathrm{r}^i_H(B)) \neq 0$.[47]

H-contrariness yields A-unexpectedness of B under CIP. A prominent special case of Hypothesis 4 obtains when $H = \bar{B}$ (cp. [3]). Note here

Theorem 9 *For any* A, B *and context* $\langle i, H\rangle$: *if* $H = \bar{B}$, *then* $(A \perp_i B| \pm H)$.

Likely instances of $H = \bar{B}$ are the stock example or the 'aspersions on Crowford' reading for 'A but B' of footnote 46. $H = \bar{B}$ is apt as a default interpretation, not least in linguists' out-of-the-blue uses. For one, it avoids introducing new atomic H into the propositional universe of discourse, i.e. into the minimal field $\mathcal{F}$ of $P^i(\cdot)$. And it yields all basic (a,b), prototypical (c), and formally elegant (d) properties of 'but':

Theorem 10 *If* $\mathrm{r}^i_H(A) > 0$ *and* $H = \bar{B}$ *and* $P^i(AB) > 0$, *then*
(a) $\mathrm{r}^i_H(B) < 0$;
(b) $\mathrm{r}^i_{\bar{H}}(B), \mathrm{r}^i_{\bar{H}}(AB) > \mathrm{r}^i_H(A)$;
(c) $P^i(B|A) < P^i(B)$; *and*
(d) $\mathrm{r}^i_H(AB) = \mathrm{r}^i_H(A) + \mathrm{r}^i_H(B)$.

The relevance analysis also accounts for an apparent language universal (Merin [55]). This is the following asymmetry between NPs, which prototypically denote 'particulars', and VPs, which typically denote 'universals': While for

Kim$_k$ walks but Kim$_k$ talks

a felicitous context can always be found (e.g. Shall we hire Kim as a confidential messenger?), we have

**Kim walks but Sandy walks*
Kim walks but Sandy { *a l s o walks* / *walks too*}

[47] Established in all essentials by Reichenbach [68] in discussion of causal relations. Cp. remarks on Def. 9 and check intuitions suggesting $(A \not\perp_i B| \pm H)$ for the context $\langle i, H\rangle$ of fn. 46.

for all contexts.[48] On present evidence,[49] any language having bona fide, specific translation equivalents of 'but' and 'also' obeys the pattern, first noted for English, and classified as a brute syntactic fact, by Harris [29].

The apparent asymmetry of, simply speaking, subject (particular-like) and predicate (universal-like) calls for an explanation.[50] Let $P^i(\cdot)$ be defined on a monadic first order quasi-English language fragment. Consider now, adapted from Carnap [12], the following

Definition 10 $P^i(\cdot)$ satisfies *non-negative instantial relevance* [NNIR] when, for arbitrary Q, a, b: $P^i(Qb|Qa) \geq P^i(Qb)$.

Under NNIR a proposition Qb cannot be Qa-unexpected in i.[51] Now, as we saw above, A-unexpectedness of B is not generally necessary for acceptability of 'A but B'. But recall also that the special case $H = \overline{B}$ of H-contrariness makes B A-unexpected and is a prime candidate for an interpretive default (Theorem 10). Let us embed this fact into a more general setting and adopt as a

Hypothesis 6 Every natural language L obeys a *Default Satisfiability Principle* [DSP]: Let L have a default condition Γ on preferred interpretation of a class $\Sigma \subseteq L$ of expressions. Let $\sigma \in \Sigma$. If σ satisfies Γ for no context $\langle i, H \rangle$ of non-degenerate use, then σ is unacceptable [write: $^*\sigma$].

Suppose now that Γ has for an instance the Conditional Independence Presumption, CIP (Hypothesis 2). Recalling Theorems 8 and 9, we have as a

Corollary 11 *Given* DSP, *instantiated either by* CIP *or more narrowly by the antecedent of Theorem 10: if the probability functions $P^i(\cdot)$ of contexts $\langle i, H \rangle$ of use for sentence schema 'Qa but Qb' generally satisfy* NNIR, *then* *'*Qa but Qb*'.

Will NNIR obtain merely as a brute, albeit evolutionarily plausible fact? Not necessarily so. For the linguistically pertinent class of probability functions, it can be motivated discourse-semantically. Here is how. Argumentation involving H-contraries uses an inference rule of Universal Probabilistic Instantiation [UIP]. To see its workings recall the earlier example sentence

[48] The *ed sentence is single-speaker and has concomitant default prosody.

[49] Latest: see LinguistList 9.240 (1998) for the result of an Internet query.

[50] Cf. Merin [56], Ch. 5 for a much fuller version.

[51] If a, b designate individual events of tossing a coin, NNIR thus tells you not to let an observation of 'heads' lower your probability for 'heads' on the next toss of a coin if you have no extra evidence (say, Tb, of a trick about to be played). By contrast, NNIR does not rule out, say, Qb being Rb-unexpected for logically independent Q and R, nor, say, its being Ra-unexpected.

Kim is a doctor but she lives in Mozambique. As before, grant as background knowledge in typical i that Mozambique has a very low GNP; as United Nations statistics say. The presumption then must be that a resident of Mozambique is unlikely to be rich. But doctors, presumably, are likely to be rich. The formal argument then includes steps of which the following one is representative:

1. $\forall x[P^i(\text{rich}(x)|\text{doctor}(x)) = \alpha \ (1 > \alpha \gg 0.5)]$ [Ass.]
2. $P^i(\text{rich}(Kim)|\text{doctor}(Kim)) = \alpha$ [1,UIP].

Is the instantiation of x to *Kim* indeed valid? Recall that probability creates an intensional context. A reminder of this fact is that doxastic necessity may be interpreted as doxastic probability 1 (see fn. 3). To state conditions for validity, we first need, appropriately specialized and phrased for brevity of argument, a standard technical

Definition 11 A probability function $P(\cdot)$ on sentences with first-order models is *symmetric with respect to a sequence* $(\pm Qa_j)_{j\in J}$ *of sentences* (Q a fixed predicate, possibly differing in polarity Q vs. $\overline{Q}$ among sequence elements) iff the probability $P(\bigwedge_j \pm Qa_j)$ of the conjunction of sequence elements stays invariant under all finite permutations of the set of individual constants a_j. $P(\cdot)$ is *symmetric* iff symmetric with respect to any such sequence in its domain. [52]

The crucial theorem is here stated briefly and informally:

Theorem 12 (Skyrms [70]) *Validity of* UIP *requires that no $P(\cdot)$-relevant information be conveyed by the individual constant. Names must be treated as properly arbitrary and thus $P(\cdot)$ must be symmetric.*

Present use of it rests on something I shall adopt as an empirical

Hypothesis 7 Natural language semantics satisfies an *Embeddability Desideratum* [ED]. Models (or Discourse Representation Structures) generated by a small, finite universe of individuals (or their discourse referents) should be embeddable in larger models of arbitrary finite or of infinite cardinality.[53]

---

[52]Symmetry extends the permutation-invariance that characterizes deductive logical relations to that of inductive relations; appropriately so to those of a relatively aprioristic kind.

[53]Unless, as Tim Fernando of U. Texas at Austin pointed out, the candidate structures for embedding correspond to theories that have only infinite models to start with or only models of bounded finite cardinality. Such cases would naturally arise in mathematical discourse. Discourse Representation Structures are due to Kamp [43].

This hypothesis, and a reminder that a countable first-order model may be represented by a set of first-order sentences (a 'state description'), in turn, lend linguistic import to the following pair of

Theorem 13 (Savage-Kemeny-Gaifman-Humburg) *A symmetric probability function $P(\cdot)$ on finite first order models m extends to symmetric $P'(\cdot)$ on extensions m' of m of infinite size only if $P(\cdot)$ satisfies* NNIR.[54]

Corollary 14 ED *(cf. Hypothesis 7) and validity of* UIP *imply* NNIR.

One might now think that such appeal to deep results is a case of theoretical overkill. And one should indeed do well to think so if ready to accept the following thesis, rarely professed though frequently implicit in casual pragmatico-semantic argumentation: The structure of natural languages and attached bits of mind is very much simpler than that of, say, the hydrogen bond. I see no compelling evidence for the truth of such a thesis.

To test the stochastic approach empirically, we must now ask what explains the healing power of *also*. A serious answer requires an independently motivated account of *also*. Hence, I shall propose one. This takes us into the phenomenon of presupposition. I treat some of its traditional instances in [59]. Here I just outline what is required to sustain the argument regarding *but also*.

VI. Presupposition: 'also' and 'too'.

Current formal accounts of presupposition, following Stalnaker [75], are based on a notion of *being presupposed in a context*. The context or *common ground* (CG) is one of discourse participants' ostensible joint and firm, epistemic or other commitments: a conjunction of propositions, hence a proposition. Proposition A is presupposed iff entailed by CG. Let us say that an *utterance* of sentence S *presupposes* A iff the utterer presumes it context-presupposed. If need be, and subject to conditions left vague in the literature, it will be so presumed by 'accommodation' (Lewis [51]), i.e. by being taken for granted as context-presupposed, even if it was not actually so context-presupposed. Call A *sentence-presupposed by* S iff every felicitous utterance of S utterance-presupposes it.

Consider now the conservative probabilistic extension CP of CG. We identify, as before, an epistemic state with a probability measure, a *common prior* (CP) over a field $\mathcal{F}$ of propositions, and adopt as a

[54]Proofs: Gaifman [21], Humburg [37]. The convergence to a class of probability functions satisfying NNIR is rapid already for fairly small, finite m'. So, to all intents and purposes, one may read *infinite* as *arbitrarily large finite*.

Definition 12 Proposition A is *presupposed in epistemic state (context) i* [*i-presupposed*] iff $\forall X[P^i(X) = P^i(X|A)]$.

This implies $P^i(A) = 1$, and we can say: $\mathsf{CG}^i =_{df} \{A : P^i(A) = 1\}$. Intuitively, A has spent all its relevance in coming to be presupposed. Now consider the focus particle, *also*. Let the first, x-argument of the schema $also(x, y)$ be the focus of *also*. Traditionally, use of $also(b, B)$ as e.g. in

> *Kim a l s o talks*

utterance-presupposes that some individual a other than b (Kim) has the property B. Kripke [49] objected,[55] with counterexamples similar to

> *Kim and Sandy will come. The boss will also come.*

This discourse sequence, now, presents a context for $also(b, B)$ in which

$$\exists x[\text{come}(x) \wedge x \neq \text{the boss}]$$

is satisfied even when, contrary to linguistic intuition, Kim *is* the boss. For $also(b, B)$ to be appropriately utterable, some Ba must be an *anaphorically salient* part of the context.[56] The 'presupposition', says Kripke, is that, for any such a: $b \neq a$; in our example, that the boss is neither of Kim or Sandy. I will show that this follows from a much more general presupposition, and that one of the premisses of the derivation is a presupposition in the traditional sense, which, in turn, is part of what Kripke identifies as anaphoricity.[57]

First, I explicate salience and (a very special case of) anaphoricity as applying to *propositions*. Salience is, no doubt, always implied by anaphoricity. The presuppositional condition will further characterize the important special case. It will not do so fully; for one, because it makes no reference to syntactic entities.

Definition 13 Proposition D is *topic-anaphorically salient* for proposition E in context $\langle i, H\rangle$ iff **(i)** $\mathrm{r}^i_H(E) \neq 0$, and **(ii)** D is i-presupposed ($P^i(D) = 1$), and **(iii)** the last (or close to last) preceding context with index $i - k$ such that $P^{i-k}(D) < 1$ satisfies $\mathrm{r}^{i-k}_H(D) \neq 0$ [where $k = 1$ or small positive integer].

---

[55] Arguing for 'too', which differs from 'also' in ways inessential here.

[56] 'Some' here admits sets $\{a_j\}_{j \in J}$ of antecedents. For brevity, I treat formally only the '=' case $j = 1$ in what follows. I treat '$\in$' and predicate-focus quite analogously in a forthcoming paper on 'also' and 'too' that contains an earlier, full account.

[57] More generally so, van der Sandt [77]. See Beaver [8] for assessment; and for a state-of-the-art survey.

Intuitively: (i) demands $\langle i, H\rangle$-relevance of E; (ii) and (iii) jointly demand that D has just or recently come to be presupposed, and has spent non-negligible relevance to H in the process. So D is, at the very least, related to E via H. The *intuition* that D is nonetheless relevant to H at a point of utterance indexed to i is, so to speak, ambicontextual.[58]

Hypothesis 8 *also*(b, B) is felicitous only if there is a (proposition expressible as) Ba topic-anaphorically salient for (the proposition expressed by) Bb and the appropriate one of the following specific felicity conditions is met:

$$\text{For } (\textit{and})\ \textit{also}(b,B):\ \operatorname{sgn}(\mathrm{r}_H^{i-k}(Ba)) = \operatorname{sgn}(\mathrm{r}_H^{i}(Bb)).$$
$$\text{For } \textit{but also}(b,B):\ \operatorname{sgn}(\mathrm{r}_H^{i-k}(Ba)) = -\operatorname{sgn}(\mathrm{r}_H^{i}(Bb)).$$

Re *salience*: In uttering *Kim a l s o lives in New York*, one might well take for granted that millions live there (Kripke [49]). Now, is one Jane Doe's living there of relevance to the issue, H? Let H = *Do we have sufficiently many friends in N.Y.?* Then, without any more ado, it isn't. But suppose we have just come to agree that J.D. lives there (cp. Def. 13.ii,iii). Then presumably we did so for a reason, namely, that she is a friend. So quasi-perceptual 'salience' is at least in parts domesticated—as relevance—into the realm of action.

Re *presupposition*: By Hypothesis 8 and Def. 13.ii, i-felicity of *also*(b, B) implies $P^i(Bb|Ba) = P^i(Bb)$. I.e. Bb is i-irrelevant to (i.e. stochastically i-independent of) its i-presupposed, topic-anaphorically salient antecedent. This yields as a

Corollary 15 *If* $b = a$, *then* $P^i(Bb) = P^i(Bb|Bb) = 1$. *Hence,* Bb *will be* i*-irrelevant to any possible* H. *Thus, if* Bb *is to be* i*-relevant at all (whatever* H *might be),* $b \neq a$ *is required.*

The familiar non-identity constraint for *also* becomes derivable. Kripke's notion of 'presupposition' arises out of obedience to 'Be relevant!'. This maxim, now, properly extends the requirement 'Be consistent!', which might be said to motivate the naive and apparently vacuous classically semantic notion of presupposition that admits only the tautology as a presuppositum.

Re *accommodation*: When can it take place? How is it distinct from context-presupposition and assertion? A standard example type of accommodation ([32]) is the presumption arising that there is a duke of Paris,

[58] For technical options explicating it, see Skyrms [71]. Feasibility is realistic: limitation to short-term memory (small integer k of updates) characterizes our conversational phenomenon, here particularly the use of *also*.

when someone says '*The duke of Paris is not bald*', negation being deemed boolean with sentence scope (cp. [22], p. 90 on why 'duke'). Kripke mentions theoreticians' folklore observation that accommodables should be 'uncontroversial'. But what does this mean? I.e. where is a theory that gives it explanatory significance? The DTP framework suggests explicating uncontroversiality by *irrelevance to the issue.* Context-presuppositions are thus uncontroversial by Def. 12.

Partial Definition 14 A proposition Θ is *properly accommodable* at $\langle i, H\rangle$ only if $0 < P^i(\Theta) < 1$ and $\mathrm{r}^i_H(\Theta) = 0$.

The duke's existence, for example, will *in itself* presumably be irrelevant for, say, our sales prospects of pate-polish. His not being bald, materially conditional on existence, might not; and the relevance of that might—indeed would [59]—increase once his existence was properly accommodated.

Consider now an utterance of $also(b, B)$ when no instance of Bx is presupposed. Could one accommodate a salient Ba? Not properly so. The proposed constraint $\mathrm{r}^{i-k}_H(Ba) \neq 0$ (default: $k = 1$) for *anaphoric* salience at i is motivated by the semi-theoretical intuition that Ba is relevant to H at the same order of magnitude as Bb. Its current irrelevance is due to its relevance having been spent when it turned from, say, an assertion at $i - 1$ into a presupposition at i. But if Ba has to be accommodated, then it has not yet become irrelevant through having spent its relevance. Thus, to be relevant to H, it must still be so at i. Hence, proper accommodability is ruled out.

This explains the well-known observation that 'also' is reluctant to license accommodation of an antecedent. I *can* say, 'Kim also talks', and hope you will get my meaning: that our friend, Sandy, talks. But, as likely as not, you will respond with puzzlement, even hostility, instead of smug complicity. Your reaction need not be due to ignorance[59] ('Who else could there be?'), but due to a reluctance to let me take for granted what should not be granted without argument, i.e. what is really an assertion. Here is the essential, properly epistemic distinction between *accommodables* and *assertables.*

Part-explication of anaphoric salience by Def. 13 has more striking, less arcane consequences, too. Consider *Sandy walks and Kim* a l s o *walks.* Def. 13.i implies that i-felicity of '*Aa and also Ab*' implies $P^i(Ab|Aa) = P^i(Ab)$. Next, following Reichenbach [68] and Suppes [76] grant a

[59]Beaver [8], Section 6.1, reports Henk Zeevat's suggestion that anaphoricity requires explicit mention for textual coherence. This speaks against unitarian tendencies for presupposition treatment. My proposal would speak for them. I should be surprised if they could prevail, but think it useful to explore how much language structure *is* motivable by constraints on rational conduct.

Partial Definition 15 A has *prima-facie causal responsibility* for B according to i only if $P^i(B|A) > P^i(B)$.[60]

This implies

Corollary 16 *If A is i-presupposed, then A lacks prima facie causal responsibility for (any) B according to i.*

The corollary and the previous fact lead to the following

Prediction 4 In *'Kim fell and she (also) broke her arm*, 'also' removes routine intimations of causality (causal implicature).

This prediction is readily confirmed. To confirm the semantics of 'also' further, consider the following simple

Fact 17 *If $P^{i'}(\cdot) =_{df} P^i(\cdot|A)$, then $\mathrm{r}^{i'}_H(AB) = \mathrm{r}^{i'}_H(A) + \mathrm{r}^{i'}_H(B)$.*[61]

Note a similar decausalizing effect of 'plus' or stressed 'and', near-synonyms to 'and also'.[62] Thus, presuppositional independence induced by 'also' guarantees the very property which makes the characterization of 'also' and 'too' as 'additive' particles (König [48], Ch. 4) so apt.[63] For note that relevance is now additive for arbitrary A, B, H.[64] Such evidence for a relevance-semantics of 'also' now gives some weight to the following

Hypothesis 9 'also' in *'A but also B'* pre-empts the A-unexpectedness default for *'A but B'*, and thus CIP, in forcing conditioning on A prior to that on B by Hypothesis 8 and Def. 13.

This coheres well with an intuition (check!) that A *but also* B intimates or favours for DT an $H \neq \bar{B}$. The condition of Hypothesis 4 that, nevertheless,

[60] 'Prima facie cause' is Suppes' notion, label credited to Hintikka. *Secunda facie*, the cause may turn out to be spurious, i.e. a case of spurious stochastic correlation. I doubt that the language faculty deals extensively in second sight.

[61] Proof: $P^{i'}(A) = 1$, hence $\mathrm{r}^{i'}_H(AB) = \mathrm{r}^{i'}_H(B) = 0 + \mathrm{r}^{i'}_H(B) = \mathrm{r}^{i'}_H(A) + \mathrm{r}^{i'}_H(B)$.

[62] Jeff Kaplan [44] notes kinship of 'a n d' with 'too'.

[63] This result has most striking implications for intuitions attending whole-sentence focus interpretation, an otherwise occult phenomenon.

[64] The difference between plain *and* and *and also* poses a problem for the extendibility, to non-deductive inferential relations, of any dynamic semantics (e.g. Heim [31, 32] and other contemporary theories) which represents simple 'and' by way of sequential updates. Our evidence makes this representation appropriate in general for 'and also' only. Thus, updating of epistemic, 'world-information' constraints and of discourse-internal constraints will fall apart in the case of plain 'and' whenever the second clause contains an anaphoric reference to a discourse-referent introduced by the first. For on current understanding that referent must become part of the discourse-context to license anaphoric linkage to a pro-form in the second clause.

A ($= Ba$) and B ($= Bb$) have inverse signs of relevance with respect to H is explicated in modified form by the appropriate specific felicity condition of Hypothesis 8.[65] For recall that intuitions are now explicated ambicontextually, or, if you will, as being more properly kinematic.

VII. Dichotomic Issues

One might ask what motivation there is for assuming dichotomic issues $\{H, \overline{H}\}$. A socio-political motivation is the two-person bargaining framework, since dichotomies yield relevance functions that determine dual quasi-utilities or indeed cardinal utilities by indexing to a single proposition. But again, there is a trans-partisan cognitive rationale being served. Recall the use made in Sections IV and V of the conditional independence presumption, CIP, in deriving linguistic predictions. Let CIP now stand for the general, extralinguistic compositionality principle noted. This principle has a more general consequence yet.

For suppose CIP were to generalize to issue partitions $(H_j : j = 1, \ldots, n)$. R.W. Johnson [41] proved that for $n > 2$, at most one evidential proposition deemed atomic can be relevant to any given H_j if conditional independence of evidential atomic propositions is to be satisfied with respect to each element of the issue partition and its complement.

Cognitive engineering cannot always reasonably restrict itself to dichotomies. It will occasionally give up CIP thus generalized. (Recall also the 'doctor' example, Section V.) But suppose the Maker of cognitive psychology values additivity and thus freedom from paradoxes of relevance. This will militate for restriction to the case $n = 2$. The adoption of CIP would thus predict a cognitive bias towards dichotomizing issues. Anecdotal evidence speaks for such a bias being evolutionary fact. Of course, more obvious reasons for it can be thought of. That's fine by me. Any prima facie independent constraint that favours dichotomy will also favour descriptive appeal to CIP.

VIII. Conclusions

Decision-Theoretic Semantics is the label I have given, for want of a more informative one, to the pragmatic approach to meaning illustrated above. The aim has been to show some consequences of taking seriously the view of language as an instrument of rational action and attendant cognition.

The theory of action and cognition assumed is that which largely informs today's curricula in political science, economics and business studies; for better or worse. It generates more structure and thus, if you will, more

[65] I.e., $\mathrm{sgn}(\mathrm{r}_H^{i-k}(Ba)) = -\mathrm{sgn}(\mathrm{r}_H^i(Bb)) \neq 0$, for small $k \geq 1$. By Fact 17, this entails $\mathrm{sgn}(\mathrm{r}_H^i(BaBb)) = \mathrm{sgn}(\mathrm{r}_H^i(Bb))$, i.e. $\mathrm{sgn}(\mathrm{r}_H^i(AB)) = \mathrm{sgn}(\mathrm{r}_H^i(B))$.

readily falsifiable hypotheses than common sense discussion of desires, beliefs, intentions, and cognitions is wont to do. It also links up significant parts of the philosophy of language with the tradition of inductive argumentation that is central to the philosophy of science. One example of this is the semantics of 'but'. Another such point of rapprochement is the distinction drawn in terms of relevance relations between assertion, implicature, accommodatum, and presupposition. There is some reason, then, to invest in applying the theory to what is perhaps the most fundamental of social institutions.

Relevance of assertions—a directional, signed notion as defined in this framework—and the political structure of partisan negotiation are two sides of the same coin. Partisan structure is inherent in the notions of argument, debate and deliberation. It is familiar in pre-Socratic reasoning and contemporary logic alike. It is what underlies and promotes *talking to a point*, rather than merely adding information to a context, never mind to which end. A pertinent concept of relevance will, as Grice noted, explicate this latter intuition, not tacitly rely on it.

Partisan structure does generate trans-partisan structure, one flight of description up. This is not merely, as common sense plausibly suggests, a matter of cancelling partisan atavisms or heuristics by taking both points of view in sequence or in parallel. We have in fact seen a very specific 'invisible hand' type effect in 'scalar implicature' (Section III.2.A). Another, more abstract instance of such trans-partisan constraints is the multiply predicted bias towards dichotomies.

Both findings are consistent with the assumption that cognitive structures are *sociomorph* in being predicated on simple, though far from simplistic structures of social action. This, in turn, should lend some formal and predictive substance to a thesis often advanced in more general terms, though less often made to yield empirical consequences. The thesis is that intricate basic mechanisms of language and of mind are social phenomena.[66]

References

[1] Adams, E. (1973). The logic of conditionals. *Inquiry* 8, 166-197.

[2] Anscombre, J.-C. (1973). Même le roi de France est sage. *Communications* 20, 40-82.

[66] Funding under grant SFB340, Deutsche Forschungsgemeinschaft (DFG) is gratefully acknowledged.

[3] Anscombre, J.-C. and Ducrot, O. (1977) Deux 'mais' en français. *Lingua* 43, 23-40.

[4] Anscombre, J.-C. and Ducrot, O. (1978) Échelles argumentatives, échelles implicatives et lois de discours. *Semantikós* 2, (2/3) 43-66. Repr in. Anscombre and Ducrot [5].

[5] Anscombre, J.-C. and Ducrot, O. (1983) *L'argumentation dans la langue.* Bruxelles: Mardaga.

[6] Austin, J.L. (1946) Other minds. *Proceedings of the Aristotelian Society*, Suppl. Vol. 20. Repr. in J.L.A. *Philosophical papers*, Oxford, Clarendon Press, 1962.

[7] Barwise, J. (1999) State spaces, local logics, and non-monotonicity. In this volume.

[8] Beaver, D. (1997) Presupposition. In J. van Benthem and A. ter Meulen (eds.) *The handbook of logic and language.* Amsterdam: North-Holland.

[9] Birnbaum, M.H. (1983) Base rates in Bayesian inference: signal-detection analysis of the cab problem. *American Journal of Psychology* 96, 85-94.

[10] Carnap, R. (1950) *Logical foundations of probability.* (2nd edn. 1962.) Chicago: University of Chicago Press.

[11] Carnap, R. and Bar-Hillel, Y. (1952) An outline of a theory of semantic information. TR 247 Res. Lab. of Electronics, MIT. Repr. in Y.B-H., *Language and information*, Reading, MA: Addison-Wesley, 1964.

[12] Carnap, R. and Jeffrey, R.C. [eds.] (1971) *Studies in inductive logic and probability*, Vol. I. Berkeley CA: University of California Press.

[13] Carston, R. (1988) Implicature, explicature, and truth-theoretic semantics. In R.M. Kempson (ed.) *Mental representation: the interface between language and reality.* Cambridge: Cambridge University Press. Repr. in Davis [14].

[14] Davis, S. [ed.] (1991) *Pragmatics.* New York: Oxford University Press.

[15] Ducrot, O. (1972) *Dire et ne pas dire: principes de sémantique linguistique.* 3rd edn., 1991. Paris: Hermann.

[16] Ducrot, O. (1973) *La preuve et le dire.* Paris: Mame. Partially reprinted with augmentations as *Les échelles argumentatives.* Paris: Minuit, 1980.

[17] Fauconnier, G. (1975) Pragmatic scales and logical structure. *Linguistic Inquiry* 6, 353-375.

[18] Field, H. (1977) Logic, meaning, and conceptual role. *Journal of Philosophy* 74, 379-409.

[19] Francescotti, R.M. (1995) 'Even': the conventional implicature approach reconsidered. *Linguistics and Philosophy* 18, 153-173.

[20] Frege, G. (1879) *Begriffsschrift.* Trsl. in J. van Heijenoort (ed.) *From Frege to Gödel.* Cambridge MA: Harvard University Press, 1967.

[21] Gaifman, H. (1971) Applications of De Finetti's theorem to inductive logic. In Carnap and Jeffrey [12].

[22] Gazdar, G. (1979) *Pragmatics.* London: Academic Press.

[23] Gärdenfors, P. (1986) Belief revision and the Ramsey test for conditionals. *Philosophical Review* 95, 81-93.

[24] Good, I.J. (1950) *Probability and the weighing of evidence.* London: Charles Griffin.

[25] Good, I.J. (1983) *Good thinking.* Minneapolis: University of Minnesota Press.

[26] Grice, H.P. (1967) 'Logic and Conversation'. William James Lectures, Harvard U., repr. in Grice (1989).

[27] Grice, H.P. (1989) *Studies in the way of words.* Cambridge MA: Harvard University Press.

[28] Groenendijk, J. and Stokhof, M. (1991). Dynamic predicate logic. *Linguistics and Philosophy* 14, 39-100.

[29] Harris, Z. (1964) The elementary transformations. Transformations and Discourse Analysis Papers 54, University of Pennsylvania. Repr. in Z.H., *Papers in structural and transformational linguistics.* Dordrecht: Reidel, 1970.

[30] Harsanyi, J.C. (1977) *Rational behavior and bargaining equilibrium in games and social situations.* Cambridge: Cambridge University Press.

[31] Heim, I. (1982) The Semantics of Definite and Indefinite Noun Phrases in English, PhD Thesis, University of Massachusetts, Amherst. Repr. New York: Garland.

[32] Heim, I. (1983) On the projection problem for presuppositions. In M. Barlow et al. (eds.) WCCFL 2 *Proc. of the West Coast Conference on Formal Linguistics., Vol. 2.* Stanford: Stanford Linguistics Association. Repr. in Davis [14].

[33] Hirschberg, J. (1985) A theory of scalar implicature. PhD Thesis U. Penn. Publ. New York: Garland 1991.

[34] Horn, L.R. (1972) The semantics of logical operators in English. PhD Thesis Yale U., Distr. Indiana University Linguistics Club, 1976.

[35] Horn, L.R. (1989) *A natural history of negation.* Chicago: University of Chicago Press.

[36] Horn, L.R. (1992) The said and the unsaid. *Proc. SALT II*, 163-192.

[37] Humburg, J. (1971) The principle of instantial relevance. In Carnap and Jeffrey (1971).

[38] Isard, S.D. (1972) Changing the context. In E.L. Keenan (ed.) *Formal semantics of natural language.* Cambridge: Cambridge University Press.

[39] Jeffrey, R.C. (1965) *The logic of decision.* 2nd edn., 1983. Chicago: University of Chicago Press.

[40] Jeffrey, R.C. (1992) *Probability and the art of judgment.* Cambridge: Cambridge University Press.

[41] Johnson, R.W. (1986) Independence and Bayesian updating methods. *Artificial Intelligence* 29, 217-222.

[42] Kahneman, D., Slovic, P. and Tversky, A. (1982) *Judgment under uncertainty: heuristics and biases.* Cambridge: Cambridge University Press.

[43] Kamp, H. (1981) A Theory of Truth and Semantic Representation. In: Groenendijk, J. , Janssen, T. M. V. and Stokhof, M. *Formal Methods in the Study of Language.* Amsterdam: Mathematical Centre Tracts.

[44] Kaplan, J. (1984) Obligatory 'too' in English. *Language* 60, 510-518.

[45] Karttunen, L. and Peters, S. (1979) Conventional implicature. In C.-K. Oh and D.A. Dineen (eds.) *Syntax and Semantics 11: Presupposition.* New York: Academic Press.

[46] Kay, P. (1990) 'Even'. *Linguistics and Philosophy* 13, 59-111.

[47] Keynes, J.M. (1921) *A treatise on probability*. London: Macmillan.

[48] König, E. (1991) *The meaning of focus particles.* London: Routledge.

[49] Kripke, S. (n.d) Presupposition and anaphora: remarks on the formulation of the projection problem. Ms. Princeton University.

[50] Lewis, D. (1969) *Convention.* Cambridge MA: Harvard University Press.

[51] Lewis, D. (1979) Score-keeping in a language game. *Journal of Philosophical Logic* 8, 339-359.

[52] Mandelbrot, B. (1961) On the theory of word frequencies and on related Markovian models of discourse. In: Jakobson, R. (ed.) *Structure of language and its mathematical aspects.* Providence RI: American Mathematical Society.

[53] Merin A. (1992) Permission sentences stand in the way of Boolean and weaker lattice-theoretic semantices. *Journal of Semantics* 9, 95-162.

[54] Merin A. (1994) Algebra of elementary social acts. In S.L. Tsohatzidis (ed.) *Foundations of Speech Act Theory.* London: Routledge.

[55] Merin, A. (1995) Decision Theory of Language Universal. In M.-L. Dalla Chiara et. al. (eds.) Abstracts of the 10th Congress for Logic, Methodology and Philosophy of Science (Section 7: Probability, Induction, Decision-Theory), Florence, August 19-25, 1995.

[56] Merin, A. (1996a) Die Relevanz der Relevanz: Fallstudie zur formalen Semantik der englischen Konjunktion 'but'. Habilitationsschrift, University of Stuttgart. [English version available soon.]

[57] Merin, A. (1996b) Formal semantic theory and diachronic data: a case study in grammaticalization. SFB 340 Technical Report 75, University of Stuttgart.

[58] Merin, A. (1997a) If all our arguments had to be conclusive, there would be few of them. SFB 340 Technical Report 101, University of Stuttgart.

[59] Merin (1997b) Scoring high. In P. Weingartner, G. Schurz, and G. Dorn (eds.) *The role of pragmatics in philosophy.* Contributions of the Austrian Ludwig Wittgenstein Society, Vol. 5, Kirchberg: ALWS.

[60] Merin A. (in press) Negative attributes, partitions and rational decisions: why *not* speak Notspeak. *Philosophical Studies.*

[61] Miller, G.A. and Chomsky, N. (1963) Finitary models of language users. In Bush, R.R., Luce, R.D. and Galanter, E. (eds.) *Handbook of mathematical psychology*, vol. II, New York: Wiley.

[62] Nash, J.F. (1953) Two-person cooperative games. *Econometrica* 21, 140-152.

[63] Peirce, C.S. (1878) The probability of induction. *Popular Science Monthly* 12, 705-18. Repr. in *Collected Papers*, 2.669-693. Cambridge MA: Belknap Press, 1932.

[64] Ramsey, F.P. (1926) Truth and probability. Repr. in Ramsey (1990).

[65] Ramsey, F.P. (1929a) Probability and partial belief. Repr. in Ramsey (1990).

[66] Ramsey, F.P. (1929b) Theories. Repr. in Ramsey (1990).

[67] Ramsey, F.P. (1990) *Philosophical Papers*, ed. D.H. Mellor, Cambridge: Cambridge University Press, 1990.

[68] Reichenbach, H. (1956) *The direction of time.* Berkeley: University of California Press.

[69] Ristad, S.E. (1993) *The language complexity game.* Cambridge MA: MIT Press.

[70] Skyrms, B. (1980) *Causal necessity.* Newhaven CT: Yale University Press.

[71] Skyrms, B. (1983) Three ways to give a probability assignment a memory. In J. Earman (ed.) *Testing scientific theories.* Minnesota Studies in the Philosophy of Science Vol X. Minneapolis: University of Minnesota Press.

[72] Skyrms, B. (1984) *Pragmatics and empiricism.* Newhaven CT: Yale University Press.

[73] Soames, S. (1982) How presuppositions are inherited: a solution to the projection problem. *Linguistic Inquiry* 13, 483-545.

[74] Sperber, D. and Wilson, D. (1986) *Relevance: communication and cognition.* 2nd edn., 1995. Oxford: Blackwell.

[75] Stalnaker, R. (1974) Pragmatic presuppositions. In M.K. Munitz and P.K. Unger (eds.) *Semantics and philosophy.* New York: New York University Press. Repr. in Davis [14].

[76] Suppes, P. (1970) *A probabilistic theory of causality.* Amsterdam: North-Holland. [Series: Acta Philosophica Fennica].

[77] van der Sandt, R. (1992) Presupposition projection as anaphora resolution. *Journal of Semantics* 9, 333-377.

Hyperproof: Abstraction, Visual Preference and Modality

Jon Oberlander Keith Stenning
Division of Informatics
University of Edinburgh
Richard Cox
School of Cognitive and Computing Sciences
University of Sussex

1 Introduction

We have been carrying out an evaluation of the effects of teaching logic with Barwise and Etchemendy's (1994) Hyperproof, a program for teaching first-order logic (see Figure 1). Inspired by a situation-theoretic approach to heterogeneous reasoning, it uses multimodal (graphical and sentential) methods, allowing users to transfer information to and fro, between modalities (see Figure 2). One of our major findings has been that individual differences between students have a significant effect on students' responses to Hyperproof: their prior cognitive style influences both the overall effectiveness of the teaching regime, and the actual proof structures that students produce under exam conditions (cf. Monaghan 1995; Stenning, Cox and Oberlander 1995; Oberlander, Cox and Stenning 1996; Oberlander et al. 1996). In the course of this larger study, we have built up a substantial corpus of 'hyperproofs'. We believe that this corpus can provide a detailed insight into various questions concerning the paths which students follow in their pursuit of proof goals.

In particular, Barwise and Etchemendy designed Hyperproof to support *heterogeneous* reasoning, in which information from differing modalities—sentential and graphical—is combined, or transferred from one modality to

Logic, Language, and Computation, Vol. II, edited by Lawrence S. Moss, Jonathan Ginzburg, and Maarten de Rijke.

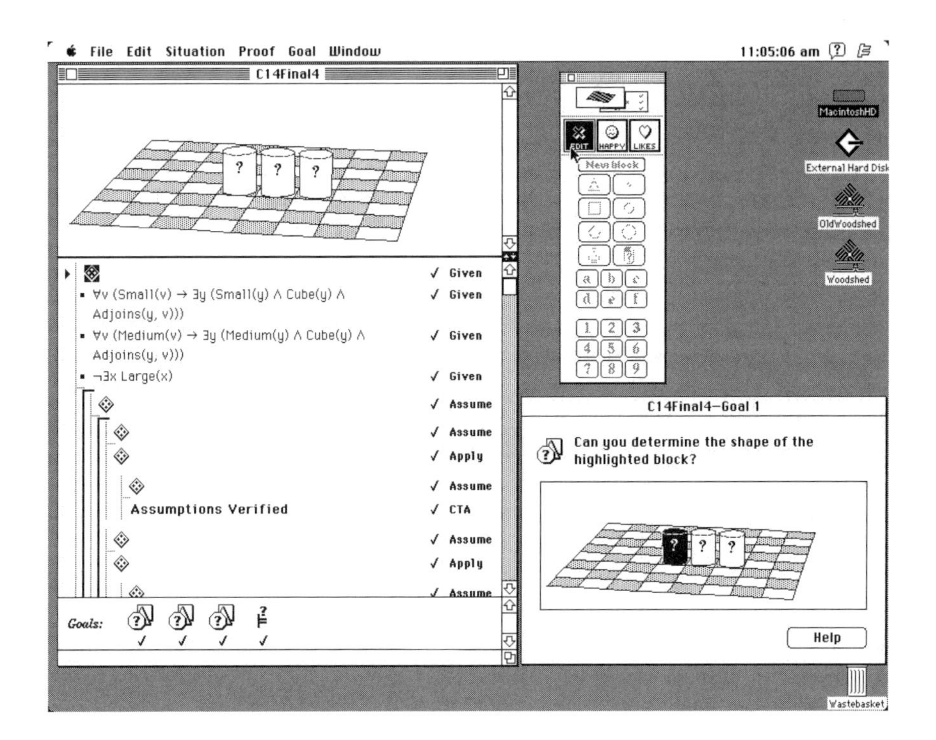

Figure 1: The Hyperproof Interface. The main window (top left) is divided into an upper graphical pane, and a lower calculus pane. The tool palette is floating next to the main window, and other windows can pop up to reveal a set of goals which have been posed.

Apply Extracts information from a set of sentential premises; expresses it graphically

Assume Introduces a new assumption into a proof, either graphically or sententially

Observe Extracts information from the situation; expresses it sententially

Inspect Extracts common information from a set of cases; expresses it sententially

Merge Extracts common information from a set of cases; expresses it graphically

Close Declares that a sentence is inconsistent with either another sentence, or the current graphical situation

CTA (Check truth of assumptions) Declares that all sentential and graphical assumptions are true in the current situation

Exhaust Declares that a part of a proof exhausts all the relevant cases

Figure 2: A set of relevant Hyperproof rules.

another. It is obviously, therefore, a multimodal system containing a visual sub-system. But given that one group of students benefits particularly from being taught with Hyperproof, we can ask: do they do well because it is a *visual* logical system, or do they do well because it is *multimodal*?

To address this question, we first frame some hypotheses concerning the relation between the individual differences in teaching outcome which we found, and the structures to be found in students' proofs; the rest of the paper then focusses on the second of these hypotheses. As background, we outline the relevant aspects of the design of the main study, indicating how it distinguishes two styles of student. We then describe (i) the way 'proofograms' are used to track the way students deal with abstractions; and (ii) the application of bigram and trigram analyses of rule use patterns in the data corpus, demonstrating that the differing styles of student end up producing multimodal proofs of distinctive types.

2 Hypotheses

The observation that graphical systems require certain classes of information to be specified goes back at least to Bishop Berkeley. Elsewhere, we have termed this property 'specificity', and argued that it is useful because inference with specific representations can be very simple (Stenning and Oberlander 1991, 1995). We have also urged that actual graphical systems do allow abstractions to be expressed, and it is this that endows them with a usable level of expressive power. Thus, Hyperproof maintains a set of abstraction conventions for objects' spatial or visual attributes. As well as concrete depictions of objects, there are 'graphical abstraction symbols', which leave attributes under-specified: the *cylinder*, for instance, depicts objects of unknown size (see Figure 1). A key step, then, in mastering an actual graphical system is to learn which abstractions can be expressed, and how.

As we describe below, our pre-tests independently allowed us to divide subjects into two cognitive style groups, on the basis of their performance on a certain type of problem item. Loosely, one group is 'good with diagrams', and the other less so. The good diagrammers turned out to benefit more from Hyperproof-based teaching than the others. Our belief is that those who benefit most from Hyperproof do so because they are better able to manipulate the graphical abstractions it offers. Call this view the *abstraction ability hypothesis*. Elsewhere, we have provided evidence in support of it (Oberlander et al. 1996).

That evidence also bears on the question in hand. Whether Hyperproof's virtue lies in its visual nature, or in its multimodality depends upon whether abstraction ability is supported by Hyperproof's visual representations —or by some other aspect of the system. One hypothesis is that the good diagrammers are simply those subjects who have a preference for the visual modality. Call this view the *visual preference hypothesis*. Another explanation would be that good diagrammers are those who are adept at translating between modalities. Call this view the *transmodal hypothesis*.

In what follows, we aim to show that the balance of evidence favours the transmodal hypothesis.

3 Distinguishing cognitive styles

In the full study, two groups of subjects were compared; one ($n = 22$ at course end) attended a one-quarter duration course taught using the multimodal Hyperproof. A comparison group ($n = 13$ at course end) were taught for the same period, but in the traditional syntactic manner supplemented with exercises using a graphics-disabled version of Hyperproof.

Subjects were administered two kinds of pre- and post-course paper and pencil test of reasoning. The first of these is most relevant to the current discussion. It tested 'analytical reasoning' ability, with two kinds of item derived from the GRE scale of that name (Duran, Powers and Swinton 1987). One subscale consists of verbal reasoning/argument analysis. The other subscale consists of items often best solved by constructing an external representation of some kind (such as a table or a diagram). We label these subscales as 'indeterminate' and 'determinate', respectively. Scores on the latter subscale were used to classify subjects within both Hyperproof and Syntactic groups into DetHi and DetLo sub-groups. The score reflects subjects' facility for solving a type of item that often is best solved using an external representation; DetHi scored well on these items; DetLo less well. For the moment, we may consider DetHi subjects to be more 'diagrammatic', and DetLo to be less so.

4 Abstraction ability results

Both the Hyperproof and Syntactic groups contained DetHi and DetLo sub-groups. All subjects sat post-course, computer-based exams, although the questions differed for the two groups, since the Syntactic group had not been taught to use Hyperproof's systems of graphical rules. Student-computer interactions were dynamically logged, permitting a full, step-by-step, reconstruction of the process of the subject's reasoning, as well as capturing the final proof produced.

Here, we discuss only the final proofs produced by the 22 Hyperproof subjects, all of whom completed the exams. The four questions that these students were set contained two types of item: determinate and indeterminate. Here, determinate problems were taken to be those whose problem statement did not utilise Hyperproof's abstraction conventions. That is: determinate problems contained only concrete depictions of objects in their initially given graphical situation, whereas indeterminate problems—such as that in Figure 1—could contain graphical abstraction symbols in the initial situation.

4.1 Proofograms

What evidence is there for the abstraction ability hypothesis? Among the Hyperproof students, do the two sub-groups—DetHi and DetLo—use graphical abstraction symbols in characteristically different ways?

We can score each step of each proof on the basis of number of concrete situations compatible with the graphical depiction; one possible scoring method is described in Oberlander, Cox and Stenning (1996). A low score always indicates more abstraction; a higher score indicates more concreteness.

We can explore the way concreteness varies through the course of a proof by graphing it against the hierarchical structure of the proof. We call such graphs 'proofograms'. Figures 3 and 5 show how subjects C2 and C14 tackle an indeterminate exam question; Figures 4 and 6 give their proofograms. The visual differences between proofograms are quite striking: one group is 'spikey'—as in Figure 4; and the other is 'layered'—as

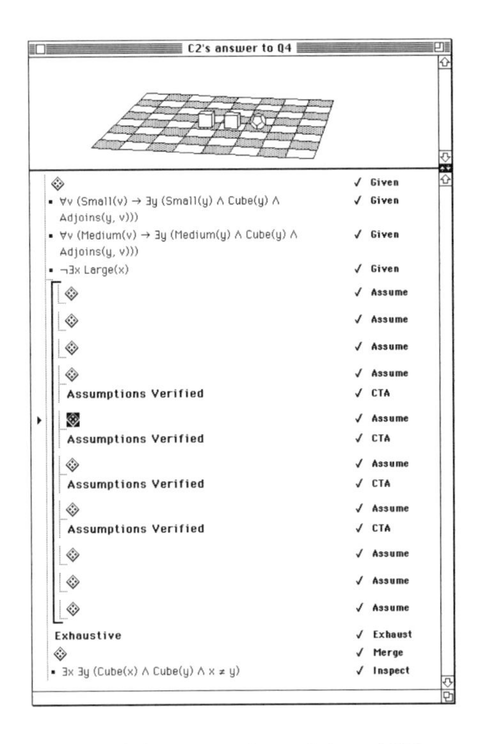

Figure 3: Submitted proof for a DetLo subject (C2) attempting an indeterminate question (Q4). The situation on view is from the 9th step of the proof.

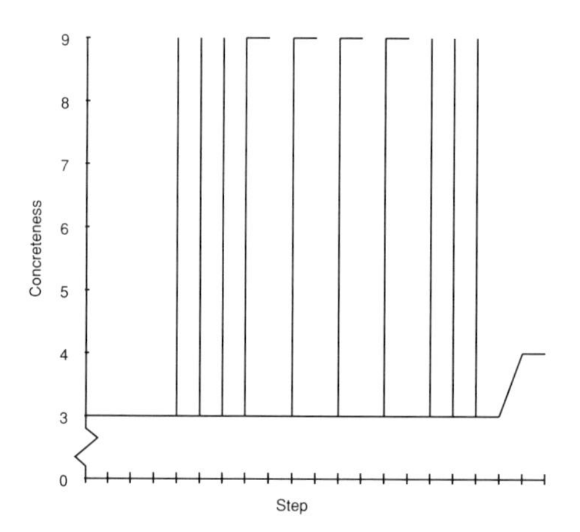

Figure 4: Proofogram for C2 attempting Q4. Proof steps are plotted on the x-axis; the concreteness of the current graphical situation is computed for each step of the proof, and is plotted on the y-axis. Horizontal lines indicate dependency structure; vertical lines indicate uses of Assume; sloping lines indicate uses of Apply or Merge. C2's proofogram is 'spikey', indicating a series of independent, concrete cases.

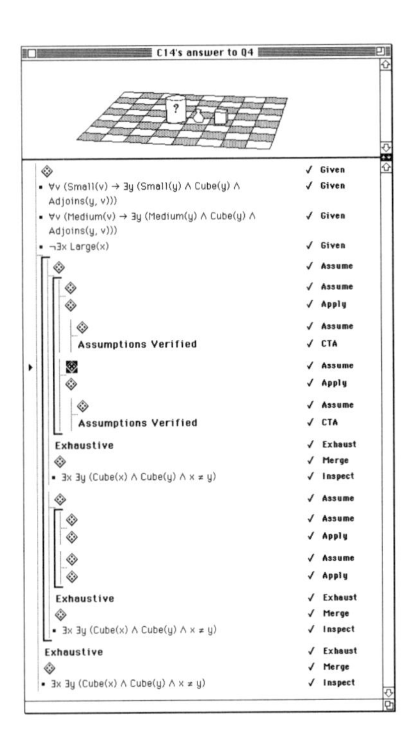

Figure 5: Submitted proof for a DetHi subject (C14) attempting an indeterminate question (Q4). The situation on view is from the 9th step of the proof.

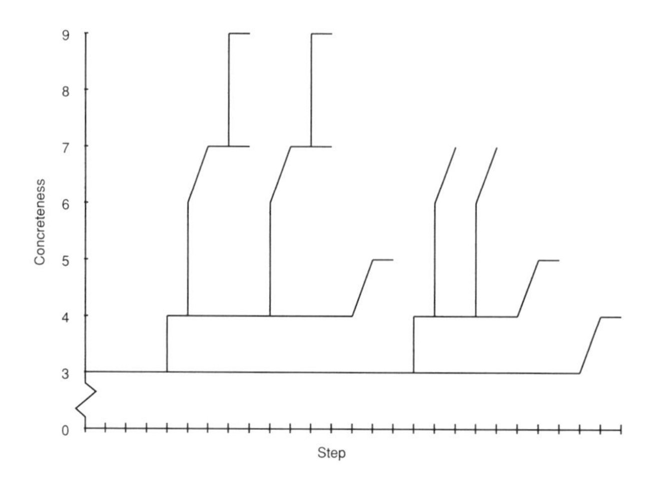

Figure 6: Proofogram for C14 attempting Q4. C14's proofogram is 'layered', indicating parallel sub-case structures with abstract superordinate cases.

in Figure 6. The differences are particularly pronounced on indeterminate questions, and Q-sort tests indicate that these questions reliably elicit layered proofs from DetHi subjects, and spikey proofs from DetLo (cf. Oberlander et al. 1996). The basic message appears to be that there is a 'staging phenomenon': DetHi introduce concreteness *by stages*, whereas DetLo introduce it more immediately. In terms of proof structure, DetHi tend to produce structured sets of cases, with superordinate cases involving graphical abstraction; DetLo tend to produce sets of cases without such overt superordinate structure. This staging phenomenon supports the abstraction ability hypothesis: the two groups are certainly using abstractions in different ways.

4.2 Corpus analysis

Of Hyperproof's rules, only Assume, Apply and Merge increase concreteness. We therefore examined the kind of patterns in which they occur through proof-corpus analysis. The proofogram results already indicate that DetHi and DetLo differ in the way they handle concreteness. Since Assume is by far the most frequent means of adding concreteness, the corpus analysis distinguishes between uses of the rule which introduce totally concrete graphical situations, and those which leave some abstractness in the graphic. The term Fullassume denotes the former type of use, and assume denotes the latter.

Using techniques developed originally for the analysis of linguistic corpora, we have carried out bigram and trigram analyses of rule use, utilising Dunning's (1993) 'Log–Likelihood Test', which can be applied to relatively small corpora. The test is designed to "highlight particular A's and B's that are highly associated in text" (p.71). Ranking the bigrams according to this test provides a good *profile* of the individual's, or the group's, rule use in the corpus. We can then compare the profiles for the sub-groups on the two question types, assessing the significance of a given bigram by using the χ^2 test on the log-likelihood value.

On indeterminate questions, we find that the bigrams assume Close, Given assume, assume Fullassume, Observe assume, NegIntro Univintro, Close assume, and Apply Observe are significant in DetHi proofs, but not in DetLo ones. Conversely, only the bigrams Inspect Merge and Given Apply are significant in DetLo proofs, but not in DetHi ones. The profiles are weakly but significantly correlated ($r = 0.167^*$).[1] When taking into account only those bigrams that are significantly associated in the profiles, the correlation is higher, but not significant ($r = 0.443, ns$).

On determinate questions, the bigrams assume Apply, CTA Observe, Close Fullassume, Observe CTA, Apply Fullassume, Inspect Implyelim, and Fullassume Fullassume are significant in DetHi proofs, but not in DetLo ones. Conversely, as with the indeterminate questions, there are only two bigrams significant in DetLo proofs, but not in DetHi ones: Inspect Merge and Exhaust Inspect. Here, the two subject groups' profiles are significantly correlated ($r = 0.537^{**}$). The correlation between significantly-associated bigrams is even stronger and still highly significant ($r = 0.809^{**}$).

This finding accords with the proofograms' indication that it is indeterminate questions which best discriminate the two subject groups. Recall that these are the questions in which the initial graphical situation is abstract, so that all concreteness must be introduced explicitly by the subjects.

The proofogram and corpus analyses therefore support the abstraction ability hypothesis. On questions where the subject must construct the concrete graphic, it seems that DetHi subjects exhibit staging behaviour, and build their graphics incrementally, whereas DetLo subjects are prone to construct their concrete graphics in one go. The abstraction ability hypothesis seems plausible, since the 'stagers' are exactly those whom our main study showed benefit most from teaching with Hyperproof (Stenning, Cox and Oberlander, 1995).

[1] Correlations reported here are non-parametric (Spearman's ρ). Significance at the $p < .05$ level is denoted by *; significance at the $p < .001$ level by **.

5 Visual preference results

But why do the subject groups diverge under these circumstances? As we have suggested, one tempting hypothesis comes from identifying our DetHi—DetLo distinction with the traditional visualiser—verbaliser distinction. If it's a matter of visual preference, then the diagrammatically capable DetHi subjects are just the visualisers, and therefore, they prefer to use the graphical modality when it is available. The diagrammatically less capable DetLo are the verbalisers, and hence prefer the sentential modality—or at least, they do not show a strong preference for the graphical.

The alternative transmodal hypothesis is that DetHi subjects are better at multimodal reasoning, mixing sentential and graphical information. On this account, DetLo might be perfectly happy in the graphical modality, so long as they do not have to translate information back and forth between the graphical and the sentential.

One way of testing these competing views is to look at the overall networks of bigram transitions, for the two subject types. The transition networks in Figures 7 and 8, represent DetHi and DetLo behaviour on indeterminate questions. In the networks, the area of a node represents the frequency with which a rule is used, while the thickness of links represents the probability of taking that exit arc, given that one is in the state denoted by the node.

First, consider the left-hand parts of the networks. Any proof must start from a Given step; now, it is clear that there are several ways in which the use of assume and Fullassume varies between the DetHi and DetLo groups. First, DetHi make more use of assume than DetLo, while the latter around twice as much use of Fullassume than the former. DetLo subjects' favouring of Fullassume over assume certainly confirms that they are not 'stagers'; but in a sense, it also suggests that it is *they* who exhibit a preference for the graphical modality, moving straight into it and working entirely within it, rather than gradually transferring information into it from the sentential pane.

Notice also that around two-thirds of DetHi transitions from Given are to assume, and the rest go to Fullassume. By contrast, just one-third of DetLo transitions from Given go to assume; some 22% go to Apply, and 44% go straight to Fullassume. So, as well as favouring Fullassume over assume, DetLo subjects also often commence proof contruction by the use of Apply. This also helps to reduce subsequent interaction between the modalities, with case construction being performed only within the graphical window. Looking at Apply in more detail, it is apparent that DetHi are more likely to use it after assume (it accounts for 28% of their transitions out of assume, as

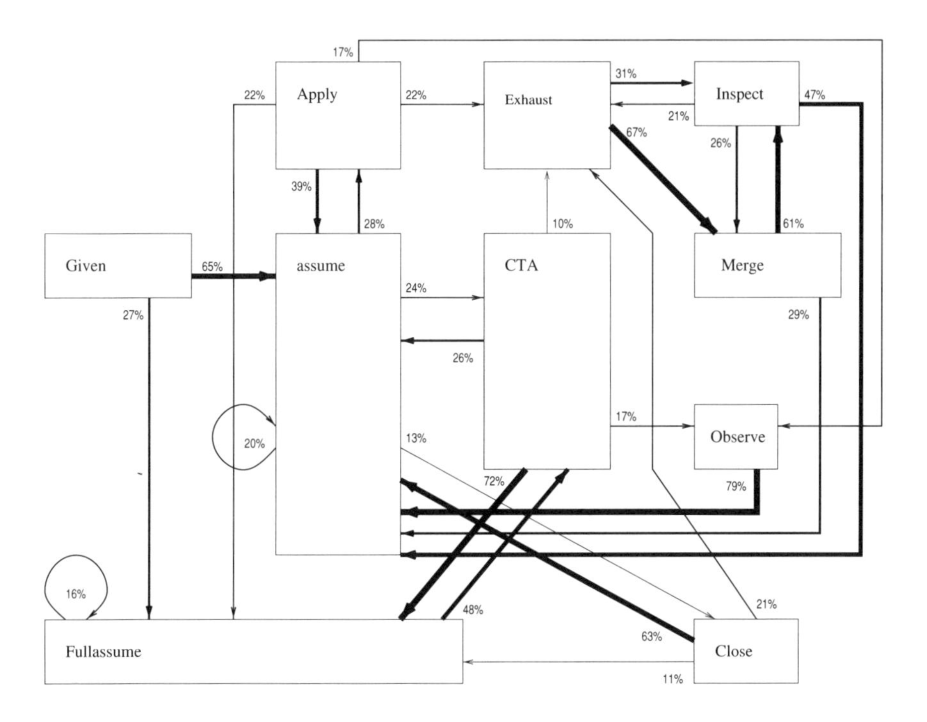

Figure 7: Transition network for DetHi behaviour on indeterminate questions. Nodes represent rules, and their areas represent the frequency at which that rule was invoked. Links represent the probability of transition from one rule to another; transitions at 10% probability and below are not shown.

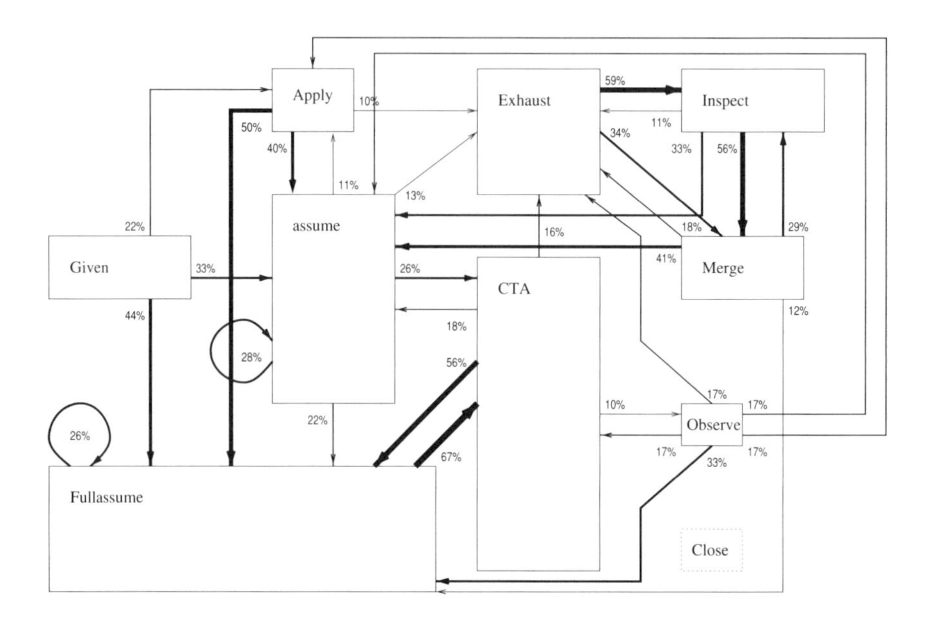

Figure 8: Transition network for DetLo behaviour on indeterminate questions. Note that Close is not visited at all.

opposed to just 11% amongst DetLo). And while both groups are as likely to use assume after Apply, DetLo are more than twice as likely as DetHi to go from Apply direct to Fullassume.

The assume Apply pattern confirms that DetHi subjects tend to add information graphical window pane gradually, either by assumption, or by transfer from the sentential pane (via Apply).

Secondly, consider the top right portions of the networks. From Exhaust, DetLo are most likely to move to Inspect, and from there to Merge. By contrast, DetHi are most likely to move to Merge, and from there to Inspect. Both Inspect and Merge find common information from the set of cases declared exhaustive by Exhaust. The difference is that Inspect provides this information sententially, and Merge does it graphically. It seems from the networks that DetHi find the graphical before the sentential, while DetLo find the sentential first, and only then carry out the graphical operation.

Taking these two parts of the network together, it should be clear by now that this is not a simple matter of DetHi preferring the visual modality, not least since DetLo move to that modality more directly. Instead, the difference seems to be that the DetHi group *operate over* the graphical situations, frequently using a graphic as input, or guide, to further stages of proof construction. The DetLo, on the other hand, seem just to *output* graphics, without subsequently using them.

Finally, consider the bottom right hand corners of the networks. There is one very striking fact: DetLo subjects *never* use the Close rule on these indeterminate questions. Close is used in proofs of inconsistency: students use it show that some assumptions contradict given information. And this means that while DetLo make considerable use of proofs of consistency—evidenced by the high frequency of use of CTA—they never proceed by showing that certain cases can be explicitly ruled out as inconsistent with existing premises and assumptions. This aversion to proof by contradiction is intriguing, because it may ultimately be related to findings concerning people's ability to verify or falsify general propositions (as in the four card selection problem, discussed, for instance, in Wason 1977).

For the time being, however, it suffices that DetHi subjects do *not* show a simple preference for the visual–graphical modality. Rather, what distinguishes them is their greater tendency to *translate* between graphical and sentential modalities in *both* directions. Suppose we call people who prefer the visual modality 'artists', and people who like to switch back and forth between representation systems 'translators'. Then it seems that abstraction ability—and hence success with Hyperproof—lies not with the artists, but with the translators.

Acknowledgements

The support of the Economic and Social Research Council for HCRC is gratefully acknowledged. The work was supported by UK Joint Councils Initiative in Cognitive Science and HCI, through grant G9018050 (Signal); and by NATO Collaborative research grant 910954 (Cognitive Evaluation of Hyperproof). The first author is supported by an EPSRC Advanced Fellowship. The paper is based on that published as Oberlander et al. (1996), but includes additional material eventually to be submitted for journal publication. Special thanks to Dave Barker-Plummer, Chris Brew, Tom Burke, John Etchemendy, Mark Greaves, Padraic Monaghan, and Richard Tobin.

References

Barwise, J. and Etchemendy, J. (1994). *Hyperproof*. Stanford: CSLI Publications.

Dunning, T. (1993). Accurate methods for the statistics of surprise and coincidence. *Computational Linguistics* **19**, 61–74.

Duran, R., Powers, D. and Swinton, S. (1987). Construct Validity of the GRE Analytical Test: A Resource Document. ETS Research Report 87–11. Princeton, NJ: Educational Testing Service.

Monaghan, P. (1995). A corpus-based analysis of individual differences in proof-style. MSc Thesis, Centre for Cognitive Science, University of Edinburgh.

Oberlander, J., Cox, R. and Stenning, K. (1996). Proof styles in multimodal reasoning. In Seligman, J. and Westerståhl, D. (Eds.) *Language, Logic and Computation: Volume 1*, pp403–414. Stanford: CSLI Publications.

Oberlander, J., Cox, R., Mongahan, P., Stenning, K. and Tobin, R. (1996). Individual differences in proof structures following multimodal logic teaching. In *Proceedings of the 18th Annual Meeting of the Cognitive Science Society*, pp201–206, La Jolla, Ca., July 1996.

Stenning, K., Cox, R. and Oberlander, J. (1995). Contrasting the cognitive effects of graphical and sentential logic teaching: reasoning, representation and individual differences. *Language and Cognitive Processes*, **10**, 333–354.

Stenning, K. and Oberlander, J. (1991). Reasoning with Words, Pictures and Calculi: computation versus justification. In Barwise, J., Gawron, J. M., Plotkin, G. and Tutiya, S. (Eds.) *Situation Theory and Its Applications*, Volume 2, pp607–621. Chicago: Chicago University Press.

Stenning, K. and Oberlander, J. (1995). A cognitive theory of graphical and linguistic reasoning: Logic and implementation. *Cognitive Science*, **19**, 97–140.

Wason, P. C. (1977). Self-contradictions. In Johnson-Laird, P. N. and Wason, P. C. (Eds.) *Thinking: Readings in Cognitive Science*, pp114–128. Cambridge: Cambridge University Press.

96307

237-65

Structured Argument Generation in a Logic-Based KB-System

Denise Aboim Sande e Oliveira
Clarisse Sieckenius de Souza[1]
Edward Hermann Haeusler

Departamento de Informática
PUC-Rio, Brazil
{denise,clarisse,hermann}@inf.puc-rio.br

Abstract This paper presents a method for the transformation of Natural Deduction proofs into hypertext-like structured explanations to be provided by knowledge-based systems. First-Order Logic and a variant of the Natural Deduction system are taken as a basis for representing knowledge and reasoning over a generic classification domain. Information selection and restructuring are used to convert proofs into argumentation schemata. Our goal is to produce structured arguments that can be converted into Natural Language explanatory hypertexts which satisfy some pragmatical constraints. Thus, pragmatically motivated rules are incorporated to the semantic component of a friendly user interface language in which explanations are requested and conveyed.

1 Introduction

Today's booming information technology brings about a renewed interest in the construction of information systems which not only store and retrieve information but also deal with them in intelligent (or rational) ways. Knowledge-based systems (kb-systems) are being used in a great variety of fields to enhance the processing of information. The impact of their role in this scenario is conditioned by their ability to present and justify their reasoning to the human user. As a result, an increasing research effort is being placed in the development of intelligent multimodal interfaces [11] and explanatory text generation components [2, 14].

Logic, Language, and Computation, Vol. II, edited by Lawrence S. Moss, Jonathan Ginzburg, and Maarten de Rijke.

[1]C. S. de Souza would like to thank CNPq, the Brazilian council for the development of science and technology, for supporting her research.

In an effort to improve the legibility of artificial reasoning patterns, we have explored the idea that a logic proof can be used as an argument structure [8]. Elaborations on proofs should produce human-readable arguments [20]] that are adequate for a set of domains. We have chosen First Order Logic (FOL) as a flexible and powerful representation language to account for generic classification domains. Logic representation and deductive reasoning, having been derived from actual human argumentation practice, present some attractive features to support the automatic generation of quality explanatory texts. By the same token, Natural Deduction (ND), derived from human mathematical proof practice, produces bidimensionally-structured tree-like proofs that are arguably easier to understand than other methods such as resolution. ND tends to produce direct proofs, notwithstanding the maintenance of full structural integrity of the original sentences. It is also important to note that, although we have chosen FOL to begin with, any other logic that can be adapted to ND reasoning would be equally acceptable as long as it had a set of ND rules that users would be able to take as "natural."

In conversational computer applications such as KB-systems, a number of pragmatic rules should be observed for interaction to proceed productively. They have been first organized by Grice into what he has called the "Co-operative Principle" [7]. His four maxims that characterize the dimensions of logic in conversation have been summarized by McCawley as: "you should assert neither more nor less than is appropriate for the purpose at hand" (Quantity); "what you assert should be true, and you should have adequate grounds for holding it to be true" (Quality); "what you mention should be relevant for the purpose at hand" (Relation); and "you should use linguistic means no more elaborate than what is needed to convey what you are asserting" (Manner) ([12], p.217).

Direct observation of logic proofs, whatever the method used to generate them, gives clear evidence for much needed transformations on them so that they can be used to provide support in human-computer explanatory dialogues. Grice's maxim of Quality is rigorously accounted for in logic-based systems, but proofs contain lots of distracting details and apparently irrelevant detours compared to "natural reasoning". The maxims of Quantity and Relevance interact in complicated ways, especially when disjunctions are used throughout the proof. Converting a ND proof into an argument schema thus involves a considerable amount of focal and topical evaluations, along with pruning machinery that contributes to maintain the structural elements within highly cohesive boundaries. The requirements on Manner are self-evidently numerous in that the linguistic means to convey assertions must be something other than FOL for the vast majority of KB-systems' users.

Moreover, conventional implicatures entailed by lexical items alone, sentences, or text spans, represent a major obstacle to meeting the requirements of Quantity and Relevance. The examples provided in this paper (such as the explanation of why cats don't live in the ocean, in sections 4 and 6) refer us to common sense knowledge residing in the misty frontier between language and cognition. They illustrate that there are implicatures in the textual explanations which suggest that some kind of logic reasoning should be present in the linguistic interface of KB-systems, and not only in the knowledge bases. For example, the text:

Cats don't live in the ocean, because they have paws and walk.

is better than the automatically generated text:
Cats don't live in the ocean, because they have four paws. Anyone who has four paws walks, so cats walk. Since anyone who walks does not live in the ocean, then cats don't live in the ocean (either).
The last phrase is implied by the use of the two previous phrases. In this case, we feel that there are strong indications of a regularity which can be mapped in the text generator, but which cannot be accounted for at the level of the reasoning module, since without the last phrase the argument is incomplete.

Our research is aimed at designing a user interface language that is easy to use and understand, and applicable to KB-systems in generic classification domains. Its formal specification pairs lexical [5], syntactic [19], semantic, and discourse [6] components, but no independent pragmatic component. Following patterns of typical computer language interpreter design, our processor additionally includes (a) a generation module, and (b) a discourse control module, to account for continuous interaction between users and system.

In this paper we explore aspects of the semantic component of such a user interface language, from a generation point of view. We show the method we have used to project some pragmatic principles of linguistic interactions onto the semantic structures manipulated by the reasoning machinery, in order to structure proof trees and select information to be conveyed to users. Most of the ideas presented here are implemented in a working prototype which is now in its final integration stage. LINX [4] is an interactive environment for logic-based KB-systems which allows users to converse with a theorem prover in natural language and/or by means of other typical widgets of graphical user interfaces. It includes all the components mentioned above, and many of the algorithms and pragmatical criteria presented in this article. The prototype is currently being used to test the various hypotheses involved in the research, using a small knowledge base. Textual interaction is based on Brazilian Portuguese. User input can be directly typed ahead or gradually built from menu-selection steps. Both forms are analysed by a grammar consistently derived from models of domain objects and relations. Users' questions trigger proof processes which produce argument structures. Such structures result from transformations over ND proofs, and are converted into recursive rhetorical schemata, further translated into hypertext-like Natural Language explanations.

The interactive model is designed so as to maintain language consistency throughout the whole interaction. It is also intended to clearly show to the user all of the system's limitations and regularities, both in terms of reasoning and communicative abilities. This should dispell the impression that the system is a replica of a human processor, and emphasize the notion that the theorem prover and its interface are not but software artifacts [3]..

In the following sections we will briefly examine the question of using pragmatical principles for the structuring of argumentative texts, make some considerations about the knowledge base and the logical proof, and then proceed to show our proof transformation methods. We finish with a detailed example of proof transformation and corresponding explanatory texts.

2 The Generation of Cooperative Explanations

Our objective is to convert a large and complex logical proof into an amenable, human-oriented, easy-to-understand, organized explanation. By considering Grice's Maxims, we arrive at four principles that we can use to design explanations:

1. *The exact way in which knowledge is coded and structured in the knowledge base is irrelevant for the user.*

2. *All information accidental to the proof process should be omitted from the explanation.*

3. *The user should be left to achieve deduction steps in a simple and direct inferential process, provided that he/she knows the premises, in their correct order, and the conclusion.*

4. *The amount of information contained in any explanation step should be limited to the amount that can be simultaneously visualized and processed by a human being without great effort.*

All the above principles translate into a set of transformations which convert a proof into a human-oriented argument structure. By masking the exact coding of the base whenever possible and adequate, by pruning extraneous references in the proof, by resuming deductive steps, and by restructuring them into smaller-sized proof "chunks", we aim to try and satisfy the general principles stated above, without additional domain-specific information.

3 Some Considerations about the Knowledge Base and Natural Deduction

We use the FOL with the Natural Deduction system [18] as the basis for the system's inference engine. Although it has the advantage that in answering questions the system generates justifications as a kind of by-product, which minimizes the problem of having to deal with two separate knowledge bases, we still have to provide it with a way of computing inferences in ND. In practice, it amounts to designing and implementing an automated theorem prover for it, built from a variant of Natural Deduction so defined as to be computable. Our automated theorem prover [15] (cf. also [17]) not only generates valid Natural Deduction proofs, but also generates them in Normal Form, which gives us a minimal formula at every proof branch, separating the elimination and introduction instances of rules in the branch. This characteristic is key to our treatment, as the Normal Form offers a simple pattern for all proofs. Any logic that can be expressed in ND in a compatible computable form could be used instead of FOL, although the impact of their new connectives and ND rules on the human understanding of system's reasonings should also be taken into account.

The quality of the knowledge base also has a direct impact on the quality of the generated proof. There are many ways of imposing structure onto logical representations of a given domain in order to improve the quality of the resulting proofs. However, we have chosen a single structure over the whole knowledge base, which is convenient when converting proofs into argument schamata and imposes no real restriction upon knowledge base engineering. We make a distinction between general axioms and particular facts in knowledge representation. General axioms are those logical formulas that, relative to the main categories of the domain, express general characteristics, attributes or relations. On the other hand, a particular fact is a formula that expresses a characteristic, attribute or relation of a particular individual instance of one of the main categories of the domain. In the absence of a previous distinction made during the knowledge engineering process, we may suppose that general axioms are all those formulas that have a quantifier as the main operator, whereas all the others are particular facts. This distinction is important when considering the premises over which we would like to generalize (see Section 6).

Moreover, we have also slightly modified the concept of "logical proof" in order to satisfy our general cooperative principles stated in the previous section. In order to hide the actual formulas in the knowledge base from explanatory texts provided to users, we extend the concept of valid proof to accept as proofs the deductions which have as undischarged top-formulas not only axioms or facts of the knowledge base, but also theorems deducible from it, as long as proofs for them are already available in the system. In this way, as we shall see, the first principle is already partially met. The proofs of those theorems can then be linked to their occurrences in the main proof, creating what we will call a Proof-Like Structure (PLS - see below). We have also extended some of the ND rules, namely accepting sets of discharged hypotheses instead of a single hypothesis whenever it is a conjunction, and accepting extended versions of the conjunction and disjunction rules to deal with conjunctions or disjunctions of more than two subformulas. We also allow deduction steps that are more complex than the basic ND introduction/elimination rules, as long as they remain logically valid (that is, that can be reconstructed as a sequence of basic ND rules).

Finally, we would like to introduce some definitions and notations. We use either $\wedge$-Intro or $\wedge$I as notation for the ND rule for the introduction of the conjunction, and similarly with all other logical connectives. A top-formula of a proof is a formula which is not a conclusion of any rule, but either an axiom, a hypothesis or a theorem (anchor for a secondary proof in a PLS). A branch of a ND proof tree is a sequence of formulas which begins with a top-formula and follows from rule premise to conclusion ending on either the proof root (the proved formula) or a secondary goal: either the minor premise of a $\rightarrow$-Elim (modus ponens) rule or the major premise (the one containing the eliminated connective) of a $\vee$-Elim or $\exists$-Elim rule.

With these considerations in mind, we proceed on to the transformation methods.

4 Eliminating Irrelevant Information

As the first principle does not have an immediate impact on the proof, let us consider the second one. In fact, in a generic local knowledge base, the axioms could well express many facts in a single formula, for example: Cats have tail and four paws, which simultaneously says two attributes of the same class *cat*. Let us suppose that this axiom would be used in a proof which however requires only the fact that a cat has four paws, as in the following example:

Working Example 1: **Axioms:**

$\forall x$ $\mathsf{cat}(x) \rightarrow \mathsf{has}(x, \mathsf{paws}) \wedge \mathsf{has}(x, \mathsf{paws}(4))$

$\forall x$ $\mathsf{has}(x, \mathsf{paws}(4)) \rightarrow \mathsf{locomotion}(x, \mathsf{walks})$

$\forall x$ $\mathsf{locomotion}(x, \mathsf{walks}) \vee \mathsf{locomotion}(x, \mathsf{flies}) \rightarrow \neg\mathsf{lives}(x, \mathsf{ocean})$

Question:

$\forall x$ $\mathsf{cat}(x) \rightarrow \mathsf{lives}(x, \mathsf{ocean})$

Answer:

No. $\forall x$ $\mathsf{cat}(x) \rightarrow \neg\mathsf{lives}(x, \mathsf{ocean})$. (See Figure 1 for the corresponding proof.)

A possible explanatory text:

Cats don't live in the ocean, because they have tail and four paws. Anyone who has four paws walks so, cats walk. Since anyone who either walks or flies does not live in the ocean, then cats don't live in the ocean.

We notice that the fact that it has tail is discarded along the way, by an elimination rule. However, it is present in the proof before that point. If the proof is translated directly into text, this fact would be expressed in the explanation and would mislead the reader into supposing that the fact that cats have tails is somehow important to the conclusion, although the relation is nowhere clear. This happens because this information introduces a new theme in the explanation text, so humans suppose that it is somehow relevant to the conclusion. The logical relation in the proof, however, derives from the precise way in which information was coded in the knowledge base, and this is irrelevant to the human user. So, according to the second principle, that information should never be mentioned.

Examining the standard ND rules for FOL, we can verify that there are two rules that either eliminate or introduce information locally irrelevant to the deduction itself, and which are more related to the actual form of the logical sentences. These two rules are the elimination of the conjunction (as in the example above) and the introduction of the disjunction (as when concluding, from cats are felines, that cats are felines or herbivores). These two rules should be avoided in general. One way could be the modification of the base by splitting all the formulas. But we consider highly inconvenient to change the form of the sentences in the base, because in the event that the user wants to know exactly what was asserted by the domain specialist, it would be very hard to find out. The other is the editing of the proof to eliminate all unnecessary occurrences of these rules, with a corresponding propagation throughout the proof tree to eliminate the vestiges of the extraneous information (cf. Figures 2 and 3).

Occurrences of these rules are either linked to occurrences of disjunctions and/or conjunctions in the axiom base or in the user question itself. In the first case, the exclusion of the extraneous information can be ignored by the reader at this level, and is in accordance with our first principle. In

$$\dfrac{\dfrac{\dfrac{\dfrac{[cat(a)] \quad \dfrac{\forall x\, cat(x) \rightarrow has(x, paws(4)) \wedge has(x,tail)}{cat(a) \rightarrow has(a, paws(4)) \wedge has(a,tail)}}{has(a, paws(4)) \wedge has(a,tail)}}{has(a, paws(4))} \quad \dfrac{\forall x\, has(x, paws(4) \rightarrow locom(x,walks)}{has(a, paws(4)) \rightarrow locom(a,walks)}}{\dfrac{locom(a,walks)}{locom(a,walks) \vee locom(a, flies)}} \quad \dfrac{\forall x\, locom(x,walks) \vee locom(x, flies) \rightarrow \neg lives(x,ocean)}{locom(a,walks) \vee locom(a, flies) \rightarrow \neg lives(a,ocean)}}{\dfrac{\dfrac{\neg lives(a,ocean)}{cat(a) \rightarrow \neg lives(a,ocean)}[cat(a)]}{\forall x\, cat(x) \rightarrow \neg lives(x,ocean)}}$$

Figure 1: A formal Proof.

$$\dfrac{\dfrac{\dfrac{[cat(a)] \quad \dfrac{*\forall x\; cat(x) \rightarrow has(x, paws(4))}{cat(a) \rightarrow has(a, paws(4))}}{has(a, paws(4))} \quad \dfrac{\forall x\; has(x, paws(4)) \rightarrow locom(x, walks)}{has(a, paws(4)) \rightarrow locom(a, walks)}}{locom(a, walks)} \quad \dfrac{*\forall x\; locom(x, walks) \rightarrow \neg\, lives(x, ocean)}{locom(a, walks) \rightarrow \neg\, lives(a, ocean)}}{\dfrac{\dfrac{\neg\, lives(a, ocean)}{cat(a) \rightarrow \neg\, lives(a, ocean)}[cat(a)]}{\forall x\; cat(x) \rightarrow \neg\, lives(x, ocean)}}$$

Figure 2: The transformed proof.

the second, the modification of the conclusion can be used by the dialogue system to identify and correct mistaken presuppositions, and to generate answers which are more informative, even without the need of presenting the explanation (as in: "Are cats felines or herbivores? Yes, cats are felines.").

$$\wedge E\,\frac{\begin{array}{c}\forall x\, cat(x) \rightarrow has(x, paws(4)) \wedge has(x, tail)\\ \Pi_1 \\ has(a, paws(4)) \wedge has(a, tail)\end{array}}{\begin{array}{c}has(a, paws(4))\\ \Pi_2 \\ \forall x\, cat(x) \rightarrow \neg lives(x, ocean)\end{array}} \quad \Rightarrow \quad \begin{array}{c}\forall x\, cat(x) \rightarrow has(x, paws(4))\\ \Pi_1' \\ has(a, paws(4))\\ \Pi_2 \\ \forall x\, cat(x) \rightarrow \neg lives(x, ocean)\end{array}$$

Figure 3: An edited proof

The validity of our transformation is stated in the next theorem:

Theorem 1 *Given a normal ND proof of a formula F, it's possible to build an extended proof with similar structure but containing no $\wedge$-Elim or (unnecessary) $\vee$-Intro rules, whose top-formulas are either the same or implied by the original ones, and whose conclusion F' implies F.*

We'll describe the algorithm for the construction of the extended proof (or, rather, the tranformation of the original proof) that satisfies the above theorem.

We are supposing that all hypotheses-discharging rules are modified to discharge sets of hypotheses whenever the original hypothesis is a conjunction, so there is no hypothesis that has as main connective a conjunction. This helps to reduce the number of $\wedge$-Elim cases.

When the proof is complete, we can proceed to eliminate the irrelevant information rules from the proof. Each undesired rule can be removed, and its branch reconstructed, by the application (or reverse application) of the same original rules (except the one being removed) to the formula in question, while the other branches remain the same. In the case of an elimination rule (the $\wedge$-Elim) the branch has to be reconstructed upwards from the rule conclusion to its top-formula, while in an introduction rule (the $\vee$-Intro) it has to be reconstructed from the single premise downwards to the branch conclusion. Anyway, we need to examine only the resulting top-formula(s) or conclusion. All the rule applications are symmetrical and can be inverted to generate secondary proofs of the modified top-formulas from the original ones, or, conversely, of the original conclusion from the new one. So the new top-formula can be used as an extended-proof anchor for a secondary proof where it's derived from the original one. The conclusion is still valid (since the validity of the deduction was unaffected) and the original conclusion can be obtained from the new one.

Note that some rules (the $\vee$-Elim rule) propagate a downward transformation to other parallel branches upward from the rule, and if one of those other branches has at least one introduction rule then the rule removal that originated the transformation (necessarily the introduction rule: $\vee$-Intro) is not possible. In this case, we can try to push the $\vee$-Elim rule upwards through identical parallel branch sequences, until the propagation is possible in all branches. If it's not possible, this means that the information being eliminated may not be irrelevant after all.

If the original top-formula was a hypothesis, and the rebuilding occurred in all its instances in exactly the same way, then the transformation can be propagated to the discharge point, by rebuilding the corresponding branch, in the same way we described above. If it succeeds, that eliminates the need for the secondary proof. If the conclusion of a branch is rebuilt, the transformation should be propagated to this next branch, in the same way. In this case the affected rule has to be a $\rightarrow$-Elim rule, in which case the rebuilding is done upwards towards its own top-formula (the rule can't be a $\vee$-Elim or $\exists$-Elim rule, since in these cases the branch ending on the major premise contains only elimination rules, which don't propagate downwards). In this way the proof rebuilding can affect large portions of the original proof. The propagation through hypotheses links is optional but recommended whenever possible, and force the reconstruction upwards in elimination rules ($\vee$-Elim and $\exists$-Elim, to the major premise branch) and downwards in introduction rules ($\rightarrow$-Intro and Classical Absurdity).

Finally, when a given hypotheses-discharging rule discharges a hypothesis (from a set of conjuncts) that is not used throughout the proof, the corresponding conjunction can be modified through the exclusion of the unused hypothesis and the modification propagated upwards or downwards through the branch, in the same way as above. In this way large conjunctions can be modified to satisfy our goal.

By reconstructing the whole proof, we arrive on a similar valid deduction. This may differ from the former on some top-formulas and the conclusion, but these top-formulas can be derived from the former ones by proofs which can be easily constructed from the original rule sequences, and the conclusion implies the former conclusion in the same way. So this derived deduction is in a sense *smaller* although logically equivalent to the former. As the new premises are theorems of the knowledge base, this new deduction fits in the extended proof definition above, and so is acceptable from the point of view of the first principle, at the same time that it satisties the second.

The above proof, transformed accordingly, would be as below, where the top-formulas marked with (*) are the ones modified as above (which have secondary proofs that derive them from the original top-formulas is shown in Figure 2.

Here is a possible explanatory text:

Cats don't live in the ocean, because they have four paws. Anyone who has four paws walks, so cats walk. Since anyone who walks does not live in the ocean, then cats don't live in the ocean (either).

5 Simplifying Proof Steps

From the point of view of a Normal Form Natural Deduction proof, the third principle can be restated as saying that in a sequence of rule applications of the same character (introduction or elimination) of a given proof, you may state the premises of the uppermost rule(s) and the conclusion of the last, because all the intermediate rules are syntactical reduction or merging rules. You may also do this with rules of different characters and even different branches. In the end you can reduce the whole proof to a single deduction step. But this is not what we are looking for, and it may result in deduction steps whose complexity can be very large. What we want to do is group rule applications in composite proof steps while keeping them simple enough to be easily understood, as in principle 4.

We can start by considering the size of the formulas in the proof. But the largest formulas in the proof are either axioms of the base, or the question submitted to the prover, since we are dealing only with Normal proofs. So the syntactical complexity of the sentences necessary to articulate the argument is at most the same as the complexity of the largest formulas in the knowledge base or the question itself, and normally smaller. Therefore we can assume that the knowledge representation language, the system interface and the knowledge-base builder only produce formulas of "acceptable" complexity, that is, that can be expressed in a way the normal human user can understand without (great) effort. If that's not so, the problem lies in the system itself, and no manipulation of the proof will be able to completely eliminate it.

Having established this, we consider the proof itself. Since we don't want to collapse the whole proof into a trivial text with no argument, we want to divide the proof into parts that can be considered separately. We choose to divide a proof branch at its minimal formula, since this formula is relevant to the transformation of the next section and shouldn't disappear. So the proof branches are divided into their elimination and introduction segments.

We then look at the relative complexity of each basic ND rule. Here we don't have experimental data to support our suggestions, so our method is subject to future verification. We consider the elimination rules with a single premise (the $\wedge$-Elim, which is removed by the previous section method, and the $\forall$-Elim) as the simplest and most obvious rules. The enunciation of the premise already announces its conclusion. So all elimination rules with a single premise should be collapsed with the subsequent rule in their branch, with the premise taking the place of the conclusion. The case of the other elimination rules ($\rightarrow$-Elim (modus ponens), $\vee$-Elim and $\exists$-Elim) involve either more than one premise or a variable substitution, both of which need to be examined in the proof to ascertain its validity, and which require an equivalent care in the textual argument, or we may end with an unclear argument. Even so, a sequence of identical rules coming from a single major premise (a large disjunction $A \vee B \vee C$ or a causal chain $A \rightarrow B \rightarrow C$, for example) can be collapsed into a single rule application, as long as the text generator is able to take advantage of the collapsed rule by expressing it in a resumed way.

On the introduction rules, the case is different. The $\forall$-Intro rule is very sensitive to the interface language, to the way it is able to express the difference between specific and generic individuals and classes. In our current knowledge base, since there are no individuals, only classes, the $\forall$-Intro rule can be mostly ignored in the text generation, with the system using the parameter that replaces the variable to identify the generic case. In such bases, a sequence of $\forall$-Intro rules can be collapsed into a single instance, with few if any problems to the text generator. In other bases, however, it may be necessary to keep the $\forall$-Intro rules separated. A similar but simpler case happens with the $\exists$-Intro rule. In general, we assume that both quantifier introduction rules can be collapsed into a single rule instance whenever they occur in sequences.

The case of the $\rightarrow$-Intro rule is also special because this rule discards a hypothesis. Depending on the kind of knowledge representation language chosen for the domain, there can be large sequences of these rules which can be collapsed without problems, or almost no case of a sequence of two $\rightarrow$-Intro rules. Since this rule (and the Absurdity) discards one or more hypotheses, and the hypotheses links are one of the main rhetorical structures to be considered when building the text, we assume that those rules should not be collapsed when they appear in sequence, but that they can be collapsed with a subsequent quantifier introduction rule (producing derivations of the form "supposing $A(x)$ we derive $B(x)$, so $\forall x A(x) \rightarrow B(x)$").

There remains only the $\wedge$-Intro rule, since the $\vee$-Intro rules are (almost always) removed by the algorithm of section 4. Sequences of $\wedge$-Intro rules can be very common, and they present no problem for comprehension. So we assume a sequence of $\wedge$-Intro rules is always collapsed into a single step, either in the proof process or here in the transformation process. More, we can safely assume that the step itself is redundant, and collapse the step into the one below it in the proof tree, by writing the premises in the place of the conclusion.

We now discuss the transformations that can take place with each of the basic ND rules, regarding sequences of rules of the same type and the collapse of different rule types. All but two of the tranformations are shown in Table 1. For example, for $\forall$-Intro, a sequence of rules can collapse into a single step.

The transformations for $\vee$-elimination are shown below:

$$\frac{\overset{\times}{A_1 \vee \ldots \vee A_i \vee \ldots \vee A_n} \quad \overset{\Uparrow}{C} \; \cdots \; \overset{\Downarrow}{C} \; \cdots \; \overset{\Uparrow}{C}}{\underset{\Downarrow}{C}} \; [A_1], \ldots [A_i], \ldots, [A_n] \; \times \quad (*)$$

$$\frac{\overset{\times}{A_1 \vee \ldots \vee A_i \vee \ldots \vee A_n} \quad \overset{\Uparrow}{C} \; \cdots \; \overset{\Uparrow}{C} \; \cdots \; \overset{\Uparrow}{C}}{\underset{\Uparrow}{C}} \; [A_1], \ldots [A_i], \ldots, [A_n] \; \times$$

The $*$ in the first transformation indicates a restriction: if this propagation upwards in the parallel branches is not possible (when one of the other branches has at least one introduction rule) then the optional propagation or rule removal which originated the process is invalid and should not be done.

Note that in all cases, a rule always collapses with the one below it, that is, replacing its conclusion with its premises.

In this simple way, a great reduction in size can be achieved in the proof, which also puts in evidence the kind of structural information that is really relevant in an argument the way in which many different pieces of information are combined and lead to the conclusion at hand. The reduction could take place at the text generator, but by transforming the proof at the logical level, we can ensure its consistency and the systematicity of the process.

6 Proof Restructuring

The proof restructuring is the most extensive proof transformation within our method. The result is what we call a Proof-Like Structure (PLS) which reflects an *argumentative structure*. To achieve it we follow a sequence of steps, indicated in the algorithm in Figure 4 below.

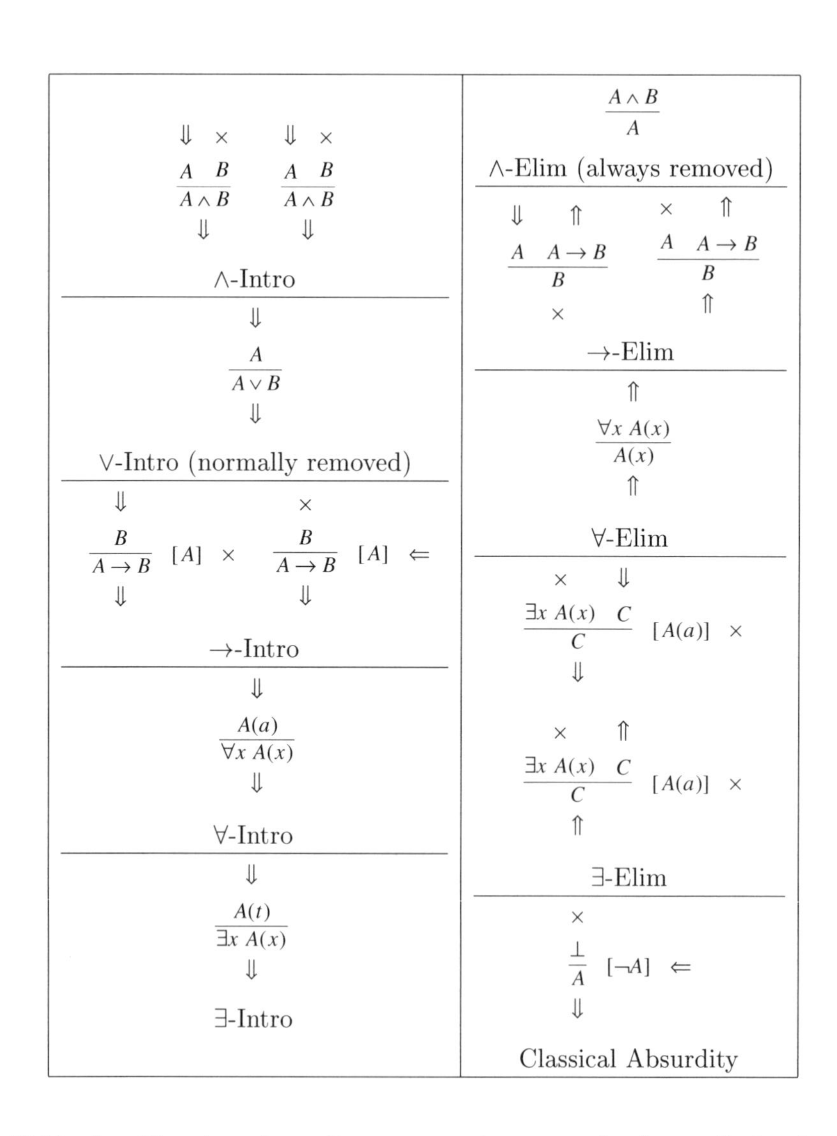

Table 1: The transformation propagates upwards from elimination rules, and downwards from introduction rules. The arrows indicate the direction of the propagation of the rebuilding process.

1. Apply the algorithm for the removal of Irrelevant Information to P;

2. For each Minimal Formula F_i in P do:

   (a) If it is the Absurdity ($\perp$), consider instead its conclusion;

   (b) Mark it as a breaking point and select its subtree TF_i;

   (c) Apply recursively Proof-Restructuring to TF_i;

   (d) Identify hypotheses and select top-formula facts and anchors to be discharged in TF_i, following pragmatical criteria to limit their number;

   (e) Undo the unifications bounding F_i and TF_i to the rest of the proof, by rebuilding the deduction from the hypotheses with freed variables;

   (f) Discharge hypotheses and top-formula anchors generating a new conclusion $F_{i'}$, and a complete proof $PF_{i'}$ of $F_{i'}$, from TF_i;

3. Select the set of Breaking Points F_j whose substitution would satisfy the pragmatical ccriteria;

4. For all the Breaking Points F_j in the selected set, do:

   (a) replace the subtree with the deduction of F_j from $F_{j'}$,

   (b) place an anchor for $PF_{j'}$ from $F_{j'}$;

5. Apply the algorithm for Simplifying Proof Steps to P.

Figure 4: The Proof Restructuring Algorithm

First, we remove the irrelevant information from the proof, possibly generating some secondary proofs in the process. Then we need to identify proof pieces that represent inference chunks whose internal structure is less important to argumentation than the relations maintained between them. Since we are talking about normal proofs in a ND system, the minimal formulas constitute natural candidates for breaking points between the chunks. We prefer to have as the conclusion of a proof piece the smallest possible formula, as they represent information 'atoms' that are reorganized in order to reach the conclusion. Minimal formulas are then our choice of breaking points, except in one case. The case is when the minimal formula is the absurdity ($\perp$); we then take as breaking point the subsequent formula in the branch, which is proved through contradiction. Note that in the case of the hypothesis-discharging elimination rules (of the existential $\exists$ and the disjunction $\vee$), whose conclusions are identical to one or more of the premises, the major premise (which differs from the rule conclusion) is the minimal (and last) formula of the branch, which has no introduction rule (in this we differ from the definitions in [15]). The proof piece is thus the whole subtree determined by a given breaking point. In this way all proof pieces have as their conclusion meaningful formulas (not the absurdity) and are valid deductions (although maybe not complete proofs) in themselves.

The second step is to recursively apply the proof restructuring to each proof piece. This can be optimized by a good selection of the first breaking points, by considering the lowest minimal formulas and checking whether the remaining deduction (the proof without their subtrees) satisfies our pragmatical criteria on the size of the resulting explanation. We consider that the pragmatical criteria have to be tailored to the specific knowledge-representation language chosen and the characteristics of the text generator. But, as a general rule, we suggest counting the number of predicates occurring in the deduction and the number of "objects" referred to by the terms in those predicates. Both those numbers (and even their sum) should be lower than the "magical number" (5 to 9) [13] of items a human being can simultaneously consider. However, this pragmatical criterium has to be validated and more experimentation should be done before we arrive on a way to optimize the criteria to specific knowledge-based systems. We are here more interested in proposing mechanisms that can be used to implement specifical criteria than concluding exactly whose criteria should be applied in each case.

Then we make each proof piece a complete proof in itself, thus generating a new conclusion for the proof piece, which will be a theorem of the knowledge base. This is done by first undoing the unification steps necessary to link the breaking point formula to the rest of the proof, which turns the proof piece into a more general deduction. The idea here is to convey all information stored in the structure of the deduction, derived from its top-formulas, regardless of the possible instantiations that could be necessary to apply it to the proof. The links to the main proof are done through unification between terms in the main proof and free variables in the conclusion, and also through the use of (undischarged) hypotheses in the deduction. Since the unification is done to the whole deduction, we must rebuild it from its top-formulas. The undischarged hypotheses will be substituted for local hypotheses, so their terms can also be turned into free variables. We should however substitute only terms that appear in the conclusion of the deduction (the breaking point), since it will be the focus of the argumentative text, and the point where the generalization should take place.

So before undoing the unification steps we must select the hypotheses in the proof piece, and identify in them the terms which appear both in a hypothesis and in the proof piece conclusion (the breaking point). These terms can be replaced with free variables, in all the top-formulas in which they appear, that is, all the hypotheses and theorem anchors have to be examined. Axioms shouldn't have free variables, so they would never need a substitution. The deduction is then rebuilt so that the conclusion has as many free variables as possible given the argument structure. The result is what we call an independent deduction, which retains no links to the previous proof.

We take a further step in the generalization of the proof piece by allowing the consideration of any particular fact or "factual theorem" (a top-formula anchor to another proof which has the same structure as a fact) as an (undischarged) hypothesis, and substituting its terms for free variables when they coincide with the conclusion, before rebuilding the deduction. The number of predicates and terms in the hypotheses and the conclusion are subject to limitation from our pragmatical criteria, that is, they should be limited by our "magical number", since they would all be mentioned in the proof conclusion. An independent deduction with a conclusion that is too complex should be avoided, in general. The characteristics of the knowledge representation language and the domain can also be used to identify terms that should (or should not) be generalized, helping the identification of the facts that can or should be considered as hypotheses. Another consideration is that it is often undesirable to include additional terms in the conclusion, by choosing as hypothesis a fact which has another term that is not being generalized. This would introduce in the main discourse another object that was previously not mentioned, and unless the domain asks for it, it should be avoided. A possible exception is when the proof structure is complex enough (through the use of "difficult" rules, for example, as the classical absurdity, or an irreducible sequence of rules with too many top-formulas and intermediate results) to warrant an alternate presentation. Then the identification of those "hypotheses-like" facts can indeed provide an explanation without using the complex axioms that originated the proof piece structure (our example in section 7 has one such piece that is more complex than the average argument structure).

We can then discharge the not-yet-discharged hypotheses, through an implication ($\rightarrow$) introduction rule discharging their conjunction. Universal quantification introduction rules are also applied to bound the variables that are free in the conclusion. The result is a new formula which is a theorem. It is as general as we can (or want to) make it, which is advantageous from the point of view of explanations. Normally, it is more informative to make a general assertion than an individualized one, because by making a general assertion, the system avoids implying to the user that the assertion would not be generalizable, and states exactly what the proof implies (that is, that the argument is valid in that range of cases). However, pragmatical criteria, as well as the coherence of system behavior, may suggest otherwise, and the generalization of facts and "factual theorems" should be subject to these considerations. Anyway, it is also of logical interest, because it enables the system to extract and identify frequently used intermediate results in the form of theorems that can be stored and reused.

For each proof piece there is now an associated theorem and its proof. We can then substitute the whole proof piece in the original proof by a new deduction of the breaking point from the associated theorem, which is a simple enough direct deduction recurring to locally available hypotheses and particular facts and anchors. For example, when the breaking point is $F(t)$, the undischarged hypothesis is $H(t)$ and the particular fact is $A(t)$, the proof piece can be substituted by the following deduction, where the main top-formula is a theorem of the base:

$$\cfrac{\cfrac{A(t) \quad [H(t)]}{A(t) \wedge H(t)} \qquad \cfrac{\forall x\, A(x) \wedge H(x) \rightarrow F(x)}{A(t) \wedge H(t) \rightarrow F(t)}}{F(t)}$$

This small deduction can replace the original proof piece while maintaining the correctness of the proof. This is an extended proof, and associated to the theorem used in the deduction there is its own proof, which can be examined to assert the validity of the argument. So we have a recursive structure, the PLS, which may be used to structure the textual justification of the main conclusion.

The last step in the algorithm is to apply the simplification of proof steps to the resulting PLS, so that the number of formulas and rules (and the maximum size of the resulting text) is reduced while preserving the clarity of the argument. In fact, the simplification of proof steps can also be made before the proof structuring, since the relevant characteristics (minimal formulas, top-formulas, conclusion) are preserved.

What remains to be done is to establish the criteria through which we will choose the appropriate breaking points in the original proof. Although every subset of breaking points would result in a PLS, these may not satisfy our cooperativity principle above. The PLS is an instrument to structure the proof, which must be applied according to pragmatical criteria, in order to produce satisfactory results.

We identify a series of criteria which are domain-independent, based on syntactical data with pragmatic considerations in mind, that should be verified when choosing the breaking points. These can be adapted to a particular domain, if neccessary, as well as being applied together with additional domain-specific criteria.

Each breaking point should be weighted according to the size of its theorem, as well as the resulting proof size, as we said above. A small theorem is preferable. This may conflict with the generalization over facts and top-formula anchors described above. These criteria can be tailored to the average formula complexity of the domain, which is also a purely syntactical information, and be used to weight the many breaking point candidates, while guiding the generalization choices in each proof piece and excluding the breaking points with unnacceptably large theorems.

In general, the largest the proof piece the better the breaking point, as long as the resulting proof retains a manageable amount of information to be presented. The number of predicates and terms making up the proof should be within the acceptable limits (around 7, cf. [13]), whenever possible.

A structural criteria to identify desirable breaking points is to look at the terms that appear in the proof conclusion. If a breaking point contains one or more of these, and none (or only a few) of other ones, it bears a more or less direct relation, in content, to the conclusion of the proof, and should be preferred to the ones that present too much new information.

In this way we strive to limit the amount of information contained in each proof making up the PLS. As the structure is recursive, each proof subsumed in a theorem should also be transformed into a PLS, so the result allows for a flexible dialogue between the user and the system, each time focusing on a specific point of the argument, until the user is satisfied.

It is evident that, in the presence of additional cognitive information related to a specific domain, new criteria could be drawn that would enhance the breaking point selection, allowing for a better argumentative focus. We therefore emphasize the flexibility of this tool for the generation of structured explanations in logic-based knowledge systems.

7 A Working Example

In this section we present an example chosen to be complex enough to show the techniques developed in this article. The knowledge base over which it is based is still under construction, and will be used for future experimentation of the algorithm together with the whole inteface. The text generator is not ready yet for this base, so the textual explanations are hand-generated.

Suppose we have the following axioms in our knowledge base (in fact, a possible future section of our Brazilian Fruits KB, in development), where "aceroleira" is a Brazilian tree of the Malpighia genus whose fruit is a small red berry, and Paranagua is a city in the Brazilian state of Parana:

Facts:

1. Malpighia(aceroleira)
2. city(Paranagua, Parana)
3. part-of(Parana, SouthRegion)
4. in(Paranagua, CoastalArea)

Axioms:

1. $\forall x$ Malpighia$(x) \rightarrow$ tropical-cycle$(x) \wedge$ MinAdTemp$(x, 10C) \wedge$ MinAdRain$(x, 1000)$
2. $\forall x$ tropical-cycle$(x) \rightarrow \neg$dormant-state(x)
3. $\forall x \forall t$ MinAdTemp$(x, t) \wedge t > 0C \wedge \neg$dormant-state$(x)$ $\rightarrow$ sensitive-to$(x,$ frost$)$
4. $\forall x \forall y \forall z$ (sensitive-to$(x, y) \vee$illness$(y, x)) \wedge$occurs$(y, z) \rightarrow \neg$do-well$(x, z))$
5. $\forall x \forall y$ city$(x, y) \rightarrow$ in(x, y)
6. $\forall x \forall y$ part-of$(x, y) \rightarrow$ in(x, y)
7. $\forall x \forall y \forall z$ in$(x, y) \wedge$ in$(y, z) \rightarrow$ in(x, z)
8. $\forall x$ in$(x,$ SouthRegion$) \rightarrow \exists t$MinAdTemp$(x, t) \wedge t < 0C$
9. $\forall x$ in$(x,$ CoastalArea$) \rightarrow$ humid(x)
10. $\forall x \forall t$ MinTemp$(x, t) \wedge t < 0C \wedge$ humid$(x) \rightarrow$ occurs(frost$, x)$

Then the question: do-well(aceroleira, Paranagua) has a negative answer, but the conventional ND proof of the negation is large (55 formula instances and 36 rule applications), and the textual explanation would require a series of complex paragraphs.

Applying the irrelevant information pruning algorithm to the proof we eliminate three rule instances arriving in a reduced proof with 52 formula instances and 33 rule applications, together with 3 secondary proofs deriving simplified instances of the axioms 1 and 4.

After applying the algorithm for simplified proof steps, we arrive at the following proof, with only 31 formula instances and 12 rule applications, with the same 3 secondary proofs (we have broken the proof and abbreviated predicate names to reduce the proof size):

(a): Proof of in(Paranagua, SouthRegion): see Figure 5(a).

(b): Proof of occurs(frost, Paranagua) using (a): see Figure 5(b).

(c): Proof of sensitive-to(aceroleira, frost) with anchors to secondary proofs ⟨1⟩ and ⟨2⟩: see Figure 5(c).

Secondary proof ⟨1⟩ (proof ⟨2⟩ is similar to ⟨1⟩, from the same axiom):

$$\dfrac{\dfrac{[M(a_1)] \quad \forall x\, M(x) \rightarrow tc(x) \wedge MAT(x,10C) \wedge MAR(x,1000)}{tc(a_1) \wedge MAT(a_1,10C) \wedge MAR(a_1,1000)}}{\dfrac{MAT(a_1,10C)}{\forall x\, M(x) \rightarrow MAT(x,10C)}[M(a_1)]}$$

(d): Proof of ¬do-well(aceroleira, Paranagua) using (b) and (c) and secondary proof ⟨3⟩:

$$\dfrac{\overset{(c)}{st(a,f)} \quad \overset{(b)}{occ(f,P)} \quad \overset{\langle 3\rangle}{\forall x \forall y \forall z\, st(x,y) \wedge occ(y,z) \rightarrow \neg dw(x,z)}}{\neg dw(a,P)}$$

Secondary proof ⟨3⟩:

$$\dfrac{\dfrac{[st(a_1,b_1)]}{st(a_1,b_1) \vee ill(b_1,a_1)} \quad [occ(b_1,c_1)] \quad \forall x \forall y \forall z\, (st(x,y) \vee ill(y,x)) \wedge occ(y,z) \rightarrow \neg dw(x,z)}{\dfrac{\neg dw(a_1,c_1)}{\forall x \forall y \forall z\, st(x,y) \wedge occ(y,z) \rightarrow \neg dw(x,z)}[st(a_1,b_1)],[occ(b_1,c_1)]}$$

This breakdown of the main proof at the minimal formulas, resulting in smaller chunks, can be reflected in an explanation text divided in many paragraphs. In fact, here we have a selection of breaking points that serve to begin the proof-restructuring process.

The next step is the generalization over the proof pieces. The proof piece in (a) could be generalized by considering city(Paranagua, Parana) as a hypothesis, but this would introduce an additional term (Parana) in the

(a)

$$\dfrac{\dfrac{city(P,Parana)\quad \forall x\forall y\; city(x,y)\rightarrow in(x,y)}{in(P,Parana)}\qquad \dfrac{part-of(Parana,SR)\quad \forall x\forall y\; part-of(x,y)\rightarrow in(x,y)}{in(Parana,SR)}\qquad \forall x\forall y\forall z\; in(x,y)\wedge in(y,z)\rightarrow in(x,z)}{in(P,SR)}$$

(b)

$$\dfrac{\dfrac{\overset{(a)}{in(P,SR)}\quad \forall x\; in(x,SR)\rightarrow \exists t\; MT(x,t)\wedge t<0C}{\exists t\; MT(P,t)\wedge t<0C}\qquad \dfrac{[MT(P,t_1)]\quad [t_1<0C]\quad \dfrac{in(P,CA)\quad \forall x\; in(x,CA)\rightarrow h(x)}{h(P)}\quad \forall x\forall t\; MT(x,t)\wedge t<0C\wedge h(x)\rightarrow occ(f,x)}{occ(f,P)}}{occ(f,P)}\;[MT(P,t_1)],[t_1<0C]$$

(c)

$$\dfrac{\overset{\langle 1\rangle}{\dfrac{M(a)\quad \forall x\; M(x)\rightarrow MAT(x,10C)}{MAT(a,10C)}}\qquad 10C>0C\qquad \dfrac{\overset{\langle 2\rangle}{\dfrac{M(a)\quad \forall x\; M(x)\rightarrow tc(x)}{tc(a)}}\quad \forall x\; tc(x)\rightarrow \neg ds(x)}{\neg ds(a)}\qquad \forall x\forall t\; MAT(x,t)\wedge t>0C\wedge \neg ds(x)\rightarrow st(x,f)}{st(a,f)}$$

Figure 5: Three proofs: (a) in(Paranagua, SouthRegion), (b) occurs(frost, Paranagua), and (c) sensitive-to(aceroleira, frost)

conclusion, resulting in the theorem

$$\forall x \ \mathsf{city}(x, \mathsf{Parana}) \rightarrow \mathsf{in}(x, \mathsf{SouthRegion}).$$

This alternative is generally unfavored, and the context provides nothing to support it. Moreover, the predicate in normally does not require elaboration or explanation, so in this knowledge base we have a pragmatical criterium that prevents the complexification of this breaking point. So the proof piece (a) is left as it is.

The proof piece in (b) is very complex, and the alternative is to generalize it to try to identify the relevant pieces of information. It can be generalized by taking as minimal formula the conclusion itself, and selecting as hypotheses the facts

$$\mathsf{in}(\mathsf{Paranagua}, \mathsf{CoastalArea}) \ \text{and} \ \mathsf{in}(\mathsf{Paranagua}, \mathsf{SouthRegion})$$

(the latter is an anchor to (a)). In this case, the generalization results in the following formula:

$$\forall x \ \mathsf{in}(x, \mathsf{CoastalArea}) \wedge \mathsf{in}(x, \mathsf{SouthRegion}) \rightarrow \mathsf{occurs}(\mathsf{frost}, x).$$

This is not prevented in our base. So we have two versions of the same proof piece: one the original (b) with $\mathsf{occurs}(\mathsf{frost}, \mathsf{Paranagua})$ as anchor, and the new one with a theorem linking the occurrence of frost to the location in the coastal area of the south region of Brazil.

However, when we try to apply the new theorem as anchor in the main proof, the new facts and terms mentioned make it too complex to be desirable, so the simple ungeneralized version is preferred. For the explanation of the secondary proof of $\mathsf{occurs}(\mathsf{frost}, \mathsf{Paranagua})$, in its turn, the new generalized version can be used, to try to convey useful information without having to follow all the complex details of the original proof piece.

The third proof piece, (c), lends itself to a straightforward generalized version, with

$$\forall x \ \mathsf{Malpighia}(x) \rightarrow \mathsf{sensitive\text{-}to}(x, \mathsf{frost})$$

as theorem. Another minimal formula ($\neg\mathsf{dormant\text{-}state}(\mathsf{aceroleira})$) results in an additional secondary proof with $\forall x \ \mathsf{Malpighia}(x) \rightarrow \neg\mathsf{dormant\text{-}state}(x)$ as theorem.

The fourth proof piece is the remaining main proof, and the resulting PLS is as below:

i) Main Proof, with 4 predicates and 3 objects:

$$\cfrac{\cfrac{M(a) \quad \overset{\langle 4 \rangle}{\forall x \, M(x) \rightarrow st(x,f)}}{st(a,f)} \qquad \overset{\langle 5 \rangle}{occ(f,P)} \quad \overset{\langle 3 \rangle}{\forall x \forall y \forall z \, st(x,y) \wedge occ(y,z) \rightarrow \neg dw(x,z)}}{\neg dw(a,P)}$$

A corresponding text would be as following (in parenthesis a textual implicature that could be removed by the text generator, as long as the anchor in this case can be considered unnecessary):

The Aceroleira does not do well in Paranagua, because it's from the Malpighia genus, Malpighia genus plants are sensitive to frost, frost occurs in Paranagua (and plants sensitive to something that occurs in a given place don't do well there).

Underlined we see the anchors for the secondary proofs:

ii) Secondary proof corresponding to Anchor 4 (see Figure 6).

A corresponding text would be:

Plants of the Malpighia genus are sensitive to frost because their Minimum Adequate Temperature is 10 degrees Celsius, which is above 0, and they don't have a dormant state (and plants whose Minimum Adequate Temperature is above 0 degrees Celsius and which do not have a dormant state are sensitive to frost).

iii) Secondary proof corresponding to Anchor 5:

$$\frac{in(P,CA) \quad \overset{\langle 8\rangle}{in(P,SR)} \quad \overset{\langle 7\rangle}{\forall x\, in(x,CA) \wedge in(x,SR) \rightarrow occ(f,x)}}{occ(f,P)}$$

A corresponding text would be:

Frost occurs in Paranagua because Paranagua is a place in the Coastal Area and is on the South Region, and places in the Coastal Area of the South Region are subject to frost.

Note that here we must present explicitly the anchor for ⟨7⟩, so the implicature can't be used to reduce the text.

iv) Secondary proof corresponding to Anchor 6:

$$\cfrac{\cfrac{\cfrac{[M(a_1)] \quad \overset{\langle 2\rangle}{\forall x\, M(x) \rightarrow tc(x)}}{tc(a_1)} \qquad \forall x\, tc(x) \rightarrow \neg ds(x)}{\neg ds(a_1)}}{\forall x\, M(x) \rightarrow \neg ds(x)}\,[M(a_1)]$$

A corresponding text would be:

Plants of the Malpighia genus don't have dormant states, because they are tropical-cycle plants (and tropical-cycle plants don't have dormant states).

v) Secondary proof corresponding to Anchor 7 (see Figure 7):

$$\frac{occ(f,p_1)}{\forall x\, in(x,CA) \wedge in(x,SR) \rightarrow occ(f,x)}\,[in(p_1,CA)],[in(p_1,SR)]$$

A corresponding text would be:

Frost occurs in the Coastal area of the South Region, because in every place in the South Region the Minimum Temperature is below 0 degrees

$$\dfrac{\dfrac{\overset{\langle 1\rangle}{[M(a_1)] \quad \forall x\, M(x) \rightarrow MAT(x,10C)}}{MAT(a_1,10C)} \quad 10C > 0C \quad \dfrac{\overset{\langle 6\rangle}{[M(a_1)] \quad \forall x\, M(x) \rightarrow \neg ds(x)}}{\neg ds(a_1)} \quad \forall x \forall t\, MAT(x,t) \wedge t > 0C \wedge \neg ds(x) \rightarrow st(x,f)}{\dfrac{st(a_1,f)}{\forall x\, M(x) \rightarrow st(x,f)}[M(a_1)]}$$

Figure 6: Secondary proof corresponding to Anchor 4

$$\dfrac{\dfrac{[in(p_1,SR)] \quad \forall x\, in(x,SR) \rightarrow \exists t\, MT(x,t) \wedge t < 0C}{\exists t\, MT(p_1,t) \wedge t < 0C} \quad \dfrac{[MT(p_1,t_1)] \quad [t_1 < 0C] \quad \dfrac{[in(p_1,CA)] \quad \forall x\, in(x,CA) \rightarrow h(x)}{h(p_1)} \quad \forall x \forall t\, MT(x,t) \wedge t < 0C \wedge h(x) \rightarrow occ(f,x)}{occ(f,p_1)}}{} [MT(p_1,t_1)],[t_1 < 0C]$$

Figure 7: Secondary proof corresponding to Anchor 7

Celsius, and if it's in the Coastal Area it's humid too, and frost occurs in places where the Minimum Temperature is below 0 degrees Celsius and where it's humid.
vi) Secondary proof corresponding to Anchor 8: it's the same deduction as in (a) above.
A corresponding text would be:

Paranagua is in the South Region because it's a city of Parana, (so it's in Parana), and Parana is part of the South Region (so it's in the South Region).

The transitivity rule should not be written since it's part of the meaning of the word "in", and the phrases inside parenthesis should also be omitted, since they represent part of the meaning of "city" and "part of".

The only proofs remaining correspond to anchors $\langle 1 \rangle$, $\langle 2 \rangle$, and $\langle 3 \rangle$. We think it's unnecessary to explain such proofs, the explanation in those cases would simply be the enunciation of the corresponding axiom.

The combination of the above explanation texts allow the interface to establish an explanation dialogue to the user, in which the system uses small paragraphs whenever possible, and the user can ask for further detailing of the argument when necessary. The use of small paragraphs of text is very attractive for most knowledge-based system applications, and the ability to present the complete argument with all the details whenever necessary has the ability to make explanations useful in those systems. The problem then moves from the explanatory system to the knowledge base itself, since the quality of the system's explanations is now limited by the quality of the domain representation.

8 Concluding Remarks

In this paper we have presented a method for transforming normalized Natural Deduction proof trees into proof structures adequate to be mapped onto text structures to render good quality Natural Language explanations for knowledge-based systems. The method is inspired in Toulmin's definition of what an argument is and in Grice's Maxims, both being renowned attempts at providing a formal basis for mutual intelligibility.

One of the main features of the resulting explanatory structure from our method is a hypertext-like structure, in which chunks of sensible information are provided to users and expanded on demand. This is an important attemp at pragmatic adequacy [10] that we do; all transformations upon ND proofs are so done that the remaining logical formulas contain information that users can bind together in an inferential process that is natural to them. Breakdowns in this process can be handled by expanding proof nodes in accordance to proof-like structures that account for the argumentative thread in the system's reasoning.

Another noteworthy feature is our attempt to keep only focal information in the proofs. This is achieved through an elimination of all information that is not directly related to the ones present in the conclusion. Cohesive and coherent knowledge representation in multimodal interfaces and textual explanations are a mandatory quality without which users are easily led astray in their interpretation of artificial intelligent agents. Although cohesion-oriented transformations can be carried out at text-structure level, our decision for performing them at proof level guarantees coherence and soundness of conclusions (while merely textual manipulations have been historically observed to produce sophisms rather than logical arguments).

We are using only the standard First-Order Logic and a Natural Deduction theorem prover, with minimal assumptions over the structure of the knowledge base and other components of the system. The overall results are encouraging enough to allow the conclusion that simple but systematic approaches can be successful in generating good quality interactions that satisfy the cooperativity goals while not attempting the perfection of human-like dialogues.

The Natural Deduction theorem prover and the proof transformation algorithm presented here are currently parts of a larger system, LINX (cf. [4]), a prototype integrating research results in many different areas. It uses a subset of Portuguese for the textual interaction. The system is in the final phases of implementation and will be used to test the many theories involved in its development.

Our ongoing research aims to furthering logic knowledge representation processing so that optimal textual and hypertextual explanations can be generated to improve the usability of knowledge-based systems in classificatory domains. Our prototype will be used to test the use of pragmatical criteria in an extended knowledge base being built, which should be large enough to provide a number of interesting proof examples. We are also investigating the demands our methods impose over the knowledge base, such as the restrictions in the complexity of the axioms, the increased demand for cohesion and coherence in the knowledge representation, and the different levels of detail in the logical relations between the concepts. We also intend to produce a stand-alone version of the theorem prover, together with the proof transformation module, to be made available to other researchers in the same field, and to be used for systematical experimentation and validation of the pragmatical criteria.

Recent results include a method for interpreting the results of the theorem prover in the case of partial or negative answers, with an analysis of the impact of the use of logical connectives in the context of cooperative interaction ([16]). We observe that the exact meaning of the logical connectives has to be taken into account when generating natural language answers from formal logical systems in order to avoid misleading the user. We have also investigated how multimodal reasoning and explanation could be implemented within the basic framework of our system, taking as reference the Hyperproof system (cf. [1]).

Long-range goals are directed towards ensuring automatic knowledge acquisition and usage as independently as possible from human engineering, the use of automated knowledge acquisition to guarantee an acceptable quality on the knowledge base, the investigation of alternative and/or expanded logics and the integration of multiple representation systems to account for complex domains with a classification aspect.

References

[1] Barwise, J. & Etchemendy, J. Hyperproof. CSLI, Stanford, distributed by Cambridge University Press, 1994.

[2] Cawsey, A. Explanation and Interaction. Cambridge,Ma. MIT Press, 1992.

[3] de Souza, C.S. The Semiotic Engineering of User Interface Languages. International Journal of Man-Machine Studies no.39, pp.753-773. November 1993.

[4] de Souza, C.S., Garcia,L.S., Dias,M.C.P. & Quental,V.S.D.B. LINX: An Integrated Interface Environment for Knowledge-Based Systems (in Portuguese). Proceedings of the II Meeting on Computational Processing of Written and Spoken Portuguese, of the XIII Brazilian Symposium on Artificial Intelligence, pp. 29-39. Curitiba, Brazil, 1996.

[5] Dias,M.C.P. The Lexicon in Automated Text Analysis and Generation Systems in Portuguese Language (in Portuguese). Ph.D. Dissertation. Departamento de Letras, PUC-Rio. Rio de Janeiro, Brazil, 1994

[6] Garcia, L.S. An Integrated Interface Environment for Knowledge-Based Information Systems (in Portuguese). Ph.D. Dissertation. Departamento de Informática. PUC-Rio. Rio de Janeiro, Brazil, 1995

[7] Grice, H. P. Logic and Conversation. In Cole, P. & Morgan, J. L. (eds) Syntax and Semantics, vol. 3: Speech Acts. Academic Press, New York, 1975.

[8] Huang, X. Planning Argumentative Texts. In Proceedings of COLING94 Kyoto, Japan, 1994.

[9] Huang, X. Human Oriented Proof Presentation: A Reconstructive Approach. Ph.D. Dissertation, DISKI 112, Infix, Sankt Augustin, Germany, 1996.

[10] Leech, G. The Principles of Pragmatics. Longman, New York, 1983.

[11] Maybury, M.T. (Editor) Intelligent Multimedia Interfaces. AAAI Press. Menlo Park, 1993.

[12] McCawley, J.D. (1981) *Everything that linguists have always wanted to know about logic.* Chicago, Illinois. The University of Chicago Press. 1981.

[13] Miller, G. The Magical Number, 7 plus or minus two: Some limits on our capacity for processing information. Psychological Review no. 63 pp. 81-97. 1956

[14] Moore, J.D. Participating in Explanatory Dialogues. MIT Press. Cambridge, Ma. 1995

[15] Oliveira, D.A.S. A Natural Deduction Theorem Prover Capable of Complementing its Knowledge (in Portuguese). Master Dissertation. Departamento de Informática. PUC-Rio. Rio de Janeiro, Brazil, 1992.

[16] Oliveira, D.A.S, de Souza, C.S., Haeusler, E.H. Dealing with Negative Answers in Logic. Communication presented at Logic in Natural Language, LINGUA'98, Recife, Brazil, 1998.

[17] Haeusler, E.H. A Generator of Interactive Natural Deduction Theorem Provers. Proceedings of the XI Brazilian Symposium on Artificial Intelligence pp.557-571, Fortaleza, Brazil, 1994.

[18] Prawitz, D. Natural Deduction - A Proof-Theoretical Study. Stockholm, 1965.

[19] Quental, V.S.D.B. A Grammar for Automatic Text Comprehension and Generation in Portuguese Language (in Portuguese). Ph.D. Dissertation. Departamento de Letras, PUC-Rio. Rio de Janeiro, Brazil, 1995.

[20] Toulmin, S. E. The Uses of Argument. Cambridge University Press, 1964.

266-78

Beliefs, Belief Revision, and Splitting Languages

Rohit Parikh

Computer Science Department
Brooklyn College and City University Graduate Center

Abstract When a theory is updated with new information, few problems arise if the new information is consistent with the theory. If, however, the new information is inconsistent with the theory, then some adjustment is needed. The AGM axioms provide guidelines for the *form* of the adjustment though not for its exact nature. What we would like to do is to maximize the information retained and also to adjust only in those areas of the old theory to which the new information is pertinent. We use the notion of language splitting to this end. This notion allows us to carve up a theory into disjoint pieces about different subject matters. We show that such a splitting is unique. Tools that we use include a notion of mixing truth assignments and Craig's lemma.

Much current work in the study of belief revision goes back to a now classic paper due to Alchourron, Gärdenfors and Makinson [AGM]. The central issue is how to revise an existing set of beliefs T to a new set of beliefs $T * A$ when a new piece of information A is received. If A is consistent with T, then it is easy: we just add A to T and close under logical inference to get the new set of beliefs. The harder problem is how to revise the theory T when a piece of information A *inconsistent* with T is received. Clearly, as Levi has suggested, T must first be contracted to a smaller theory $T' = T \dot{-} \neg A$ which *is* consistent with A and then A added to T'. However, it is not clear how $T \dot{-} \neg A$ should be obtained. The mere deletion of $\neg A$ from T will clearly not leave us with a theory and there is in general no unique way to get a theory T' which is contained in T and does not contain $\neg A$.

Logic, Language, and Computation, Vol. II, edited by Lawrence S. Moss, Jonathan Ginzburg, and Maarten de Rijke.

[1]Also affiliated with the departments of Mathematics and Philosophy at CUNY Graduate Center. I thank Sam Buss, Samir Chopra, Horacio Arlo Costa, Konstantinos Georgatos and Bernhard Heinemann for comments. Research supported in part by NSF grant CCR-9208437 and a grant from the Research Foundation of CUNY.

Suppose, for example, that I believe that country Saturnia is hot and country Urania is cold. Now I discover that the two countries have very similar climates. Do I drop my belief that Saturnia is hot or that Urania is cold? Clearly I cannot retain them both.

The AGM approach does not actually tell us what to think about the two lands in question. What it does tell us is *if* we do have some procedure for updating, what logical properties such a procedure should satisfy. These properties (the AGM axioms) have been widely studied and model theoretic results proved for them (see [G], [KM]. [S]). Yet some issues remain.

Notation: In the following, L is a finite propositional language.[2] We assume that the constants *true, false* are in L. We shall use the letter L both for a set of propositional symbols and for the formulae generated by that set. It will be clear from the context which is meant. $A - B$ means that A and B are logically equivalent, i.e. that $A \leftrightarrow B$ is a tautology, i.e. true under all truth assignments. Similarly, $A \Rightarrow B$ means that $A \rightarrow B$ is a tautology. If X is a set of formulae then $Con(X)$ is the logical closure of X. In particular, X is a theory iff $X = Con(X)$. We shall use letters T, T' etc. for theories. $T * A$ is the revision of T by A, and finally, $T \dot{+} A$ is $Con(T \cup \{A\})$, i.e. the result of a brute addition of A to T (followed by logical closure) without considering the need for consistency.

AGM have proposed the following widely accepted axioms for the revision operator $*$:

1. $T * A$ is a theory.
2. $A \in T * A$
3. If $A - B$, then $T * A = T * B$.
4. $T * A \subseteq T \dot{+} A$
5. If A is consistent with T, i.e. it is not the case that $\neg A \in T$,
   then $T * A = T \dot{+} A$.
6. $T * A$ is consistent if A is.

Remark: Sometimes two more axioms having to do with revision by conjunctions are also included. Since we do not find a strong intuitive reason behind them we have omitted them. Please see the end of the paper for a discussion.

Unfortunately, the AGM axioms are consistent with the trivial update, which is defined by:

If A is consistent with T, then $T * A = T \dot{+} A$, otherwise $T * A = Con(A)$.
Thus in case A is inconsistent with T, under this update, all information in T is simply discarded. Clearly this is unsatisfactory because we would

[2] This restriction is only made for convenience. The results continue to hold for a countably infinite first order language without equality.

like to keep as much of the old information as is feasible. Hence the AGM axioms need to be supplemented to rule out the trivial update. However, various actual proposals have run into trouble, either by being too flexible and allowing implausible update operators, or worse, by allowing *only* the trivial update (see [DP]; [L] and [AP] give examples of some difficulties with the approach of [DP]).

We propose axioms for update operators which are consistent with the AGM axioms and which block the trivial update. The axioms are based on the notion of splitting languages. We shall explain first the intuitive idea behind this. The existing set of beliefs T may contain information about various matters. E.g. my current state of beliefs contains beliefs about the location of my children, the state of health of my teeth, and beliefs about the forthcoming election in India. In case one of my beliefs about the location of my children turns out to be false, it surely ought not to affect my beliefs about the election, since the subject matters of the two beliefs do not interact in any way. In order to model this intuition mathematically, we need to define in a rigorous way what it means to say that some given set of beliefs can be split among various unrelated matters. The notion of splitting languages does this for us[3]. Intuitively a theory T in language L *splits* if L is a union of two or more disjoint sub-languages, and the beliefs in T are generated by separate beliefs in the various sub-languages.

Definition 1: 1) Suppose T is a theory in the language L and let $\{L_1, L_2\}$ be a partition of L. We shall say that L_1, L_2 *split* the theory T if there are formulae A, B such that A is in L_1, B is in L_2 and $T = Con(A, B)$. Similarly we say that (mutually disjoint) languages $L_1, L_2, .., L_n$ split T if there exist formulae $A_i \in L_i$ such that $T = Con(A_1, ..., A_n)$. We may also say that $\{L_1, ..., L_n\}$ is a T-splitting.

2) If $L_1 \subset L$ then we say that T is *confined* to L_1 if $T = Con(T \cap L_1)$. (Note that in that case T also splits between L_1 and $L - L_1$, with the $L - L_1$ part being trivial, i.e. any formula of $L - L_1$ which is a theorem of T will be a tautology.)

In part 1 of the definition, we can think of T as being generated by the various T_i in languages L_i. Then the condition implies that T contains no "cross-talk" between L_i and L_j for distinct i, j. Part 2 of the definition says that T knows nothing about the part $L - L_1$ of L.

Remark: If P and P' are partitions of L, P is a T-splitting and P refines P' then P' will also be a T splitting.[4] For example suppose that

[3] This is a very natural notion and indeed we suspect that the notion of splitting has a wider application than just in belief revision.

[4] *P refines P'* if every element of P is a subset of some element of P'. Equivalently, the equivalence relation corresponding to P extends the equivalence relation corresponding

$P = \{L_1, L_2, L_3\}$ is a T-splitting and let $P' = \{L_1 \cup L_2, L_3\}$. Then P' is a 2-element partition, P is a 3-element partition which refines P' and P' is also a T-splitting. For let $T = Con(A_1, A_2, A_3)$ where $A_i \in L_i$ for all i. Then $T = Con(A_1 \wedge A_2, A_3)$ and $A_1 \wedge A_2 \in L_1 \cup L_2$ so that P' is also a T-splitting.

Example: Let $L = \{P, Q, R, S\}$, and $T = Con(P \wedge (Q \vee R))$. Then $T = Con(P, Q \vee R)$, and the partition $\{\{P\}, \{Q, R\}, \{S\}\}$ will be (the finest) T-splitting. $\{\{P, Q, R\}, \{S\}\}$ is also a T-splitting, but not the finest. Also, T is confined to the language $\{P, Q, R\}$ and knows nothing about S.

Lemma 1: Given a theory T in the language L, there is a unique *finest* T-splitting of L, i.e. one which refines every other T-splitting.

Lemma 1 says that there is a unique way to think of T as being composed of disjoint information about certain subject matters.

Lemma 2: Given a formula A, there is a *smallest* language L' in which A can be expressed, i.e., there is $L' \subseteq L$ and a formula $B \in L'$ with $A - B$, and for all L'' and B'' such that $B'' \in L''$ and $A - B''$, $L' \subseteq L''$.

Although A is equivalent to many different formulas in different languages, lemma 2 tells us that nonetheless, the question, "What is A actually *about*?" can be uniquely answered by providing a smallest language in which (a formula equivalent to) A can be stated.

All proofs are at the end of this paper.

The axioms:

The general rationale for the axioms is as follows. If we have information about two subject matters which, as far as we know, are unrelated (are split) then when we receive information about *one* of the two, we should only update our information in that subject and leave the rest of our beliefs unchanged. E.g. suppose I believe that Barbara is rich and Susan is beautiful and only that. Later on I meet Susan and realize that she is not beautiful. My beliefs about Barbara should remain unchanged since I do not connect Susan and Barbara in any way. If on the other hand I had initially believed that Barbara had made her money *as the agent* for Susan who was a beautiful model, then the two beliefs would be *connected* and finding that Susan was not beautiful could require an adjustment also of my beliefs about Barbara.

In fact, the notion of language splitting seems intrinsic to any attempt to form a theory of anything at all. Any observation or experiment gives us an enormous amount of information. E.g. when we are dealt a hand

to P'. P will have smaller members than P' does and more of them.

of cards, we are dealt them in a certain order, either by the right hand or the left hand of the dealer, who may have grey or brown or blue eyes. We usually ignore all this extra information and concentrate on the *set* of cards received. There is a tacit assumption, for instance, that the color of the dealer's eyes will not affect the probability that the hand contains two aces. This assumption that we can ignore some aspects while we are considering others is inherent in almost all intellectual activity.[5]

Now we give our new axioms P1–P3, giving intuitive justification for each. We also give a single axiom P which implies all of P1–P3.

Axiom P1: If T is split between L_1 and L_2, and A is an L_1 formula, then $T * A$ is also split between L_1 and L_2.

Justification: The two subject areas L_1 and L_2 were unconnected. We have not received any information which *connects* these two areas, so they remain separate.

Axiom P2: If T is split between L_1 and L_2, A, B are in L_1 and L_2 respectively, then $T * A * B = T * B * A$.

Justification: Since A and B are unrelated, they do not affect each other and so it should not matter in which order they are received.

Axiom P3: If T is confined to L_1 and A is in L_1 then $T * A$ is just the consequences in L of $T *' A$ where $*'$ is the update of T by A in the sub-language L_1.

Justification: Since we had no information about $L - L_1$ and have received none in this round, we should update as if we were in L_1 only. $L - L_1$, about which we have no prior opinions and no new information, should simply not have any impact.

All these axioms follow from axiom P, below.

Axiom P: If $T = Con(A, B)$ where A, B are in L_1, L_2 respectively and C is in L_1, then $T * C = Con(A) *' C \dot{+} B$, where $*'$ is the update operator for the sub-language L_1.

Justification: We have received information only about L_1 which does not pertain to L_2 so we should revise only the L_1 part of T and leave the rest alone.

It is easy to see that P implies P1. To see that P3 is implied, we use the special case of P where the formula B is the trivial formula *true*. To see that P implies axiom P2, suppose T is split between L_1 and L_2, and A, B are in L_1 and L_2 respectively. Let $T = Con(C, D)$ where $C \in L_1$ and $D \in L_2$. Then we get $T*A*B = (T*A)*B =_P [(Con(C)*' A) \dot{+} D]*B =_P (Con(C)*' A) \dot{+} (Con(D)*'' B)$. The two occurrences of $=_P$ indicate where

[5]The third chapter of Cherniak's book [C] gives arguments why beliefs must thus be divided into subsets, and cites supporting statements from Quine and Ullian [QU] as well as from Herbert Simon [Si].

we used the axiom P. Now the last expression $(Con(D) * B) \dot{+} (Con(C) * A)$ is symmetric between the pairs (C, A) and (D, B) and calculating $T * B * A$ yields the same result.

Remark: The trivial update procedure cannot satisfy P2 (or P), though it does satisfy P1 and P3. It follows that any procedure that does satisfy P cannot be the trivial procedure.

Justification: Let $T = Con(P, Q)$ and let $A = P$ and $B = \neg Q$. Then the trivial update yields $T * A * B = T * B = Con(\neg Q)$ and $T * B * A = Con(B) * A = Con(\neg Q, P)$. This violates P2. Also, $T * B = Con(\neg Q)$ which violates P. Thus P, or P2 alone, rules out the trivial update.

In theorem 1 we shall restrict AGM 1–6 to the case of those updates where both T and A are individually consistent and only their union might not be. This is because we can suppose that our current state of belief T about the subject matter of L originates in a state T_0 where we only believe tautologies (or if not, T_0 is at least consistent) and T is obtained from T_0 through zero or more revisions. Suppose that at some stage we are told an inconsistent formula A. Then axiom 3 tells us that this is equivalent to being told a blatant contradiction like $P \wedge \neg P$ and we would simply not believe A in that case. Hence if A is inconsistent, then $T * A$ should be just T. The AGM axioms 1–6 in their original form *force* that $T * true = T$ for consistent T and *disallow* it for an inconsistent T. Our restriction has the fortunate consequence that $T * true$ always equals T.

Definition 2: Given a theory T, language L and formula A, let L'_A be the smallest language in which A can be expressed and L^T_A be the smallest language containing L'_A such that $\{L^T_A, L - L^T_A\}$ is a T-splitting. Thus L^T_A is a union of certain members of the finest T-splitting of L, and in fact the smallest in which A can be expressed.

Example: Let $L = \{P, Q, R, S\}$, and $T = Con(P \wedge (Q \vee R) \wedge S)$. Then $T = Con(P, Q \vee R, S)$ and $\{\{P\}, \{Q, R\}, \{S\}\}$ will be the finest T-splitting. If A is the formula $P \vee \neg Q$, then L'_A is the language $\{P, Q\}$. But L^T_A, the smallest language *compatible* with the T-splitting, will be the larger language $\{P, Q, R\}$ which is the union of the sets $\{P\}$ and $\{Q, R\}$ of the finest T-splitting.

Theorem 1: There is an update procedure which satisfies the six AGM axioms and axiom P.

Proof: We define $T * A$ as follows. Given T and A, if A is consistent with T then let $T * A = T \dot{+} A$.

Otherwise, if A is not consistent with T, then write $T = Con(B, C)$ where B, C are in L_A^T, $L - L_A^T$ respectively.[6] Then let $T * A = Con(A, C)$. B, C are unique up to logical equivalence, hence this procedure yields a unique theory $T * A$. To see that it satisfies axioms 1–6 of AGM is routine. For the proof that it satisfies axiom P, see the proofs section of this paper. □

Example: Let, as before, $T = Con(P, Q \vee R, S)$. Then the partition $\{\{P\}, \{Q, R\}, \{S\}\}$ is the finest T-splitting., Let A be the formula $\neg P \wedge \neg Q$, then L_A^T is the language $\{P, Q, R\}$. Thus B will be the formula $P \wedge (Q \vee R)$ and $C = S$. B represents the part of T incompatible with the new information A. Thus $T * A$ will be $Con((\neg P \wedge \neg Q), S)$. The update procedure of theorem 1 notices that A has no quarrel with S and keeps it. As we will see, axiom P requires us to keep S.

Remark: In this update procedure we used the trivial update on the sub-language L_A^T, but we did not need to. Thus suppose we are given certain updates $*_{L'}$ for sub-languages L' of L. We can then build a new update procedure $*$ for all of L by letting $T * A = (B *_{L'} A) \dot{+} C$ in the proof above, where $L' = L_A^T$. What this does is to update B by A on L_A^T according to the old update procedure, but preserves all the information C in $L - L_A^T$.

Georgatos' axiom: K. Georgatos has suggested that axiom P2 be strengthened to require that in fact we should have $T * A * B = T * B * A = T * (A \wedge B)$. Thus axiom P2 would be revised as follows:
Axiom P2g: If T is split between L_1 and L_2, A, B are in L_1 and L_2 respectively, then $T * A * B = T * B * A = T * (A \wedge B)$.
But this is easily achieved. Suppose we are given a current theory T with its partition P_1 and a new piece of information C, and the theory $Con(C)$ has its own partition P_2. Now let P be the (unique) finest partition such that both P_1 and P_2 are refinements of P. E.g. if P_1 is $\{\{P, Q\}, \{R\}, \{S\}, \{T\}\}$ and P_2 is $\{\{P\}, \{Q, R\}, \{S\}, \{T\}\}$, then P will be $\{\{P, Q, R\}, \{S\}, \{T\}\}$. Now P is also *a* partition for both T and C though not necessarily the finest partition for either. Say $P = (L'_1, L'_2, L'_3)$ where T is axiomatized by (A_1, A_2, A_3) in (L'_1, L'_2, L'_3) respectively, and C by (C_1, C_2, C_3), also in (L'_1, L'_2, L'_3). Then let $T * C$ be $Con(D_1, D_2, D_3)$ where $D_i = A_i * C_i$. Now we get the AGM axioms, axiom P, and also the Georgatos axiom. We skip the proofs which are similar to the other proofs earlier.

---

[6] We allow that $L_A^T = L$, in which case $L - L_A^T$ will be the empty language with no non-logical symbols. L_A^T will still contain the constants *true* and *false*, and we take C to be the trivial formula *true*.

Computational Considerations: The following are rather trite observations, but may be of value. If we have a theory which has a large language, but which is split up into a number of small sub-languages, then the revision procedure outlined above is going to be computationally feasible. This is because when we get a piece of information which lies in one of the sub-languages (or straddles only two or three of them) then we can leave most of the theory unchanged and revise only the affected part. All of this is just common sense and very likely how we humans actually think. The results above give us a formal framework which shows that this is actually doable in a precise way.

Other bases for L: We recall that the set of truth assignments over some language $L = \{P_1, ..., P_n\}$ is just an n-dimensional vector space over the prime field of characteristic 2. Hence there is more than one basis of atomic predicates for this language. We regard the P_i as basic, but this is not necessary. For example, the language $L = \{P, Q\}$ is also generated by R, Q where R is $P \leftrightarrow Q$. To see this we merely need to see that P can be expressed in terms of Q, R. But this is easy, for $P - (Q \leftrightarrow R)$. Such a way of formalizing a language may be natural at times. For example, if my children Vikram and Uma have gone to a movie together, then it is very likely that both have come home or neither has. So it may be natural for me to formalize my belief in the language $\{R, U\}$ where R is $V \leftrightarrow U$. I may originally think that they are together but not home, so my theory will be $Con(R, \neg U)$. Later, if I find that Uma is home, I will revise to $Con(R, U)$, retaining my belief that they are together. It is obvious that the results of this paper will continue to hold for such adjustments of the "atomic" symbols.

The Proofs:

The proof of Lemma 1 depends on lemma A below.

Definition: Let $\{L_1, ..., L_n\}$ be a partition of L, and $t_1, ..., t_n$ be truth assignments. Then by $Mix(t_1, ..., t_n; L_1, ..., L_n)$ we mean the (unique) truth assignment t on L which agrees with t_i on L_i.

Example: Suppose that $L = \{P, Q, R\}$, $L_1 = \{P, Q\}$ and $L_2 = \{R\}$. Let the truth assignment t_1 be (1,1,1) on P, Q, R respectively, where 1 stands for *true*. Similarly, t_2 is (0,0,0). Then $Mix(t_1, t_2, L_1, L_2)$ will be (1,1,0) and $Mix(t_2, t_1, L_1, L_2)$ equals (0,0,1). Now the formula $A = (Q \rightarrow R)$ does not respect the splitting L_1, L_2. Hence both t_1 and t_2 satisfy A, but $Mix(t_1, t_2, L_1, L_2)$ does not (although $Mix(t_2, t_1, L_1, L_2)$ does).

Lemma A: $\{L_1, ..., L_n\}$ is a T-splitting iff for every $t_1, ..., t_n$ which satisfy T, $Mix(t_1, ..., t_n; L_1, ..., L_n)$ also satisfies T.

Proof of lemma A: ($\Rightarrow$) Suppose $T = Con(A_1, ..., A_n)$ where $A_i \in L_i$. If $t_1, ..., t_n$ satisfy T, let $t = Mix(t_1, ..., t_n; L_1, ..., L_n)$. Then, for each i, since t agrees with t_i on L_i, t satisfies A_i. Hence it satisfies T.

($\Leftarrow$) Write $t \models T$ to mean that T is true under t and let $Mod(T) = \{t|t \models T\}$. Now let X_i be the projection of $Mod(T)$ to L_i. I.e. $X_i = \{t'|(\exists t)(t \models T \wedge t' = t \uparrow L_i\}$, where $t \uparrow L_i$ is the restriction of t to L_i. If $t \in Mod(T)$ then for all i, $t \uparrow L_i \in X_i$, i.e. $t \in X_1 \times X_2 \times ... \times X_n$. Hence $Mod(T) \subseteq X_1 \times X_2 \times ... \times X_n$.

But the reverse inclusion is also true. For if $t \in X_1 \times X_2 \times ... \times X_n$, then for each i, there must exist t_i which agree with t on L_i and such that $t_i \models T$. But then $t = Mix(t_1, ..., t_n; L_1, ..., L_n)$ so $t \models T$ also. Thus $Mod(T) = X_1 \times X_2 \times ... \times X_n$.

Let $B_i \in L_i$ be such that $X_i = Mod(B_i)$. Then it is immediate that $Mod(T) = X_1 \times X_2 \times ... \times X_n = Mod(B_1) \times Mod(B_2) \times ... \times Mod(B_n)$. Hence $T = Con(B_1, ..., B_n)$ and so $\{L_1, ...L_n\}$ is a T-splitting. $\Box$

Proof of lemma 1: Suppose that $P = \{L_1, ...L_n\}$ is a *maximally fine* T-splitting. Such a P must exist but there is no a priori reason why it should also be *finest*; there might be more than one maximally fine partition. However, we will show that this does not happen and that a maximally fine partition is actually *finest*, i.e. refines every other T-splitting. If it does not, then there must exist a T-splitting $P' = \{L'_1, ..., L'_m\}$ such that P does *not* refine P'. Then there must be i, j such that L_i overlaps L'_j but is not contained in it. By renumbering we can take $i = j = 1$, so that L_1 overlaps L'_1 but is not contained in it. Consider $\{L'_1, (L'_2 \cup ... \cup L'_m)\}$ which is also a (2-element) T-splitting which P does not refine.

So without loss of generality we can take $m = 2$ and P' to be a two-element partition $\{L'_1, L'_2\}$. We now show that $\{L_1 \cap L'_1, ..., L_n \cap L'_1, L_1 \cap L'_2, ..., L_n \cap L'_2\}$ must also be a T-splitting. But this is a contradiction, for even if we throw out all empty intersections from these $2 \times n$ intersections, we still have at least $n + 1$ non-empty ones. For L_1 gives rise to two non-empty pieces, and all the other L_i must give rise to at least one non-empty piece. We thus get a *proper* refinement of P which was supposedly *maximally* refined.

To show that $\{L_1 \cap L'_1, ..., L_n \cap L'_1, L_1 \cap L'_2, ..., L_n \cap L'_2\}$ is also a T-splitting, we use lemma A. Let $t_1, ..., t_n, t'_1, ..., t'_n$ be any truth assignments which satisfy T. Let

$$t''' = Mix(t_1, ..., t_n, t'_1, ..., t'_n; L_1 \cap L'_1, ..., L_n \cap L'_2)$$

We have to show that t''' satisfies T. Let $t' = Mix(t_1, ..., t_n; L_1, ..., L_n)$ and let
$t'' = Mix(t'_1, ..., t'_n; L_1, ..., L_n)$. Since $\{L_1, ..., L_n\}$ is a T-splitting, both t', t''

satisfy T. Also, $\{L'_1, L'_2\}$ is a T-splitting and hence $t^{iv} = Mix(t', t''; L'_1, L'_2)$ does satisfy T.

To see that t''' satisfies T, it suffices to show that $t''' = t^{iv}$. We note now that for example, t^{iv} agrees with t' on L'_1 and the latter agrees with t_1 on L_1. Hence t^{iv} agrees with t_1 (and hence with t''') on $L_1 \cap L'_1$. Arguing this way we see that t^{iv} agrees with t''' everywhere so that $t^{iv} = t'''$. Thus $t''' \in Mod(T)$. Now use lemma A.

Thus $\{L_1 \cap L'_1, ..., L_n \cap L'_1, L_1 \cap L'_2, ..., L_n \cap L'_2\}$ is a T-splitting which properly refines P, which was maximally fine. This is a contradiction. Since assuming that P was not finest led to a contradiction, P is indeed the finest T-splitting. □.

Proof of lemma 2: Let us say that A is *expressible* in L' if there is a formula B in L' with $A - B$. We want to show that there is a smallest such L'. So let L_1 (with B_1) and L_2 (with B_2) be *minimal* such languages. Then $A - B_1$ and $A - B_2$ and so we have that $B_1 \Rightarrow B_2$. By Craig's lemma there is a B_3 in $L_1 \cap L_2$ such that $B_1 \Rightarrow B_3 \Rightarrow B_2$. But then we must have $A \Rightarrow B_1 \Rightarrow B_3 \Rightarrow B_2 \Rightarrow A$ and all are equivalent. Hence A is expressible in $L_1 \cap L_2$. By the minimality of L_1 and L_2 we must have $L_1 = L_1 \cap L_2 = L_2$ and L_1, L_2 were in fact not just minimal but actually smallest. □

For the proof that the procedure of theorem 1 satisfies axiom P, we need lemma B. The language L^T_A is as in definition 2.

Lemma B: Suppose that C is inconsistent with T. Let $\{L_1, ..., L_n\}$ be the finest T-splitting and let $T = Con(A_1, ..., A_n)$ where $A_i \in L_i$. Given a formula C, $L^T_C = \bigcup L_i : L_i$ overlaps $L'_C = \bigcup L_i : L_i \subseteq L^T_C$.
Moreover, if $*$ is the update procedure of theorem 1, then

$$\begin{aligned} T * C &= Con(C\ ,\ A_j : L_j \text{ does not overlap } L^T_C) \\ &= Con(A_j : L_j \text{ does not overlap } L^T_C) \dot{+} C. \end{aligned}$$

The proof of the lemma B is quite straightforward and relates updates to the finest splitting of T. During the update, those A_j such that L_j does not overlap L^T_C are exactly the A_j that remain untouched by the update. C has no quarrel with *them*. The other A_j are dropped and replaced by C.

Proof of theorem 1 (continued): It is sufficient to show that axiom P holds.

To see that P holds, suppose that $T = Con(A, B)$ where A, B are in L_1, L_2 respectively and C is in L_1. Let $\{L'_1, ..., L'_n\}$ be the finest T-splitting of L (and therefore refines $\{L_1, L_2\}$). Let $T = Con(A_1, ..., A_n)$ with $A_i \in L'_i$. Note that both L^T_A and L^T_C will be subsets of L_1 and indeed they will each be a union of some members, contained in L_1, of some of the

L'_i. A will be a consequence of some of the A_i, each from some member of this finest T-splitting and contained in L^T_A.

Under the update of theorem 1, those A_i which lie in L^T_C will be replaced by C. The others will remain. Hence, $Con(A) * C = Con(A_i : L_i \subseteq (L^T_A - L^T_C), C)$.

Also $B = \bigwedge A_i : L_i$ does not overlap L_1. So $Con(A) * C \dot{+} B$ equals $Con(A_i : L_i \subseteq (L^T_A - L^T_C), C, \bigwedge A_i : A_i$ does not overlap $L_1) = T * C$. □

Remark: The notion of splitting languages and the lemmas can easily be extended to first order logic without[7] equality. We use the fact that if a first order theory (without equality) is consistent then it has a countably infinite model, in particular, a model whose domain is the natural numbers. Call such models standard. Now given a standard first order structure $\mathcal{M}$ which interprets a language L and a sub-language L' of L, we can define a reduct $\mathcal{M}'$ of $\mathcal{M}$ which is just $\mathcal{M}$ restricted to L'.[8] Given a partition $L_1, ..., L_n$ of a language L and standard L-structures $\mathcal{M}_1$,...,$\mathcal{M}_n$ we can define $Mix(\mathcal{M}_1, ..., \mathcal{M}_n, L_1, ..., L_n)$ to be that standard structure $\mathcal{M}'$ which agrees with $\mathcal{M}_i$ on L_i. Then lemma 1 can be generalized in the obvious way and all our arguments go through without any trouble.

Comment on AGM axioms 7 and 8: Axiom 7 of AGM says that $T * (A \wedge B) \subseteq (T * A) \dot{+} B$ and axiom 8 says further that if B is consistent with $T * A$ then the two are equal. We do not feel that these axioms are consistent with the spirit of our work for the following reason. Suppose that $A = (\neg P \vee Q)$ and $B = (P \vee Q)$, then $A \wedge B$ is equivalent to Q and says nothing about P. Now revising a theory T first by A could cause us to drop some P-related beliefs we had, and revising after that with B we might not recover them. But revising with $A \wedge B$ should leave our P beliefs unchanged, provided that our beliefs about P and Q were not connected. Thus contrary to 7, revising with the conjunction may at times preserve more beliefs than revising first with A and then with B. This is why it does not seem to us that axioms 7 and 8 *should* hold in general.

To give a somewhat different, concrete example, suppose you believe that to *reach* a certain place, the *path* should be clear. Write this as $R \to P$. You also believe (strongly) that your grandmother is afraid of flying and therefore is *not taking flying lessons.* Call this (latter) $\neg F$. You now

[7]Sam Buss has pointed out that problems arise if equality is present. Once we have equality present, one can express the formula for instance which says that the model has cardinality 2. Another formula in a disjoint language can say that the cardinality is at least 3. These two formulae will conflict though they have no non-logical symbols in common.

[8]For instance suppose $L = \{P, R\}$ and $L' = \{P\}$. If $\mathcal{M}$ is $(N, \mathcal{P}, \mathcal{R})$ then $\mathcal{M}'$ would be $(N, \mathcal{P})$.

receive the news $A = (\neg P \vee F)$, that either the path is not clear or your grandmother is taking flying lessons. You conclude that the path is not clear and that you will not reach in time. Later you are told $B = (P \vee F)$, that either the path is clear or that your grandmother is taking flying lessons. You will conclude that you will reach your destination after all. But you will never acquire the belief that your grandmother is taking flying lessons. But you would have acquired that belief if you had been told $A \wedge B$, i.e. F.

Postscript: Subsequent to our submission of this paper to ITALLC, a paper by del Cerro and Herzig [CH] has appeared which also mentions dependence in the context of belief revision. However, the sort of dependence they discuss is from probability theory and not the one that we have discussed above.

References

The technical material depends only on the reference [AGM] and then only on the definitions. The other references are given for background. [QU], [C] and [Si] discuss motivation. The others consist mostly of technical developments arising from [AGM].

[AGM] Carlos Alchourron, Peter Gärdenfors and David Makinson, "On the logic of theory change: partial meet contraction and revision functions", *J. Symbolic Logic* 50 (1985) 510–530

[AP] Horacio Arlo Costa and Rohit Parikh, "On the inadequacy of C2 (research note)", typescript, May 1995.

[C] Christopher Cherniak, *Minimal Rationality*, MIT Press 1986.

[CH] Luis Farinas del Cerro and Andreas Herzig, "Belief change and dependence", proceedings of *Theoretical Aspects of Rationality and Knowledge*, Ed. Y. Shoham, Morgan Kaufmann 1996, 147-161.

[DP] Adnan Darwiche and Judea Pearl, "On the logic of iterated theory revision", *Theoretical Aspects of Reasoning about Knowledge*, Edited by R. Fagin, Morgan Kaufmann 1994, 5–22

[G] Adam Grove, "Two modellings for theory change", *J. Phil. Logic* 17 (1988) 157–170.

[KM] Hirofumi Katsuno and Alberto Mendelzon, "Propositional knowledge base revision and minimal change", *Artificial Intelligence*, 52 (1991) 263–294.

[L] Daniel Lehmann, "Belief Revision, revised", *Proceedings of 14th International Joint Conference on Artificial Intelligence*, Morgan Kaufmann 1995, 1534–1541.

[QU] Willard van Orman Quine and Joseph Ullian, *The Web of Belief*, 2nd edition, Random house 1978.

[S] Krister Segerberg, "Belief revision from the point of view of doxastic logic", *Bull. of the IGPL* 4 (1995) 535–553.
[Si] Herbert Simon, *Administrative Behaviour*, Macmillan 1947.

Prolegomena to A Theory of Disability, Inability and Handicap

John Perry
Center for the Study of Language and Information
Stanford University

Elizabeth Macken
Center for the Study of Language and Information
Stanford University
and
David Israel
Artificial Intelligence Center
SRI International

1 Introduction

Underlying the political activism that led to the Americans with Disabilities Act (ADA) was what Ron Amundsen has called the *environmental conception of disability.* In [6] we called this the *circumstantial conception of disability and handicap*, and contrasted it with the *intrinsic conception.* We use *disability* to mean loss of a function, such as moving the hands or seeing, that is part of the standard repertoire for humans. *Handicap* is a species of inability, in particular, the inability to do something that one wants to do and most others around one can do.[1] The intrinsic conception imagines a tight connection between disability and handicap; the circumstantial conception loosens and relativizes that connection. The circumstantial conception reminds us that we all depend on various tools and structures—in particular, on cultural artifacts—to enable us to do what we want to do. In many cases it is the design of these tools and structures

Logic, Language, and Computation, Vol. II, edited by Lawrence S. Moss, Jonathan Ginzburg, and Maarten de Rijke.

[1] An *impairment* is a physiological disorder or injury; impairments may be the ground or cause of a disability.

that prevents disabled people from accomplishing what they want, rather than anything intrinsically connected to the disability. For example, very few people can get from the first floor to the second floor of a building without the assistance of some structure such as stairs, ramps, or elevators. If no such structures are available, everyone is handicapped; if stairs are available, but not a ramp or an elevator, people with various disabilities are handicapped; if ramps or elevators are available, very few people are. Disabled people, like everyone else, are handicapped in the absence of the structures and tools that enable them to perform the tasks they need and want to do.

The ADA, in requiring that employers and others *reasonably accommodate* disabled workers, reflects this circumstantial conception of disability and handicap. The underlying idea is simply that many of the tasks that are necessary for getting to a job site and then accomplishing what the job requires can be done by individuals with disabilities, given the proper equipment and facilities. The way the disabled worker accomplishes these tasks may differ from the way other workers do. She may, for instance, use a wheelchair rather than walk to get to the job site; she may use a voice-recognition tool rather than typing on a keyboard to input to a computer.

Accommodation can be brought about in two ways. Where situations have been designed without consideration for individuals with disabilities, retrofitting is required; e.g. installing ramps or elevators, widening hallways, etc. Far better is the second way: to design with an eye more toward enabling the accomplishments required to satisfy the demands of the task rather than toward enabling a small range of (even widely employed) ways of satisfying those requirements. Providing stairs enables people who can walk to locomote between flights by (something akin to) walking—though it's still difficult for people who walk with crutches, say. Providing ramps enables both walkers and a wide range of non-walkers to locomote between flights—the former by walking, the latter, in other ways—and it makes it easier for walkers with crutches.

In developing and applying the circumstantial conception of disability, the following basic concepts are clearly central:

- Ability, inability, disability.
- Doing the same thing in different ways.
- Accommodating and enabling.

In this paper we extend the *Content Theory of Action* (CTA)[2], to try to elucidate these concepts. This attempt at elucidation is itself at most a

[2]Presented in [4, 5], as the "IPT" account.

prolegomena to a study that can usefully feedback into the moral and legal issues involving disabilities.

In §2, we review the CTA and introduce some new concepts to help capture more adequately the structure of abilities and inabilities. In §3, we define ability and inability, and use the CTA to motivate and explain the notion of *enabling strategies.* In §4 , we show how this notion in turn motivates design principles that are in accord with the circumstantial conception. In §5, we conclude with a brief discussion of epistemic disabilities.

2 The Content Theory of Action

2.1 The Meaning of movements

The strategy of CTA is to develop a theory of action that is modeled in important ways after the relational theory of meaning developed in [2].

- In [2], utterances were viewed as particulars that involve speaking or writing sentences of various types, in virtue of which various things get said, depending on the circumstances of utterance (the context).

  According to CTA, acts or movements are particulars that involve the execution of movements of various types, in virtue of which various results occur, depending on the circumstances of the act.

- In [2] the content of an utterance of a sentence is a collection of described situations; roughly, a proposition.

  In CTA, the results of movements of particular types in particular circumstances are modeled by propositions. (Hence the "Content" theory of action.)

- We distinguish between direct and indirect discourse descriptions of utterances. Direct discourse identifies (more or less) the type of the expression uttered, while indirect discourse characterizes an utterance by way of its contents.

  In CTA we distinguish two ways of characterizing acts, as executions of movements of particular types, and as accomplishments, that bring about various states of affairs. Describing acts in terms of the movements executed is analogous to direct discourse description; describing them in terms of the results accomplished is analogous to indirect discourse.

- In [2], Barwise and Perry associate relations between contexts and contents with types of expressions; they take these relations to be the *meanings* of the expression types.[3]

  In CTA, we associate relations between circumstances and results with types of movement; we take these relations to be the *meanings* of the movement types.

Consider, for example, the type of movement one makes when pushing an elevator button. In different circumstances, a movement of this type will bring about different results. Standing in front of an elevator, it will call the elevator to one's floor; standing inside the elevator it will go to a different floor; standing in front of an angry brute, it will cause one to get beat up, and perhaps to lose the offending digit. Think of the circumstances as the context and replace the described situation with the resulting situation, and the familiar pattern will emerge. Then think of both the circumstances and the resulting situation as characterized by propositions, and you have the basic concept of CTA, the meaning of a type of movement. In [2], the context of an utterance was analyzed into two components, the discourse situation and connective situation. Below we will see an analogue of this decomposition.

Where $\mathcal{C}$ is a set of basic constraints, $[\![M]\!]_{\mathcal{C}}$ is a relation between circumstances C and a result P. $[\![M]\!]_{\mathcal{C}}(C, \mathrm{P})$ obtains just in case according to $\mathcal{C}$, when a movement of type M occurs in circumstances C, P results.

More explicitly, we define: $[\![M]\!]_{\mathcal{C}}(C, P)$ iff

- any movement m that is of type M, made in circumstances of type $C(x_1, \ldots, x_n, m)$, will have as a result that $P(x_i, \ldots, x_l)$ $(1 \leq i \leq l \leq n)$,
- where the x_i are additional parameters for objects involved in C.[4]

2.2 Two types of actions: executions and accomplishments

In CTA, movements are acts. Acts are particulars, actions are properties of agents at times. An act is identified by an agent, a location, a time and a type of movement. Agents have action-properties in virtue of being the agents of the acts. We recognize two kinds of actions. *Executions* are properties that agents have locally and non-circumstantially, in virtue of the

[3]Barwise and Perry used the term "interpretation," but "content" has become generally accepted.

[4]Henceforth we suppress explicit mention of the relativity to constraints.

type of movement that they produce. Where M is a movement type, $\mathcal{E}[M]$ is the property of producing a movement of type M. *Accomplishments* are properties that agents have in virtue of the results they bring about. These results depend on the circumstances in which their acts occur, as well as the type of movement involved. The execution/accomplishment distinction is analogous to the direct/indirect discourse distinction, in characterizing utterances. Where P is a proposition, we use $\mathcal{B}[P]$ to denote the property of bringing it about that P.

2.3 Relations between and among actions

A key concept in CTA, and one of particular importance in thinking about disabilities, is that of one action being a *way to* perform another action. In CTA we have factored this into two relations. MO (for *mode of*) is a relation between executions and accomplishments, and WO (for *way of*) is a relation between accomplishments. WTO (for *way to*) is the general concept, which holds between a pair of actions if the first is either a mode of or a way of performing the second.

- Executing M is a **mode of** bringing it about that P in circumstances C iff any movement of type M in circumstances C will have the result that P:

  $MO(\mathcal{E}[M], \mathcal{B}[P], C)$ iff $[\![M]\!]$(C, P).

- Bringing it about that P is a **way of** bringing it about that Q in C if any M whose execution is a mode of bringing it about that P in C is also a mode of bringing it about that Q in C:

  $WO(\mathcal{B}[P], \mathcal{B}[Q], C)$ iff
  $(\forall M)$ if $MO(\mathcal{E}[M], \mathcal{B}[P], C)$ then $MO(\mathcal{E}[M], \mathcal{B}[Q], C)$.

- An action A_1 is a **way to** perform another action A_2, if either A_1 is an execution and a mode of performing A_2 or A_1 is an accomplishment and a way of performing A_2:

So moving one's left ring finger is a *mode of* bringing it about that an "s" appears on one's computer screen. Sending a linefeed to the computer may be a *way of* sending a message. Intuitively, sending a linefeed is something that can be done by executing a wide variety of movements, so it is an accomplishment, not an execution. The line between executions and accomplishments can be drawn at different places depending on the analytical project at issue, however. Of course getting the "s" to appear, or the linefeed sent, doesn't just depend on which movement one executes. It also depends on the circumstances an agent is in. One has to have one's hands poised over a keyboard, the keyboard has to be connected to a computer, which has to be turned on, etc. Note that more must be said about the circumstances in the second case than in the first; the agent must be in a mail program of a certain sort, etc. All of these relations are not just relations between actions, but relations between actions and circumstances.

In this paper we introduce the concept of a *path.* A path is a sequence of actions, each of which is a way to perform the next in certain circumstances. These actions form a sequence in the set theoretical sense, not the temporal sense.

We distinguish between *execution paths* and *distal paths.* This distinction is analogous to that between *mode of* and *way of.* Execution paths start with an execution, distal paths do not. We think of execution paths as starting with an agent executing some movement. We think of distal paths as leading to some goal. The terminology comes from the possibility that the distal path depends on circumstances at some distance from the agent. What inventors, engineers, builders, repair persons and others do is connect execution paths to distal paths, giving us paths from execution to goal, that is methods for doing new things, or old things in new ways.

Moving one's hand, thus turning the key, thus completing the circuit, thus energizing the starter motor is an execution path. Energizing the starter motor, thus bringing it about that the starter turns, thus bringing it about that the flywheel turns, thus bringing it about that the engine turns over, thus bringing it about that the car starts, is a distal path from engaging the starter motor to starting the car. When we put the execution path and the distal path together, we have a an execution path all the way to the likely goal of our action, starting a car. Logically, we put paths together by creating a circumstance in which all the needed *way-to* relations hold. In this example, a mechanic might quite literally put things together. The path may have been broken because of a change in circumstance: the spring in the starter motor wears out, so that it does not engage with the flywheel. All the way-to links on the path from moving the hand to engaging the starter motor still obtain in these circumstances, as do the ones from turning the flywheel to starting the engine. But bringing it about that the starter turns is no longer a way to bring it about that the flywheel turns. The repairman changes the circumstances back to what they should be, restoring that link in the path.

- Let $\mathcal{R}$ be a sequence of actions, either executions or accomplishments:

  $\mathcal{R} = \langle\langle A_1 \ldots A_n \rangle\rangle$

  $\mathcal{R}$ is a *path* to A_n in C iff $(\forall i, 1 \leq i \leq n-1), \mathrm{WTO}(A_i, A_{i+1}, C)$.

$\mathcal{R}$ is an *execution path* if A_1 is an execution, otherwise it is a *distal path.*

To avoid misunderstanding, we need to emphasize that the theory here pertains to a single movement (however complex) and the "upshots" it has as a result of wider and wider circumstances. When such movements are complex, they will decompose into a temporal structure of movements. The upshots of each of these movements will depend on the other movements and their upshots. That is part of our picture, but not part of the present theory. Moreover, paths, as so far conceived, are not intended even to extend to cases in which an agent does first one thing, via an execution, and then does another thing, via another execution. Our paths are the simplest cases of a much wider class of *action-structures.*

3 Needed Concepts

3.1 Ability and inability

The CTA analysis suggests a conception of ability.

First, we say in what circumstances an agent can perform an execution:

- An agent α has the ability to perform $\mathcal{E}[M]$ in C if α can form a volition to execute M and this volition reliably causes M in C.[5]

Then we give the general definition:

- An agent α has the ability to bring it about that P in C if there is a path $\mathcal{R}$ for bringing it about that P in C, and α can perform the first action in $\mathcal{R}$ in C.

An agent has an inability to perform an action A in C if he does not have an ability to do so.

There are in general two (combinable) strategies for getting rid of inabilities. One is to enlarge the executions the agent can perform; the other to build new connections between executions and distal goals. And we use both strategies. A person may exercise, stretch, etc. to become able to perform new simple executions. They may practice so as to be able to perform complex movements in a coordinated and speedy way. It is fair to say, however, that the vast majority of new inventions, technologies and products that win wide acceptance require very little in the way of learning new basic executions, and relatively little in the way of developing new skills.

[5]For more on volitions and the motivating complexes that cause and rationalize acts, see [5].

We build tools and structures that allow us to do new things with our old movements. The circumstantial conception of disability and handicap simply amounts to continuing this preferred strategy in the case of individuals with disabilities.

3.2 Doing the same thing in different ways

The CTA gives a clear understanding of the different ways in which actions can be similar and of the ways in which they can can differ.

- Actions can involve the same movements.

  Example: Henry and George did exactly the same thing, for Henry patterned his movement exactly on George's. Yet Henry went to jail and George did not.

- Actions can involve the same accomplishments and same movements.

  Example: Henry and George both signed "George Smith" on check.

- Actions that involve the same movements and accomplishments can also involve different accomplishments.

  Example: Henry forged George's name, but George did not forge his own name.

- Actions that involve different movements can involve the same accomplishments.

  Example: George and Fred each paid the mechanic $100; George did it by signing his name on one of his checks, Fred by signing his name on one of his checks.

  Example: George and Elwood each paid the mechanic $100, George by signing a check, Elwood by putting his ATM card through the machine.

Given the wide ranging results that any movement may have, discussion of whether two acts amount to doing the same thing is clearly idle in the absence of some structure providing criteria of comparison. Virtually any movement disturbs the air around it, so virtually any pair of acts will be doings of the same thing, if we allow so much latitude. No two movements will disturb just the same air in the same place at the same time, so no two acts will being doing the same thing if we allow no latitude. Any time there is a substantive discussion of whether acts amount to doing the same thing, there is some focus of interest and some limitation of methods that structures the problem.

We are typically interested in alternative ways of accomplishing some result of antecedent interest. The actions we are interested in are goal driven; it is not just any result, but some specific result, that the agent aims at. Then, within the context of having alternative ways of accomplishing the same goal, other factors will enter to determine if the acts are truly equivalent given the interests involved. Often these other parameters can be seen as matters of sharpening the goal. The task at hand, for example, may not be simply to deliver the mail, but to deliver the mail to the existing second floor mailboxes, or to deliver the mail by a certain time in the afternoon.

Any time there is sufficient clarity about what is meant by "doing the same thing," there will be conceptual resources that make it worthwhile to move to some other level of comparison that explicitly takes the various parameters of action into account.

3.3 Connecting Paths

The telephone provides another example of how inventors, engineers, architects, builders and others enable people to achieve goals by providing paths from executions people can perform or easily learn to perform to the goals they want to reach. Making noise by moving our mouths, and so creating meaningful signals, messages, that people could perceive, was an ability humans had long before the telephone. Communicating with people a long ways away by getting them to perceive messages, usually visible signals, that expressed our thoughts was a time-honored strategy for attaining various further goals. The telephone made it possible to use our ability to create messages by moving our mouths (combined with our abilities to dial or poke buttons), to get messages to people long distances away. In our terminology, the telephone changed the circumstances in a way that made producing a message with one's mouth a way to get a message to someone a long ways away. It connected an execution path and a distal path.

The change that is made in one's circumstances that links the execution path and the distal path—in this case, the provision of a telephone and a telephone system—we call an *enabling circumstance.* In this new circumstance, some accomplishment along the execution path becomes a way to perform some accomplishment along the distal path. We call these the *interfacing accomplishments.* In this case, speaking is a way to get a message to a distant person, given telephones and a telephone system. Speaking and getting a message to a person are the interfacing accomplishments. We think these concepts are very helpful in thinking about disabilities and handicaps.

Now consider α, a quadriplegic, capable of speech and movement of the head from side to side and back and forth, sitting alone in an ordinary chair in a room in front of a computer. α has various desires. He would like to input to the computer. He would like to be able to leave the room for a meeting in a conference room on the next floor. α cannot do these things, because there is no execution path from the movements he can execute to his goals. That is, there is no sequence of actions, starting with head movement or speech, that is a way to input to a computer, or move to the second floor conference room.

How do we go about creating an execution path for α to reach a particular goal?

We need two paths and an enabling circumstance. We need path from an execution α can perform and we need a distal path to α's goal and we need to create a circumstance in which an accomplishment along the former path becomes a way to accomplish a step in the latter.

Suppose, that is, that $\mathcal{P}_E = \langle\langle A_1, \ldots A_n \rangle\rangle$ is an execution path from executions our agent can perform in C, and suppose that $\mathcal{P}_G = \langle\langle B_1, \ldots \mathcal{B}(G) \rangle\rangle$ is a path to the goal $\mathcal{B}(G)$ in C. Then what needs to be done, in order to make it so that α can bring it about that G in C is to find an enabling circumstance C' we can add to C to create a new circumstance in which one of the actions, A_i, on the execution path is a way of performing one of the actions, B_j, along the path to $\mathcal{B}(G)$.

$$WTO(A_i, B_j, C \sqcap C').$$

In such a case, A_i, B_j are the interfacing accomplishments. Note that here we assume that paths are closed under suffixes:

- If $\langle\langle A_1, \ldots, A_n \rangle\rangle$ is a path in C then $\langle\langle A_i, \ldots, A_n \rangle\rangle$, $1 < i < n$, is a path in C

3.4 Enabling Strategies

On our picture, there are two things to work with when an agent has a goal he cannot reach (other than abandoning the goal). We can try to:

- *Standardize the agent*: Change the agent so that he can execute the movements that interface with existing paths to the goal.
- *Connect the paths*: Change the agent's circumstances so that so the movements he can already execute provide execution paths to the goal.

Repairing the spinal damage so that our agent can stand and move his legs in the standard way and walk to the conference room is the first strategy of enablement. There are strategies that fall short of a cure that approximate to the first strategy. For example, one might develop a body suit that permits the agent to move joints and legs through small external motors by using speech, small head movements, eye movements, or some device that would pick up minute muscle or brain wave activity.

There is nothing wrong with such strategies. But exclusive focus on standardizing the agent suggests a commitment to the intrinsic conception of disability and handicap. The circumstantial conception, on the other hand, suggests that there are a number of other ways to enable individuals with disabilities to reach their goals. And, of course, this is what we typically do in the case of non-disabled individuals to remove inabilities. We provide stairs because people cannot fly from floor to floor. We provide cars and roads because people cannot walk or run long distances in short amounts of time. We provide tools and structures that allow people to do new things with old movements.

In the case of the quadriplegic, providing a voice recognition system to operate the computer, a head operated wheelchair to move about, and installing whatever ramps are necessary to make moving forward a way of moving to the second floor, are examples of the second strategy.

The second strategy typically involves two parts:

- Providing the agent with tools so that he has execution paths to some intermediate accomplishment:
  - A head operated wheel chair, so he can move forward on a smooth surface from point A to point B.
  - A voice recognition system that allows him to input to a computer capable of receiving voice input.
- Providing infrastructure that makes the intermediate accomplishment a way of achieving the ultimate goal:
  - A ramp that leads from the office to the second floor where the conference room is.
  - A device that allows him to interact with non-voice operable computers, and local and worldwide computer networks, via the computer he can control with his voice.

4 Design Principles

In the discussion above, we assumed the existence of a great deal of technological infrastructure—the existence of computers and networks of computers of various kinds, and of certain types of software. If these weren't available, none of us would be able to send email messages. There is also, of course, the existence and well-functioning of the larger electrical systems within which the computers live. The existence of this infrastructure can be said, in turn, to be an enabling circumstance for all our computing activities. This is quite typical: much that we do is possible only in an environment in which enabling circumstances have been established by the culture and technology society has produced.

In the modern world, the things an agent can do are most often not merely a product of the movements the agent can execute, but these combined with various artifacts: structure and equipment provided by human beings. Broadly speaking, one can distinguish between the infrastructure at a location, and the equipment that an agent has. When we drive or bike to work, for example, the roads from Palo Alto to Stanford are parts of the infrastructure provided at these locations. The cars and bikes we use are part of the agent's equipment.

When infrastructure is provided at a location in order to make it possible for agents to achieve certain goals, there is a presupposition about what the agents can do, the abilities they bring with them to the infrastructure. A stairway, for example, seen as providing a way of moving from one level to another in a building, presupposes the ability (roughly) to lift oneself eight inches. This is most naturally done with the legs, but can be done with the arms, and also with some very high-tech wheel-chairs. A ramp, provided for the same purpose, only presupposes the ability to move forward, something that can be done by walking or by using an ordinary wheelchair.

The presupposed abilities are the ones that are necessary to interface with the infrastructure, to make use of it for the purpose in question.

Consider our quadriplegic from the previous section. Both a staircase and a ramp will extend the standard path for moving from point to point by moving one's legs into a path for getting to the new level. But they take off from this standard path in different places. The crucial interfacing accomplishment for the staircase is moving one's legs. The crucial interfacing accomplishment for the ramp is moving forward. The ramp interfaces at a *more generic* point than the staircase. Adding a ramp changes the circumstances in a way that extends both execution paths for moving forward, walking and moving one's head while seated in a wheelchair, to paths for getting from one level to another.

The CTA, then, in concert with the circumstantial conception of disability urges principles such as the following:

- For the Rehabilitation Engineer: Look for executions the person can perform, that have the potential to interface with existing distal paths to goals.

- For the inventor and manufacturer: Design distal paths that will provide interfacing accomplishments for the widest possible range of execution paths.

The latter design principle we call the principle of *generic interfacing.* The engineer is typically designing tools and structures that will allow people to do new things with old movements, or do old things in faster, cheaper, or more entertaining ways using old movements. The principle of generic interfacing simply means that the design should allow people to do these new things, use these new paths to old things, with as wide a variety of movements as possible.

Keyboards were designed as they were to enable typing as the standard interfacing accomplishment to distal paths for accomplishing things on a computer. The method of input envisaged was clearly depressing keys by moving fingers. But keyboards have proven a fairly generic point of interfacing. There are a great many execution paths that do not involve moving one's fingers that connect to depressing keys. One can use a head-stick, hold a pencil with one's teeth, use one's feet, etc. All of these methods and others have been used by individuals with disabilities to interface with computers.

Use of speech recognition technology, on the other hand, allows one to interface at one point further along the path, bypassing the keyboard altogether. Thus, the provision of speech recognition technology provides for a wider class of users by enabling a larger class of interfacing accomplishments.

Speech technology can interface with paths to inputting to the computer at several points. The *Total Access System* designed at Stanford's Archimedes Project interfaces with the keyboard buffer via a *Total Access Port* (TAP).[6] With other systems the point of input is further along the path to the CPU.

5 Epistemic Disabilities

The discussion so far has been limited to motor disabilities, such as quadriplegia for it is here that the approach to disabilities that flows from CTA is most intuitive. But there are other kinds of disabilities. Epistemic disabilities have to do with the inability to gain and use information using standard paths. Epistemic disabilities divide into sensory disabilities, such as blindness and deafness, and cognitive disabilities. We will conclude by

[6] The Total Access System and the TAP were invented by Neil Scott. See http://www-csli.stanford.edu/arch/arch.html for more information.

saying a bit about how the CTA, the circumstantial conception of disability, and the design principles apply in the case of epistemic disabilities.

In the case of epistemic disabilities, a situation or process that would standardly lead to an internal cognitive state does not do so, because of an internal impairment of some sort.[7]

Consider what is involved in a sighted person using vision to access the information on a printed page. The person turns towards the page, opens his or her eyes, and focuses them. In certain circumstances, the room being lit, for example, this will bring about certain results. The agent will have richly structured visual sensations and thereby pick up the information on the page—see what it says. The movements—turning towards the page, opening the eyes and focusing—were ways to bring about this succession of internal results: sensations and perceptions.

This path to getting information from a printed page depends not only on external circumstances, such as the room being lit, but also on internal circumstances. The visual system must be intact.

This provides us with a motivation to distinguish more sharply between the agent's circumstances and the circumstances at the agent's location. What we are thinking of now as the agent's circumstances includes the internal facts about brain and body, and the connections with various tools that provide the means for interfacing accomplishments.

---

[7]But first we need to make some distinctions and terminological points. Suppose that α is at location l, and it is obvious to us that it would be a good idea for α to bring it about that P, and in fact α can bring it about that P at l. But α doesn't do this. Suppose, for example, that α can get to his philosophy class on time by hopping on the bike around the corner and pedaling across campus.

One explanation may be that α doesn't have the requisite desires to do that act in question. He could care less if he gets to philosophy class on time, or ever. It would it most cases be odd to describe this as an inability. We say, "he can do it; he doesn't want to".

Another possibility is that α doesn't have the requisite beliefs; α doesn't realize that the circumstances at l are such that he can bring it about that P; α may not realize that there is a bicycle around the corner that he can use. We usually do not describe this as inability; we say, "He can do it; he just doesn't know that he can".

Another possibility is that α doesn't have the requisite know-how; he wants to bring it about that P, and knows that the circumstances are such that he could do it, in some sense, but α lacks the skills needed to bring it about that P. So in this case, perhaps α cannot ride a bike. He can execute all of the individual movements that are necessary, but has just never learned how to keep his balance. In these cases, we would typically say that α can't bring it about that P. Here there is an inability, but we wouldn't usually say there was a disability, since α can, in some sense, make all the movements he needs to. He just hasn't practiced enough to do it at will.

In each of these cases, the person can make the requisite movements, but nevertheless does not do the action, because they lack a requisite internal *cause*.

The epistemic disabilities form a fourth category in which it is something internal that prevents the goal from being reached.

Consider the need to remembers people's names at parties. For many people, the following method works. When the person is introduced, focus on them while repeating their name several times to yourself. Here we think of focusing and repeating to oneself as executing the cognitive analogue of movements, with the hoped-for result that one learns the names and acquires the ability.

For some people, this just doesn't work, presumably because of some difference in internal circumstance.

For the sake of the latter group (and lazy members of the first group), name-tags are often provided. This provides a different method for acquiring the ability to call a person by their name at a party, viz., just look at their name tag.

Finally, let us return to the case of the epistemic accomplishment of obtaining information from a display on a monitor. We were supposing our blind reader was dealing with a not quite modern system, in particular a pre-GUI system, one in which all the information to be displayed could be (and was) displayed on the screen via text. In the world after the GUI revolution, provision of screen readers and/or Braille transcribers is no longer enough to render hearing and feeling adequate interface accomplishments, relative to paths to the new, even wider range of accomplishments available to the sighted.

Interaction with computer systems is aimed, first and foremost, at interfacing with the meaning of or content in files—which we use as a maximally generic term for data structures. The principle of generic interfacing suggests the following design principles with respect to computer systems: make both input and output as device-neutral as possible, that is, bypass as much as possible the requirements (on executions) of particular peripheral devices. This is typically accomplished by providing alternative peripheral devices enabling other input/output modes. But this depends, in turn, on the form in which the information is carried being accessible to those modes. To the extent that the GUI revolution narrows the interfacing accomplishments for interacting with computers to visually picking up graphical information and pointing with a mouse, it represents a violation of the principle of generic interfacing.

References

[1] http://www-csli.stanford.edu/arch/research.html

[2] Barwise, Jon and John Perry. *Situations and Attitudes* (Cambridge: MIT/Bradford: 1983).

The Archimedes Project is a project at Stanford whose mission is to provide individuals with disabilities access to computers and access to people through computer technology. The Project is based on the philosophy encapsulated in the following six principles:

- Everyone requires help in gaining and effectively using information, not only those individuals who have disabilities.
- In itself, information is neither accessible nor inaccessible; the form in which it is presented makes it so.
- To be disabled is not necessarily to be handicapped. Handicaps can often be removed where disabilities cannot.
- Handicaps often arise from decisions to design tools exclusively for individuals with the standard mix of perceptual and motor abilities.
- Designed access is preferable to retrofitted access.
- Solutions that provide general access can benefit everyone.

Further information about the project is available on the web: http://www-csli.stanford.edu/arch/arch.html.

[3] Goldman, Alvin. *A Theory of Human Action.* (Englewood Cliffs, New Jersey: Prentice Hall, 1970.)

[4] Israel, David, John Perry and Syun Tutiya. "Actions and Movements." In *Proceedings of IJCAI-'91*, Mountain View, CA: Morgan Kaufmann, August, 1991.

[5] Israel, David, John Perry and Syun Tutiya. "Executions, Motivations and Accomplishments." *The Philosophical Review* (October, 1993): 515–40.

[6] Perry, John, Elizabeth Macken, Neil Scott and Jan McKinley. Disability, Inability and Cyberspace. To appear in Batya Friedman, editor, *Designing Computers for People—Human Values and the Design of Computer Technology* (Stanford: CSLI Publications, forthcoming).

[7] *International classification of impairments, disabilities, and handicaps: a manual of classification relating to the consequences of disease.* Geneva: World Health Organization, 1980. 205p.

Constraint-Preserving Representations

Atsushi Shimojima

Japan Advanced Institute of
Science and Technology

1 Introduction

According to Barwise and Hammer (1995), a logical system is a mathematical model of some pretheoretic notion of consequence and an existing (or possible) inferential practice that honors it. This broad view underlies the recent works of Shin (1994), Hammer (1995), Luengo (1995), and Fisler (1995), who build logical systems, with full syntax and semantics, to model inferential practices that use non-sentential representations such as Venn diagrams, Euler diagrams, Peirce's existential graphs, Harel's higraphs, geometry diagrams, state charts, circuit diagrams, and timing diagrams. As with most cases of mathematical modeling, the method of building a logical system has its upsides and downsides. On the one hand, it is very well suited to address the questions such as whether the modeled inferential practice always produces *valid* inferences with respect to its target domain and whether it is capable of producing *all* the valid inferences. These questions can be respectively studied as the matters of *soundness* and *completeness* of the system that models the inferential practice in question. On the other hand, building a logical system does not directly answer the questions concerning the *efficacy* of the inference practice, or more specifically to my concern, the questions concerning the efficacy of the *representations* used in it. Why does a particular mode of representation sometimes appear to be more efficacious than others as an aid to human problem-solving and sometimes not?

The present work is a part of a larger project aimed at answering this question. Our method is mathematical modeling in a broad sense, but it does not consist in building logical systems. At the base of our modeling is the following alternative perspective, called "the Constraint Hypothesis": (i) representations are objects in the world, and as such they obey

Logic, Language, and Computation, Vol. II, edited by Lawrence S. Moss, Jonathan Ginzburg, and Maarten de Rijke.

certain structural constraints that govern their possible formation, and (ii) the variance in inferential potential of different modes of representation is largely attributable to different ways in which these structural constraints on representations match with the constraints on targets of representation.

There are, of course, other research traditions that focus on the efficacy issues. The Constraint Hypothesis aside, our project differs from these predecessors in its strategy and focus. First, in explaining a particular kind of efficacy exhibited by a representation, we spend most of our energy in describing the property of the representation (or of the underlying representation system) that is responsible for the efficacy. Thus, we mainly study the properties of the *tool* itself, unlike the psychologists who are more interested in the *user* and her cognitive processes (Palmer 1978, Bauer and Johnson-Laird 1993, Stenning and Oberlander 1995). Secondly, our explicit call for an *explanatory* theory discerns our project from the methodological studies of statistical data presentation (Bertin 1973, Tufte 1983, 1990). Thirdly, our primary interest in *human* reasoning distinguishes ourselves from the AI researchers working on the computational modeling (and implementation) of representation-mediated reasoning (Gelernter 1959, Larkin and Simon 1987, Lindsay 1988, Funt 1980) and from those in the knowledge representation tradition (Sloman 1971, 1975, Hayes 1974, Levesque and Brachman 1985).

To substantiate the Constraint Hypothesis, we have analyzed several different properties of representation systems, namely, the abilities to provide "free ride" and "half ride," "overspecificity," and "autoconsistency" (Barwise and Shimojima 1995, Shimojima 1996a, 1996b), which are responsible for particular kinds of representational efficacy and inefficacy. This paper applies the Constraint Hypothesis to yet another kind of efficacy exhibited by representations. The issue is why certain modes of representations appear to exhibit the consequence relation holding on their targets while they do not appear to explicitly present the consequence relation. We will first present a typical example of representation with this efficacy (section 2), and then informally argues that it is due to the fact that the underlying representation system "preserves constraints on its targets in constraints on the representations" (section 3). In section 4, we will characterize this property of constraint preservation precisely, in the mathematical framework of situation theory (Barwise and Perry 1983, Barwise 1989, Devlin 1991) and its descendant information theory (Barwise 1993, Barwise, Gabbay and Hartonas 1995, Barwise and Seligman 1996). After discussing a different example of constraint-preserving system (section 5), we will introduce the notions of "surrogate inference" and "attunement" to sketch the cognitive process in which the user exploits the constraint preservation property of representation systems (section 6).

2 Representative Example

	Stem	Leaves
0|9 = 900 feet	0	98766562
	1	97719630
	2	69987766544422211009850
	3	876655412099551426
	4	9998844331929433361107
	5	97666666554422210097731
	6	898665441077761065
	7	98855431100652108037
	8	653322122973
	9	377655421000493
	10	0984433165212
	11	4963201631
	12	45421164
	13	47830
	14	00
	15	676
	16	52
	17	92
	18	5
19|3 = 19,300 feet	19	39730

Figure 1

Consider a so-called "stem-and-leaf" plot (Tukey 1977, Figure 1) that displays the heights of the world's 218 volcanos on the unit of 100 feet. This plot has twenty "stems" from the top to the bottom, labeled "0" through "19." Each digit ("leaf") in a given stem stands for an individual volcano whose height is in the range indicated by the label (the horizontal order of digits means nothing). Thus, since the stem labeled "16" consists of the numerals "5" and "2," it means that the volcanos whose heights are between 16,000 feet and 17,000 feet consist of one that is about 16,500 feet tall, one that is about 16,200 feet tall, and nothing else[1]. Thus, this entire plot carries enormous information about how many volcanos fall within each range and how tall they are. Furthermore, if we compare the lengths of several stems, we obtain information about how various height-ranges compare with each other about the numbers of volcanos falling within them. For example, if we compare the lengths of the three middle stems labeled "7," "8," and "9," we can read off the following information:

(θ_1) The number of volcanos whose heights are between 7,000 feet and 8,000 feet is greater than the number of volcanos whose heights are between 8,000 feet and 9,000 feet and the number of volcanos whose heights are between 9,000 feet and 10,000 feet.

---

[1] Here "between x and y" means "more than or equal to x but less than y."

The utility of this plot is that this information is presented in connection with the information about what makes it true. Read the three bars individually, and you obtain the following information:

(θ_2) The volcanos whose heights are between 7,000 feet and 8,000 feet consist of a volcano about 7,900 feet tall, one about 7,800 feet tall, another about 7,800 feet tall, ..., one about 7,700 feet tall, and nothing else

(θ_3) The volcanos whose heights are between 8,000 feet and 9,000 feet consist of a volcano about 8,600 feet tall, one about 8,500 feet tall, one about 8,300 feet tall, ..., still another about 8,300 feet tall, and nothing else.

(θ_4) The volcanos whose heights are between 9,000 feet and 10,000 feet consist of a volcano about 9,300 feet tall, one about 9,700 feet tall, another about 9,700 feet tall, ..., another about 9,300 feet tall, and nothing else.

Whenever the states of affairs θ_2, θ_3, and θ_4 hold, it follows that the state of affairs θ_1 holds, that is, θ_1 is a *consequence* of $\{\theta_2, \theta_3, \theta_4\}$ relative to a certain constraint governing the target situation that the plot represents. But the sheer fact that the information θ_1 is presented with the supporting information θ_2, θ_3, and θ_4 does not make the plot really unique. You can surely describe each of these four pieces of information in a sentence (as I have just done), and the set of sentences written down in this way can be said to present the information θ_1 along with the supporting information θ_2, θ_3, θ_4. Rather, what makes the plot different from such a set of sentences is that the plot presents the two sets of information *in the way that makes a consequence relation between them easier to recognize*, while the set of sentences provides no such aid.

Nevertheless, the plot does *not* explicitly present the consequence relation. There seems no definite feature of the plot that indicates the consequence relation, while we can easily identify the features of the plot that indicate the information θ_1, θ_2, θ_3, and θ_4. To wit:

(σ_1) The stem labeled "7" is longer than both the stem labeled "8" and the stem labeled "9".

(σ_2) In the stem labeled "7," there are a numeral "9," a numeral "8," another numeral "8," ..., a numeral "7," and no more.

(σ_3) In the stem labeled "8," there are a numeral "6," a numeral "5," a numeral "3," ..., still another numeral "3," and no more.

(σ_4) In the stem labeled "9," there are a numeral "3," a numeral "7," another numeral "7," ..., another numeral "3," and no more.

These are states of affairs holding in the plot, and on the semantic conventions that we adopt for the plot, they indicate the pieces of information θ_1, θ_2, θ_3, and θ_4 respectively. We read off these pieces of information from the plot by observing σ_1, σ_2, σ_3, and σ_4 in the plot.

In contrast, we can not as easily identify a state of affairs holding in the plot that indicates the consequence relation holding between $\{\theta_2, \theta_3, \theta_4\}$ and θ_1. Nevertheless, the plot seems to help us recognizing the consequence relation. If we borrow Wittgenstein's convenient terms, the plot *shows* the consequence relation without *saying* that it holds.

3 Informal Analysis

Thus, the target phenomenon of this paper is that (i) there is a kind of representation that *shows* the logical or extra-logical consequence relation among certain pieces of information $\theta_1, \ldots, \theta_n$ they present, while (ii) it does not *explicitly present* the consequence relation in the same way it presents $\theta_1, \ldots, \theta_n$, and (iii) the showing of the consequence relation is relative to the mode in which $\theta_1, \ldots, \theta_n$ are presented. What explains this phenomenon?

In my view, the crucial fact in our example is that the state of affairs σ_1 holding in the plot is a consequence of the other states of affairs $\{\sigma_2, \sigma_3, \sigma_4\}$ holding in the plot *just as* the information θ_1 about the target situation is a consequence of the information $\{\theta_2, \theta_3, \theta_4\}$ about the target situation[2]. Figure 2 represents this property of the plot, where "$\Rightarrow$" denotes the indication relation based on our semantic conventions and $\top$ denotes the consequence relation.

States of affairs that hold in the plot

$$\left\langle \begin{array}{ccc} \{\sigma_2, \sigma_3, \sigma_4\} & \Rightarrow & \{\theta_2, \theta_3, \theta_4\} \\ \top & & \top \\ \sigma_1 & \Rightarrow & \theta_1 \end{array} \right\rangle$$

Information about the target situation

Figure 2

This property gives a user of this plot two alternative ways of recognizing the consequence relation holding between the information $\{\theta_2, \theta_3, \theta_4\}$ and θ_1. First, the agent may appeal to her pre-knowledge about the target domain, and directly figure out that θ_1 is a consequence of $\{\theta_2, \theta_3, \theta_4\}$. (This

[2]Note that σ_1 is a consequence of $\{\sigma_2, \sigma_3, \sigma_4\}$ only when all the relevant numerals have nearly the same width. This syntactic stipulation is crucial to the utility of the plot, indeed.

is a case in which the agent directly recognizes the right-hand $\top$ in Figure 2, so to speak.) Alternatively, instead of relying on her knowledge about the target domain, she may inspect the plot itself. Relying on the fact that σ_1 is a consequence of $\{\sigma_2, \sigma_3, \sigma_4\}$, she may conclude that the consequence relation holds between θ_1 and $\{\theta_2, \theta_3, \theta_4\}$. (The agent is using the left-hand $\top$ as a step toward recognizing the right-hand $\top$.) She can take this second strategy safely, as long as the plot has the property schematized in Figure 2. We refer to this property of the system of stem-and-leaf plots as "constraint preservation."[3]

We propose that this mechanism of constraint preservation accounts for the stem-and-leaf plot's capacity of showing the consequence relation without explicitly presenting it. On this account, there need be no state of affairs holding in the plot that indicates the constraint $\{\theta_2, \theta_3, \theta_4\} \vdash \theta_1$. Still, the plot can be said to exhibit it because there is a matching constraint $\{\sigma_2, \sigma_3, \sigma_4\} \vdash \theta_1$ among states of affairs holding in the plot. The plot shows the target constraint via the constraint governing the plot itself (plus the semantic indication relation $\Rightarrow$).

Why then is it that the set of sentences that present the same information θ_1, θ_2, θ_3, and θ_4 do not help us recognizing the target constraint in the same way the stem-and-leaf plot does? Compare Figure 2 with Figure 3, which represents the case of presenting θ_1, θ_2, θ_3, and θ_4 in the form of sentence. Even if you write down all the sentences that present the information θ_2, θ_3, and θ_4, there need not be any fact about the set of sentences that semantically indicates the information θ_1. (You simply need *add* a sentence in order to present θ_1.) This means that there is no consequence relation between the sentential facts, say $\{\sigma_2', \sigma_3', \sigma_4'\}$, that indicate the information $\{\theta_2, \theta_3, \theta_4\}$ and the sentential fact, say σ_1', that indicates θ_1. Hence there is no $\top$ from $\{\sigma_2', \sigma_3', \sigma_4'\}$ to σ_1' in Figure 3, and the system of our sentences is not constraint-preserving. In this situation, the only way to figure out the consequence relation between θ_1 and $\{\theta_2, \theta_3, \theta_4\}$ is to rely on our preknowledge about the target domain and directly recognize the right-hand $\top$. There is nothing in the representation itself that guides us to recognize it, or confirms the correctness of our recognition.

---

[3]Note that what is at issue here is the *recognition of* the consequence relation between θ_1 and $\{\theta_2, \theta_3, \theta_4\}$, rather than the *inference* from $\{\theta_2, \theta_3, \theta_4\}$ to θ_1. Yet the system of stem-and-leaf plots indeed gives us two alternative ways of *inferring* from $\{\theta_2, \theta_3, \theta_4\}$ to θ_1, and they seem to be major sub-processes of the direct and the indirect process of recognizing the consequence relation. We will discuss this topic in section 6.

States of affairs that hold in the set of sentences	$\{\sigma'_2, \sigma'_3, \sigma'_4\}$	$\Rightarrow$	$\{\theta_2, \theta_3, \theta_4\}$	Information about the target situation
			$\top$	
	σ'_1	$\Rightarrow$	θ_1	

Figure 3

With this account in hand, we do not have to rely on the following obscure account of the phenomenon. Assume that the plot presents the information θ_1, θ_2, θ_3, and θ_4 in a "better" way than its sentential counterpart. It does not matter for the moment what exactly the word "better" means here, or what explains this superiority of the plot over the sentences. What is important is that if the plot is superior to the sentences in presenting the information θ_1, θ_2, θ_3, and θ_4, then a user of the plot probably has a better chance to recognize the constraint $\{\theta_2, \theta_3, \theta_4\} \vdash \theta_1$ than a user of the sentences. For perhaps, the recognition of this constraint involves the recognition of the individual pieces of information θ_1, θ_2, θ_3, and θ_4. On this account, the plot does not directly help us with the recognition itself; we still have to rely on our pre-knowledge about the target domain to figure out that there holds such a constraint; still, the plot presents the antecedents $\{\theta_2, \theta_3, \theta_4\}$ and the consequent θ_1 of the consequence relation so well that we feel as though it exhibited the consequence relation itself. It supplies such nice sandwich bread that it makes us feel as though a filling were also supplied.

It would take a good amount of hard work to provide precise meanings and justifications to the basic assumptions of this account. In particular, depending on how we work out the notion of "better" presentation, the postulated "sandwich" effect may or may not strong enough to make us feel as though the stem-and-leaf plot exhibited the constraint in question. It is theoretically possible that we recognize the pieces of information θ_1, θ_2, θ_3, and θ_4 as clearly as possible without realizing an elementary consequence relation holding among them. As it is, the "sandwich" account is almost empty in the absence of supplementary accounts of what "better" presentation amounts to and how it gives rise to the postulated effect.[4]

[4]If the "sandwich" account could be worked out, there would be important differences in the predictions that the two accounts make. First, since the sandwich account renders the recognition of the target constraint a matter of the user's pre-knowledge, it predicts that a representation exhibits no target constraint completely new to the user. The property of constraint preservation is independent of the user's pre-knowledge, and so our account does not make that prediction. Secondly, assuming that a representation presents the pieces of information $\delta_1, \ldots, \delta_n$ in the "better" way, the sandwich account

4 Preserving and Exhibiting Constraints

We believe that the above discussions have shown that the property that we call "constraint preservation" is *real*, namely, that some representation systems definitely have it. In this section, we will characterize the property more precisely, by embedding it in an explicit mathematical framework. The framework contains the following basic concepts: (i) "situation" and "state of affairs," which are the basic building blocks of the framework's ontology, (ii) "representation," "source domain," "target domain," "indicating relation," and "signaling relation," which are concerned with what it is for a representation to present a piece of information, and (iii) "constraint on representations" and "constraint on targets."[5] After introducing these notions one by one, we will invoke them to characterize what constitutes a "representation system" and what it is for a representation system to preserve a target constraint in a source constraint.

So, there exist situations and state of affairs. Some situations are actual while others are not actual. Situations are "sites" in which various states of affairs hold or do not hold. When a state of affairs σ holds in a situation s, we say that s *supports* σ, written $s \models \sigma$. We use English lowercases $s, s_i, t, \ldots$ for situations, actual or non-actual, and use greek lowercases $\sigma, \sigma_i, \theta, \ldots$ for states of affairs. English capitals $S, S_i, T, \ldots$ are for sets of situations, and Greek capitals $\Sigma, \Sigma_i, \Theta, \ldots$ are for sets of states of affairs.

Each state of affairs σ has a unique dual $\bar{\sigma}$ such that $\sigma = \bar{\bar{\sigma}}$. Although σ and $\bar{\sigma}$ never hold together in a single situation, it may be that neither σ nor $\bar{\sigma}$ holds in a given situation. The situation may not settle the issue of whether σ or $\bar{\sigma}$, so to speak. We say that a situation s is *determinate* relative to a state of affairs σ if and only if either σ or $\bar{\sigma}$ holds in s. For example, let s_J be the current political situation of Japan. Although the state of affairs that Ichiro Ozawa is the cleverest politician does not hold in s_J probably, s_J is determinate relative to this state of affairs, since its dual (the state of affairs that Ozawa is not the cleverest politician) holds in s_J. On the other hand, s_J is not determinate relative to the state of affairs that I love chocolate. For, although I do love chocolate, it is not the political situation of Japan that settles the issue of whether I love chocolate or not.

In our conception, a representation is a situation that we (cognitive

predicts that the representation exhibits *every* constraint, real or unreal, holding among $\delta_1, \ldots, \delta_n$ as long as it is in the user's pre-knowledge. Our account is more selective, predicting that only those constraints that are preserved by the system are exhibited.

[5]The concepts in (i) have their origins in the literature on situation theory, especially, Barwise and Perry (1983), Barwise (1989), and Devlin (1991). Many concepts in (ii) and (iii) are borrowed from more recent literature on the qualitative theory of information, especially, Barwise (1993), Barwise, Gabbay, and Hartonas (1995), and Barwise and Seligman (1996).

agents) create in order to present information about a particular object to ourselves or others. Thus, the English sentence that you are reading now is a situation that I created (or let the printer create) to present certain information about our concept of representation. The map of Kyoto that I drew yesterday to show my wife the location of Kinkakuji is a situation that I created to present the information about the temple.

Given a representation system $\mathcal{R}$, there is a range of information that the representations in the system can possibly present. We model this range of information by means of a set of states of affairs Θ'. To each state of affairs θ in Θ', we assign one or more states of affairs σ, typically one different from θ itself, with the agreement that whenever σ holds in a representation s, we can read off the information θ from s. If a state of affairs σ lets us read off a piece of information θ as the result of this agreement, we say "σ indicates θ" and write $\sigma \Rightarrow \theta$. Thus, a representation in the system $\mathcal{R}$ presents information partly by virtue of the states of affairs that hold in it and the indication relation $\Rightarrow$.

Let S be all the situations (actual or non-actual) that are determinate relative to each member of $domain(\Rightarrow)$. Let Σ be the set of state of affairs relative to which each member of S is determinate. Then, S consists of situations that may or may not support states of affairs in Σ. Let $\models_S$ be this binary relation of supporting defined on $S \times \Sigma$. We call the triple $\mathbf{S} = \langle S, \Sigma, \models_S \rangle$ the "source domain" of the given representation system $\mathcal{R}$. Likewise, let T be all the situations (actual or non-actual) that are determinate relative to each member of $range(\Rightarrow)$. Let Θ be the set of states of affairs relative to which each member of T is determinate. Letting $\models_T$ be the supporting relation defined on $T \times \Theta$, we call the triple $\mathbf{T} = \langle T, \Theta, \models_T \rangle$ the "target domain" of the system $\mathcal{R}$.[6]

In this paper, we assume that every state of affairs in $domain(\Rightarrow)$ indicates a unique state of affairs. Thus, $\Rightarrow$ is a (often partial) function from Σ into Θ. For sets of states of affairs $\Sigma_i \subseteq \Sigma$ and $\Theta_j \subseteq \Theta$, we say "$\Sigma_i$ indicates Θ_j" and write "$\Sigma_i \Rightarrow \Theta_j$" iff Θ_j is the image of Σ_i under $\Rightarrow$.

Suppose a state of affairs σ holds in a situation s and $\sigma \Rightarrow \theta$. We may say that s presents the information θ. But which situation in T does s carry the information θ about? The information θ is true of some elements of T but false of the others, and depending on which situation s is about, s may be either true or false as a representation. Thus, for s to be a really meaningful representation, it must be determined which situation in T the situation s carries information θ about—which situation s is targeting at.

[6]Thus, the source domain and the target domain of a representation system constitute "classifications" as defined in Barwise and Seligman (1996). This makes our mathematical framework easily embeddable in the general theory of information flow that Barwise and Seligman are developing.

We model this targeting relation by means of a (possibly partial) function from the set of situation S to the set of situation T, writing "$s \leadsto t$." Thus, $domain(\leadsto)$ consists of all the situations in S that have definite targets in T, and $range(\leadsto)$ consists of all the situations in T that are targeted by some situations in S.

We are in the position of characterizing what it is for a situation to present information about another situation in a representation system.

Definition 1 (presenting information) Let $\mathbf{S} = \langle S, \Sigma, \models_S \rangle$ and $\mathbf{T} = \langle T, \Theta, \models_T \rangle$ be the source domain and the target domain of a representation system $\mathcal{R}$. Let $s \in S$, $t \in T$, and $\theta \in \Theta$. We say that the situation s *presents* the information θ *about* the situation t in $\mathcal{R}$ iff there is a state of affairs $\sigma \in \Sigma$ such that:

1. $s \models_S \sigma$,
2. $\sigma \Rightarrow \theta$,
3. $s \leadsto t$.

The representation s is *true in* the system $\mathcal{R}$ iff each information that s presents about its target t in $\mathcal{R}$ is supported by t.

Central to our analysis is the notion of constraint on representations. What is it for a constraint to hold on the representations in a representation system? To answer this question, we need first determine what constitutes a representation *within* a representation system. Consider the source domain $\mathbf{S} = \langle S, \Sigma, \models_S \rangle$ of a representation system $\mathcal{R}$. Remember that the set S contains *every possible* situation that is determinate relative to the set of states of affairs $domain(\Rightarrow)$. For two reasons, S is too large to be considered as the set of representations within the system $\mathcal{R}$. First, S may contain situations whose syntactic features are "ill-formed" with respect to the syntactic formation rules stipulated in $\mathcal{R}$. Second, S may contain "abnormal" situations, namely, exceptions to the natural laws that normally regulate the domain. Given that we are interested in the efficacy of the system $\mathcal{R}$ when it is used in *normal* circumstances, we do not want to count those abnormal situations as representations of the system. For these reasons, we decide to count as proper representations of $\mathcal{R}$ only those situations in S that are both normal and well-formed. We call the set of these situations "$\hat{S}$."

With $\hat{S}$ in hand, we can now characterize what it is for a constraint to hold on the representations in the system $\mathcal{R}$. We call any pair $\langle \Gamma, \Delta \rangle$ of subsets of Σ a "source sequent" and write it as "$\Gamma \vdash \Delta$." We say, "a situation s in S *respects* a source sequent $\Gamma \vdash \Delta$" to mean that if s supports *all* members of Γ, then s supports *at least one* member of Δ. What we call *constraints on representations* are those source sequents that are respected by all representations in $\mathcal{R}$. That is, a source sequent $\Gamma \vdash \Delta$ is a *constraint on the representations of* $\mathcal{R}$ iff all situations in $\hat{S}$ respect $\Gamma \vdash \Delta$.

What is it, then, for a constraint to hold on the *targets* of a representation system? Let $\mathbf{T} = \langle T, \Theta, \models_T \rangle$ be the target domain of the representation system $\mathcal{R}$. We can think of different kinds of regularities governing the elements of T. Some regularities are due to natural laws such as mechanical laws. Some regularities are due to stipulative laws such as the U.S. laws and hotel regulations. In fact, it is these regularities that make it possible for us to make a valid inference about the target domain at all. Now, since the set of situations T consist of all the possible situations that are determinate relative to every state of affairs in *range*($\Rightarrow$), S may well contain exceptions to these natural or stipulative regularities. We exclude these exceptions, and use "$\hat{T}$" to the resulting subset of T. We take $\hat{T}$ to be the set of proper targets of the system $\mathcal{R}$.

Given any pair $\langle \Gamma, \Delta \rangle$ of subsets of Θ, we call it a *target sequent* and write it as "$\Gamma \vdash \Delta$." We call a target sequent a *constraint* on the targets of the system $\mathcal{R}$ iff all situations in $\hat{T}$ respect it.

We end up associating six elements to the intuitive notion of representation system. They are a source domain $\mathbf{S} = \langle S, \Sigma, \models_S \rangle$, a target domain $\mathbf{T} = \langle T, \Theta, \models_T \rangle$, an indicating relation $\Rightarrow$, an signaling relation $\rightsquigarrow$, a set of normal and well-formed representations $\hat{S}$, and a set of normal targets $\hat{T}$. More specifically:

Definition 2 (representation system) A *representation system* is a sextuple $\mathcal{R} = \langle \mathbf{S}, \mathbf{T}, \Rightarrow, \rightsquigarrow, \hat{S}, \hat{T} \rangle$ such that:

- $\mathbf{S}$ is a source domain $\langle S, \Sigma, \models_S \rangle$;
- $\mathbf{T}$ is a target domain $\langle T, \Theta, \models_T \rangle$;
- $\Rightarrow$ is a (possibly partial) function from Σ into Θ;
- $\rightsquigarrow$ is a (possibly partial) function from S to T;
- $\hat{S} \subseteq S$;
- $\hat{T} \subseteq T$.

Now we characterize what it is for a representation system to preserve a target constraint.

Definition 3 (preservation of constraint) Let $\mathcal{R} = \langle \mathbf{S}, \mathbf{T}, \Rightarrow, \rightsquigarrow, \hat{S}, \hat{T} \rangle$ be a representation system. The system $\mathcal{R}$ *preserves* the target constraint $\Theta_1 \vdash \theta$ *in* the source constraint $\Sigma_1 \vdash \sigma$ iff:

- $\Theta_1 \vdash \theta$ is a constraint on the targets of $\mathcal{R}$;
- $\Sigma_1 \vdash \sigma$ is a constraint on the representations of $\mathcal{R}$;[7]
- $\Theta_1 \Rightarrow \Sigma_1$;
- $\sigma \Rightarrow \theta$.

Thus, the preservation of a target constraint is a property of a representation system as a whole. In contrast, the exhibition of a target constraint is a property of a particular representation in a constraint-preserving system:

Definition 4 (exhibition of constraint) Let s be a representation of a representation system $\mathcal{R}$. Then s *exhibits* a target constraint $\Theta_1 \vdash \theta$ via a source constraint $\Sigma_1 \vdash \sigma$ iff:

- $\mathcal{R}$ preserves $\Theta_1 \vdash \theta$ in $\Sigma_1 \vdash \sigma$;
- Every member of Σ_1 (and hence σ) holds in s.

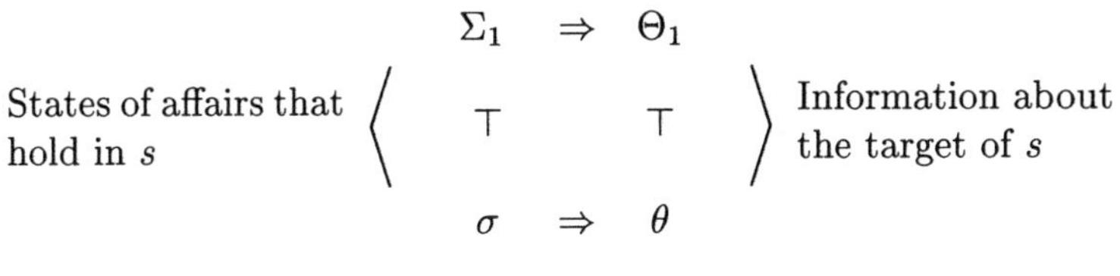

States of affairs that hold in s

$$\left\langle \begin{array}{ccc} \Sigma_1 & \Rightarrow & \Theta_1 \\ \top & & \top \\ \sigma & \Rightarrow & \theta \end{array} \right\rangle$$

Information about the target of s

Figure 4

On our characterizations of "presenting" and "exhibiting" (definitions 1 and 4), exhibiting a target constraint does not imply presenting the constraint as a piece of information. Nor presenting implies exhibiting. These are two fundamentally different ways in which a representation makes a target constraint available to the user. Thus, two representations that seem informationally equivalent in the "presentation" level may not be equivalent in the "exhibition" level. This explains some of the cases in which we feel one of two representations to be "more informative" than the other even though they seem to present the same information. The stem-and-leaf plot and its sentential counter-part are a case in question.

[7] More correctly, "$\Theta_1 \vdash \theta$" and "$\Sigma_1 \vdash \sigma$" should be "$\Theta_1 \vdash \{\theta\}$" and "$\Sigma_1 \vdash \{\sigma\}$," but we will not bother.

5 Another Example

To appreciate the diversity of cases in which representations exhibit constraints on their targets, it would be useful to consider an example that is *prima facie* very different from the stem-and-leaf plot.

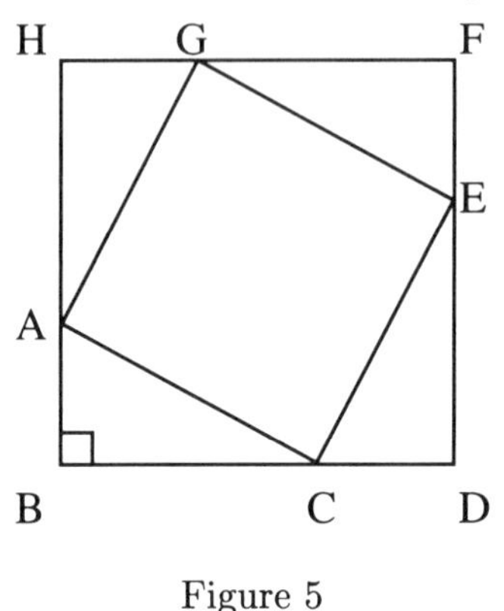

Figure 5

Figure 5 is a diagram commonly used to demonstrate the Pythagorean Theorem in geometry. A typical demonstration accompanying this diagram goes as follows. Let ABC be an arbitrary triangle with the right angle B. Let GHA, EFG, and CDE be congruent to ABC where A is between B and H, G between H and F, E between F and D, and C between D and B. Then $BHFD$ and $AGEC$ are squares. Also, $BHFD$ is decomposed into $AGEC$, ABC, GHA, EFG, and CDE. So, letting x, y, and z be the lengths of the segments AB, BC, and AC respectively, the following holds:

$$(x+y)^2 = z^2 + 4(\frac{1}{2}xy)$$

And so:

$$x^2 + y^2 = z^2$$

It is obvious that this diagram somehow helps us to understand the demonstration.[8] First of all, it helps us to comprehend what the demonstration says in each individual step—how individual points are named and

[8]Indeed, we could consider this diagram not only as an aid to understand the demonstration, but as a *part* of the underlying proof. By extending the work of Luengo (1995), we could construct a system of geometry proof that have diagrams as its "well-formed objects," and restate the above demonstration as a formal proof, where the above diagram is constructed step by step by several applications of "inference rules." But the point we are making here is independent of this issue.

how polygons are assumed to be positioned, for example. But more importantly for our purpose, the diagram also seems to helps us to *swallow* the demonstration, namely, to see *why the demonstration is valid.*

Imagine the demonstration were done without the diagram. Then, of course, we would have a harder time to simply comprehend the individual steps of the demonstration. But even after we manage to comprehend all this, we still would have a harder time to convince ourselves of the validity of the demonstration. For example, one of the crucial steps in the above demonstration is the move from the first two assumptions to the claim that begins with "Then." My claim implies that, without the diagram, we would not as easily accept this move, even when we fully comprehend what those assumptions and claim mean. That is, the diagram somehow helps us to understand the validity of the following implicit lemma:

If: (θ_5) ABC is a triangle with the right angle B,
(θ_6) triangles CDE, EFG, and GHA are congruent to ABC, and
(θ_7) C is between B and D, E is between F and D, and C is between D and B,

then: (θ_8) $BDFH$ is a square, and
(θ_9) $ACEG$ is a square.

Despite this capacity, the diagram does not seem to present this lemma explicitly. That is, there seems no definite states of affairs holding in the diagram that indicate the constraints $\{\theta_5, \theta_6, \theta_7\} \vdash \theta_8$ and $\{\theta_5, \theta_6, \theta_7\} \vdash \theta_9$ hold in the Hilbert space. The diagram does present the information θ_5, θ_6, θ_7, θ_8, and θ_9 individually. We can identify the states of affairs holding in the diagram that indicate these pieces of information. (See below.) But there seems no states of affairs that indicate the constraints in question. Again, we have a diagram that shows target constraints without presenting them explicitly.

(σ_5) The polygon icon ABC is a triangle* with the right* angle B.

(σ_6) The triangle icons CDE, EFG, and GHA are congruent* to the triangle icon ABC.

(σ_7) The point icon C is between* the point icons B and D, E is between F and D, and C is between D and B.

(σ_8) The polygon icon $BDFH$ is a square*.

(σ_9) The polygon icon $ACEG$ is a square*.[9]

[9]Here the asterisks indicate that the expressions "triangle," "congruent," "between," "square," and "decomposed" are used to describe physical, perceptually accessible properties or relations holding in the diagram, rather than abstract properties and relations holding in the Hilbert space.

We invoke the notion of constraint preservation to account for this capacity of the diagram. Let $\mathcal{G} = \langle \mathbf{S}_g, \mathbf{T}_g, \hat{S}_g, \hat{T}_g, \Rightarrow_g, \rightsquigarrow_g \rangle$ be the system of geometry diagrams underlying the above diagram, where $\mathbf{S}_g = \langle S_g, \Sigma_g, \models_{S_g} \rangle$ and $\mathbf{T}_g = \langle T_g, \Theta_g, \models_{T_g} \rangle$. Then $\sigma_5, \sigma_6, \sigma_7, \sigma_8, \sigma_9 \in \Sigma_g$, and $\theta_5, \theta_6, \theta_7, \theta_8, \theta_9 \in \Theta_g$. Notice that, as a matter of the constraints governing the diagrams in $\mathcal{G}$, σ_8 and σ_9 are consequences of $\{\sigma_5, \sigma_6, \sigma_7\}$, namely, that the constraints $\{\sigma_5, \sigma_6, \sigma_7\} \vdash \sigma_8$ and $\{\sigma_5, \sigma_6, \sigma_7\} \vdash \sigma_9$ hold on the representations of $\mathcal{G}$. But $\{\sigma_5, \sigma_6, \sigma_7\} \Rightarrow_g \{\theta_5, \theta_6, \theta_7\}$, $\sigma_8 \Rightarrow_g \theta_8$, and $\sigma_9 \Rightarrow_g \theta_9$. Thus, we can say that the system $\mathcal{G}$ preserves the target constraints $\{\theta_5, \theta_6, \theta_7\} \vdash \theta_8$ and $\{\theta_5, \theta_6, \theta_7\} \vdash \theta_9$ in the source constraints $\{\sigma_5, \sigma_6, \sigma_7\} \vdash \sigma_8$ and $\{\sigma_5, \sigma_6, \sigma_7\} \vdash \sigma_9$ respectively. (See Figure 6.)

$$\begin{array}{ccc} \{\sigma_5, \sigma_6, \sigma_7\} & \Rightarrow_g & \{\theta_5, \theta_6, \theta_7\} \\ \top & & \top \\ \sigma_8 & \Rightarrow_g & \theta_8 \end{array} \qquad \begin{array}{ccc} \{\sigma_5, \sigma_6, \sigma_7\} & \Rightarrow_g & \{\theta_5, \theta_6, \theta_7\} \\ \top & & \top \\ \sigma_9 & \Rightarrow_g & \theta_9 \end{array}$$

Figure 6

Thus, the diagram in question, which supports $\sigma_5, \sigma_6, \sigma_7$ and therefore σ_8, σ_9, can be said to exhibit the constraints $\{\theta_5, \theta_6, \theta_7\} \vdash \theta_8$ and $\{\theta_5, \theta_6, \theta_7\} \vdash \theta_9$. This fact accounts for the diagram's capacity of showing the lemma without saying it.

Given a representation system, it is determined what target constraints are exhibited by an individual representation of the system. However, the exhibition of a particular constraint may or may not be relevant to the ongoing process of reasoning, and even when it is, it may be so only marginally. For example, suppose the geometry diagram in Figure 5 is now accompanied by the following slightly different demonstration of the Pythagorean Theorem. Let ABC be an arbitrary right triangle with the right angle B. Let $ACGE$ be a square (on the hypothenuse of ABC), and GHA, EFG, and CDE be triangles congruent to ABC. Then A is between B and H, G between H and F, E between F and D, and C between D and B. So, the lengths of BH and DB are all equal to the sum of the lengths of AB and BC. Hence $BHFD$ is a square, which is decomposed into $AGEC$, ABC, GHA, EFG, and CDE. This gives us the same equations that we had in the original demonstration.

This demonstration does not appeal, either implicitly or explicitly, to the lemma we discussed above. Thus, although the diagram still exhibits the target constraints $\{\theta_5, \theta_6, \theta_7\} \vdash \theta_8$ and $\{\theta_5, \theta_6, \theta_7\} \vdash \theta_9$, this fact is not relevant to our understanding of this demonstration's validity. What is relevant in this case is the fact that the diagram exhibits the following constraint (among others):

$$\{\theta_5, \theta_9, \theta_6\} \vdash \theta_7$$

I leave the reader to see how the system $\mathcal{G}$ preserves this constraint, and why the exhibition of this constraint may help us understand the validity of the second demonstration. Note that the exhibition of this constraint was not relevant in our understanding of the first demonstration's validity.

Thus, our account does not commit ourselves to the view that whenever a representation exhibits a target constraint, the exhibition must play a significant role in the given process of reasoning. Although our account describes how a representation exhibits a target constraints, it says nothing about the overall usefulness (or uselessness) of the exhibition. The issue depends on various factors, such as how heavily the overall process of reasoning appeals to the exhibited constraint, and it is simply beyond the scope of our account.

So far, we have only focused on the diagram's function to help us understand the validity of the proofs *after* they are constructed. However, our account also predicts that the diagram should have a heuristic function, to help us *discover* a proof. For, given the diagram's capacity of exhibiting the target constraint $\{\theta_5, \theta_6, \theta_7\} \vdash \theta_8$ and $\{\theta_5, \theta_6, \theta_7\} \vdash \theta_9$ (or $\{\theta_5, \theta_9, \theta_6\} \vdash \theta_7$ in the second case), the diagram may very well help us to *come up with* the lemma supported by these constraints. We can then use this lemma in the construction of a proof either explicitly or implicitly. The diagram can have this heuristic function regardless whether we have constructed it methodically with some provision of a proof, or else accidentally by pure trial and error. For, on our characterization (definition 4), the exhibition of target constraints does not depend on how the diagram is originally drawn. This explains why many of us feel that the discovery of the diagram in Figure 5 is "more than half way through" the discovery of the proof. The same mechanism would also explain our common experiences in geometry, such as that "just adding a single auxiliary line has revealed everything." Unfortunately, we do not have enough space to fully justify these claims. We only propose them as a pointer to further applications of our analysis.

6 Cognitive Considerations

We have claimed that when a representation exhibits a target constraint $\Theta_1 \vdash \theta$ in a source constraint $\Sigma_1 \vdash \sigma$, the user has two alternative ways of recognizing the target constraint: the user may rely on her pre-knowledge about the target domain to directly figure out $\Theta_1 \vdash \theta$, or else she may derive it, via the semantic indication relation $\Rightarrow$, from the source constraint $\Sigma_1 \vdash \sigma$. Our account appeals to the existence of this second process to explain how the representation, which does not present the target constraint explicitly, nevertheless can help us to recognize it.

To complete this account, however, we need discuss this second process more fully, specifying the following points:

(a) How exactly we "derive" the target constraint $\Theta_1 \vdash \theta$ from the source constraint $\Sigma_1 \vdash \sigma$ via the indication relation $\Rightarrow$;

(b) How this indirect process facilitates our recognition of the target constraint *better than* the direct process does.

So far, we have studied particular kinds of representations, and showed what property is responsible for their remarkable function. The focus has been on the tool itself, so to speak. In contrast, (a) and (b) are questions about the user, about the cognitive process in which people exploits the property of the tool in question. Although our primary purpose is the description of the tool, we will now discuss (a) and (b) briefly, if only to sketch a plausible path to the answer.

6.1 Indirect Process of Recognizing Constraints

To answer the question (a), assume that humans have the general ability of what we call "surrogate inference," namely, inference that concerns one domain but relies on a constraint on another domain. We can use our conceptual framework to characterize the process of surrogate inference more specifically. Let $\mathcal{R} = \langle \mathbf{S}, \mathbf{T}, \Rightarrow, \rightsquigarrow, \hat{S}, \hat{T} \rangle$ be a representation system, and suppose $\mathcal{R}$ preserves a target constraint $\Theta_1 \vdash \theta$ in a source constraint $\Sigma_1 \vdash \sigma$. Thus, all the $\Rightarrow$'s and $\top$'s in Figure 4 hold (see section 4). Under these conditions, we may safely make an inference from the information Θ_1 to the information θ on the basis of the source constraint $\Sigma_1 \vdash \sigma$. We call this inference "surrogate" since normally, an inference from Θ to θ should be based on the constraint $\Theta_1 \vdash \theta$ on the target domain $\mathbf{T}$, whereas this inference is based on the constraint $\Sigma_1 \vdash \sigma$ on the source domain $\mathbf{S}$. Here, which surrogate inference on $\mathbf{T}$ is supported by $\Sigma_1 \vdash \sigma$ is determined by the indication relation $\Rightarrow$. This inference is valid since the constraint $\Theta_1 \vdash \theta$ holds on the target domain $\mathbf{T}$. Thus, if for some reason an inference based on $\Sigma_1 \vdash \sigma$ is easier for an agent than an inference based on $\Theta_1 \vdash \theta$, it is natural that she develops the ability of this particular surrogate inference.

Assuming that the agent makes a surrogate inference from Θ_1 to θ on the basis of $\Sigma_1 \vdash \sigma$, how does she proceed to recognize the target constraint $\Theta_1 \vdash \theta$? We will not dwell on this question, since it is a part of the much more general issue about how an inference from the antecedent Δ_1 to the consequent δ gives rise to a recognition of the constraint $\Delta_1 \vdash \delta$. We therefore satisfy ourselves by assuming that generally, the former is a major sub-process of the latter.

More specifically, a cognitive agent derives a target constraint $\Theta_1 \vdash \theta$ from a source constraint $\Sigma_1 \vdash \sigma$ in two steps: (i) make a surrogate inference from Θ_1 to θ on the basis of $\Sigma_1 \vdash \sigma$ and then (ii) recognize the constraint $\Theta_1 \vdash \theta$ with the support of this inference. Just what this "support" amounts to is a part of the general issue about the relationship between inferences and recognitions of constraints.

6.2 Cognitive Significance of the Indirect Process

Now for the issue (b). If an agent p derives $\Theta_1 \vdash \theta$ from $\Sigma_1 \vdash \sigma$ in this way, how does this indirect process facilitate p's recognition of $\Theta_1 \vdash \theta$ *beter than* the direct process does? Should there be no difference in the extents to which the two processes facilitate the recognition, our account could not be an explanation of why the constraint-exhibiting representation could ever be a *help* to p's recognition of $\Theta_1 \vdash \theta$.

In my view, the sheer fact that the indirect process gives an *alternative* way of recognizing $\Theta_1 \vdash \theta$ is already a partial answer to this question. However, we can pin down more acute cognitive significance of this process by appealing to the notion of "attunement to constraints" in situation theory.

When an agent lives in an ecological system in which certain constraints (such as natural laws) prevail, and those constraints have crucial importance to the agent's success and survival as an organism, the agent typically develops an ability of acting upon the constraints spontaneously, in ways to benefit her own survival. Drawing on Gibson's work (1979) on ecological psychology, Barwise and Perry (1983) characterize this ability as the "attunement" to the constraints in question, and claim that attunement is the base for our capacities of inference and other forms of "information extraction." We base our characterization of attunement on Barwise's ideas (1989, pp. 52–53): an agent p is *attuned to a constraint* $\Delta_1 \vdash \delta$ if there is a constraint (on the agent's part) according to which p's having the information Δ_1 entails p's having the information δ. Here, having information is a mental state of the agent, but it may not consist in having a mental representation that somehow presents the information.

Although there could be many alternative ways of characterizing attunement, the following points are in the core of the notion:

- An agent p infers from the information Δ_1 to the information δ more or less immediately, depending upon how strongly p is attuned to the constraint $\Delta_1 \vdash \delta$.

- The strength of p's attunement to $\Delta_1 \vdash \delta$ largely depends on what ecological environment p lives in and how crucial the constraint $\Delta_1 \vdash \delta$ is to p's survival in the environment.

With these ideas in mind, look back to Figure 4, below the definition of constraint exhibition. Suppose, for reasons related to an agent's living environment, she is attuned to the source constraint $\Sigma_1 \vdash \sigma$ more strongly than the target constraint $\Theta_1 \vdash \theta$. Then the representation s may give the agent not only an alternative way of recognizing $\Theta_1 \vdash \theta$, but also an *easier, more immediate* way of recognizing it. For, given the greater degree of attunement to the source constraint, it should be easier to infer upon the source constraint than upon the target constraint, and in some cases, this difference may make even a *surrogate* inference upon the source constraint easier than a direct inference upon the target constraint. Since an inference from Θ_1 to θ is a major sub-process of a recognition of the constraint $\Theta_1 \vdash \theta$, a recognition of the constraint via the surrogate inference may well be easier than a recognition via the direct inference.

Take the case of the stem-and-leaf plot. The plot exhibits the target constraint $\{\theta_2, \theta_3, \theta_4\} \vdash \theta_1$ in the source constraint $\{\sigma_2, \sigma_3, \sigma_4\} \vdash \sigma_1$, namely, in the constraint about *spatial* and *perceptual* properties, concerning spatial components and comparative lengths of the stems on the two-dimensional plot. Since this type of spatio-perceptual constraints are everywhere in our environment, we may presume that many of us have developed a strong attunement to them. In contrast, the target constraint is a constraint about numerical or scalar properties, concerning the heights of the volcanos and the comparison of the numbers of volcanos in various height ranges. Thus, there seems a good reason to believe that humans have developed much stronger attunement to the source constraint than to the target constraint. Indeed, the difference seems great enough to make a recognition of the target constraint easier via a surrogate inference upon the source constraint than via a direct inference upon the target constraint.

Or consider the (first) case of the geometry diagram, where the relevant source constraints are $\{\sigma_5, \sigma_6, \sigma_7\} \vdash \sigma_8$ and $\{\sigma_5, \sigma_6, \sigma_7\} \vdash \sigma_9$. These are again constraints about *spatio-perceptual* properties on a two-dimensional diagram, and we may assume that humans have come to be strongly attuned to them. What about the target constraints $\{\theta_5, \theta_6, \theta_7\} \vdash \theta_8$ and $\{\theta_5, \theta_6, \theta_7\} \vdash \theta_9$? They are constraints upon the two-dimensional Hilbert space, and hence we may perhaps say that they are about *spatial* properties. However, they are not about *perceptual* properties, and unless we are geometricians, we have not developed an attunement to these non-perceptual constraints. This contrast in attunement seems to be strong enough to account for why the recognition of the target constraints can be easier via a surrogate inference than via a direct inference.

7 Conclusion

We have pointed out that there is a kind of representations that show constraints on their targets without explicitly presenting them. To account for this capacity, we invoked the mathematical framework of situation theory and its descendant theory of information, and defined the precise sense in which the *constraint preservation property* of the underlying representation system lets the representations *exhibit* the constraints without *presenting* them. Our analysis, coupled with the notions of surrogate inference and attunement, also gives a plausible account for the cognitive processes in which the user exploits the mechanisms of constraint preservation and exhibition.

Definition 3 in section 4 shows that the constraint preservation property consists in a particular match between the constraints on representations and the constraints on targets. The analysis in this paper therefore lends a support for the main hypothesis of our overall project, namely, the Constraint Hypothesis: the variance in inferential potential of different modes of representation is largely attributable to different ways in which the structural constraints on representations match with the constraints on targets of representation.

References

Barwise, J. (1989) *The Situation in Logic.* Stanford, CA: CSLI Publications.

Barwise, J. (1993) Constraints, Channels, and the Flow of Information. *Situation Theory and Its Applications, Volume 3.* Eds. P. Aczel, D. Israel, Y. Katagiri and S. Peters. Stanford, CA: CSLI Publications.

Barwise, J., Gabbay, D. and Hartonas, C. (1995) On the Logic of Information Flow. *Language, Logic and Computation, Volume 1.* Eds. D. W. Westerståhl and J. Seligman. Stanford, CA: CSLI Publications.

Barwise, J. and Hammer, E. (1995) Diagrams and the Concept of Logical System. *What Is a Logical System?.* Ed. D. Gabbay. Oxford: Oxford University Press.

Barwise, J. and Perry, J. (1983) *Situations and Attitudes.* Cambridge, MA: The MIT Press.

Barwise, J. and Seligman, J. (1996) Information Flow in Distributed Systems. Unpublished.

Barwise, J. and Shimojima, A. (1995) Surrogate Reasoning. *Cognitive Studies: Bulletin of the Japanese Cognitive Science Society* **2**: 7–27.

Bauer, M. I. and Johnson-Laird, P. N. (1993) How Diagrams Can Improve Reasoning. *Psychological Science* **4**: 372-378.

Bertin, J. (1973) *Semiology of Graphics: Diagrams, Networks, Maps.* Translated by William J. Berg. Madison, WI: The University of Wisconsin Press, 1983.

Devlin, K. (1991) *Logic and Information.* Cambridge, UK: Cambridge University Press.

Fisler, K. (1995) Exploiting the Potential of Diagrams in Guiding Hardware Reasoning. *Logical Reasoning with Diagrams.* Eds. G. Allwein and J. Barwise. Oxford: Oxford University Press.

Funt, B. V. (1980) Problem-Solving with Diagrammatic Representations. *Diagrammatic Reasoning: Cognitive and Computational Perspectives.* Eds. J. Glasgow, N. H. Narayanan and B. Chandrasekaran. Menlo Park, CA and Cambridge, MA: AAAI Press and MIT Press, 1995.

Gelernter, H. (1959) Realization of a Geometry-Theorem Proving Machine. *Computers and Thought.* Eds. E. A. Feigenbaum and J. Feldman. New York: McGraw Hill.

Gibson, J. J. (1979) *The Ecological Approach to Visual Perception.* Boston: Houghton Mifflin.

Hammer, E. (1995) *Logic and Visual Information.* Cambridge, UK: Cambridge University Press.

Hayes, P. J. (1974) Some Problems and Non-Problems in Representation Theory. *Readings in Knowledge Representation.* Eds. R. J. Brachman and H. J. Levesque. Los Altos, CA: Morgan Kaufmann Publishers, 1985.

Larkin, J. H. and Simon, H. A. (1987) Why a Diagram Is (Sometimes) Worth Ten Thousand Words. *Diagrammatic Reasoning: Cognitive and Computational Perspectives.* Eds. J. Glasgow, N. H. Narayanan and B. Chandrasekaran. Menlo Park, CA and Cambridge, MA: AAAI Press and MIT Press, 1995.

Levesque, H. J. and Brachman, R. J. (1985) A Fundamental Tradeoff in Knowledge Representation and Reasoning (Revised Version). *Readings in Knowledge Representation.* Eds. R. J. Brachman and H. J. Levesque. Los Altos, CA: Morgan Kaufmann Publishers.

Lindsay, R. K. (1988) Images and Inference. *Diagrammatic Reasoning: Cognitive and Computational Perspectives.* Eds. J. Glasgow, N. H. Narayanan and B. Chandrasekaran. Menlo Park, CA and Cambridge, MA: AAAI Press and MIT Press, 1995.

Luengo, I. (1995) Diagrams in Geometry. Ph.D. Thesis. Department of Philosophy, Indiana University.

Palmer, S. E. (1978) Fundamental Aspects of Cognitive Representation. *Cognition and Categorization.* Eds. E. Rosch and B. B. Lloyd. Hillsdale, NJ: Lawrence Erlbaum Associates, Publishers.

Shimojima, A. (1996a) On the Efficacy of Representation. Ph.D. Thesis. Department of Philosophy, Indiana University.

Shimojima, A. (1996b) Reasoning with Diagrams and Geometrical Constraints. *Language, Logic and Computation, Volume 1.* Eds. D. Westeståhl and J. Seligman. Stanford: CSLI Publications.

Shin, S.-J. (1994) *The Logical Status of Diagrams.* Cambridge, UK: Cambridge University Press.

Sloman, A. (1971) Interactions between Philosophy and AI: the Role of Intuition and Non-Logical Reasoning in Intelligence. *Artificial Intelligence* **2**: 209–225.

Sloman, A. (1975) Afterthoughts on Analogical Representations. *Readings in Knowledge Representation.* Eds. R. J. Brachman and H. J. Levesque. Los Altos, CA: Morgan Kaufmann Publishers, 1985.

Stenning, K. and Oberlander, J. (1995) A Cognitive Theory of Graphical and Linguistic Reasoning: Logic and Implementation. *Cognitive Science* **19**: 97–140.

Tufte, E. R. (1983) *The Visual Display of Quantitative Information.* Cheshire, CN: Graphics Press.

Tufte, E. R. (1990) *Envisioning Information.* Cheshire, CA: Graphics Press.

Tukey, J. W. (1977) *Exploratory Data Analysis.* Reading, MA: Addison-Wesley Publications.

Information, Belief and Causal Role

Paul Skokowski

McDonnell-Pew Centre for Cognitive Neuroscience
Oxford University

Introduction

In an important series of books and papers, Fred Dretske has developed a theory of content to explain how information can be put to work in a system in which internal states indicate outside conditions.[1] The theory formulates an account of belief which combines information-based semantics with a naturalistic approach to learning. This combination allows for an explanation of the fundamental ability of beliefs to misrepresent, or have false content. The theory accounts for the difference between information and representation both causally and in terms of content.

Dretske's theory of belief ultimately depends on the concept of the promotion, also called the selection or recruitment, of an internal indicator type of state to a representational type.[2] Promotion is what allows learning to solve the "Design Problem" of getting a system to do M when and only when an external condition F exists. Indeed, Dretske tells us that an indicator is "... *selected* as a cause of M because of what it indicates about F. Unless this is done, the Design Problem cannot be solved. Learning cannot take place."[3] Promotion, therefore, is a cornerstone of the entire program. But the concept of promotion offered by this theory is flawed by a failure to individuate internal state types by the only tools available: content and causal role. This failure leads to a breakdown of the concept of promotion with the consequence, as we have just seen, that learning cannot take place.

This is a problem. Without learning belief – and in particular the ability of beliefs to misrepresent – can no longer be accounted for. Frankly, I do not wish to undermine the theory of content Dretske proposes. I am convinced

Logic, Language, and Computation, Vol. II, edited by Lawrence S. Moss, Jonathan Ginzburg, and Maarten de Rijke.

that he is on the right track. Therefore I propose that more explanation is needed, and so I view this paper as pinpointing a problem that can be solved with modifications. Indeed, the solutions sketched at the end are consistent with the spirit and framework of Dretske's theory of content. Nevertheless the problems examined here will need to be overcome in order to retain a viable model of the acquisition of representational states through promotion and learning.

I will begin by giving a short introduction to the key points of the theory of promotion and learning. This will include discussions of the causal relationships that internal states enter into with respect to external conditions and subsequent movements. I will then give a simple example to illustrate the causal problems with the theory of promotion, and consider counterarguments to this example. I will introduce the notions of innocent and efficacious properties, and show that Dretske is committed to efficacious properties when identifying internal state types. Finally, I will discuss how the problems raised by this example are due to inconsistent requirements for the causal roles of types of internal states involved in promotion.

In conclusion two compatible approaches will be offered for resolving the problem. The concept of promotion on offer involves tracking *types* of information carrying states through the learning phase. One approach for resolving the problem with promotion is to track concrete objects instead, while another is to hold that a new state must be installed which then plays the right causal role.

1 Background

Central to the account of promotion and learning are the concepts of *indication*, *representation* and *function of indication*, which help describe how beliefs acquire content and get hooked up to motion through *causal* encounters with the environment. I will therefore begin by discussing these concepts before turning to the details of promotion and learning which rely on them.

For Dretske an internal state of an agent indicates an external condition because of a causal regularity: a certain type of external condition F will cause a type of internal state B. The trees in front of me cause a state inside my head about trees. Generally, when a state *indicates* some condition, this means the state carries *information* about that condition. Information is different from representational content in this important way: information is content that is true, whereas representational content may be false.[4] That is, the content of a representational state may mis-represent its surroundings. Beliefs are a paradigm example of representational states. I may believe the Giants are the best team in baseball, while in fact the A's are the best team in baseball. My belief is a state carrying a content, but the content happens to be false. It is a representational state which is managing to *mis*-represent the Giants.

Further, the representational content of a system is that content which the system has the *function* of indicating. A system may indicate, that is, carry information about, many different things; but its representational content is limited to that content which it has the function to indicate. A plane's altimeter carries information about, and therefore indicates, both pressure outside the airplane *and* altitude. Nevertheless the dial only *represents* altitude. That is the function of the altimeter. It has the function of indicating altitude; it does not have the function of indicating pressure.

Beliefs, in this theory, are internal states that can cause motion. My belief that a tree is in front of me causes me to change direction during my stroll. Beliefs are arrived at from the *promotion* or *recruitment* of internal indicators through *learning.* Before learning, a type of internal state *indicates* an external condition of some type, and after learning, my internal state type *represents* that type of condition. Internal states thus acquire the ability to mis-represent due to the learning process, and become full-fledged beliefs. Using Dretske's notation, after learning, state B gets its hands on the steering wheel by becoming a cause of motions M. It becomes an executive state capable of guiding behavior. Learning is a promotion or recruitment of B to cause M in conditions F. Because of these facts, the content of B *explains* the agent's behavior, M.

A brief word on conative states and external contingencies. Usually it takes both a belief state *and* a desire state to cause a motion. I will assume a desire state for a goal exists from now on, and concentrate on beliefs only, for simplicity. This doesn't affect the analysis that follows, since promotion concentrates on representational, not conative, states.[5] I will also assume that the external contingencies required for the promotion of information states to beliefs remain more or less constant throughout the learning process. Note that individual learning results in an *internal* change in an agent rather than a change in some external contingency. I can hit a slice backhand now because I have learned to do so; nothing has changed in the ball, racquet, court, opponent, or other external circumstance that is responsible for the action. (Even my desire to use slices as a strategy needn't change throughout the learning process.) As we've just seen and will soon see again, the changes from learning are exhibited in the internal states of the agent.

The internal indicator states denoted by B are the same type of state before and after learning.[6] Types of states are what are promoted or recruited. Note that individual, particular, states cannot be promoted or recruited in any interesting way. Particular states occur at a time and so cannot re-occur at a later time as the *same* particular. But the same *type* of state can re-occur. I turned left yesterday and again today when an oak was in front of me. But yesterday my belief was caused by the oak in my yard, while today my belief was caused by the oak in the park. The very same particular belief did not cause my movements on both days, though both beliefs were of the same type. So when Dretske talks of promoting or recruiting a state, he means a *type* of state is so promoted or recruited; he does not mean the very same particular state re-occurs.[7]

We could, with Kim, call an individual occurrence of an internal state B an event.[8] Similarly for a particular external condition F or a motion M. I will follow Dretske in calling such individual occurrences *states*. Dretske and Kim also use physical properties to determine a state's type (they refer to the physical property N of a token state as the *type* this token falls under). Think of a state as an object exemplifying a property or properties at a time. Then, given a state with a physical property **A**, I will follow Dretske and Kim and say that the state is of *type* **A**. Thus it is the properties of states which determine their type.[9]

Outside conditions cause internal states; types of internal states get promoted to new roles as beliefs; beliefs cause motions. Causality is a crucial thread which ties this theory of content together, and we need to be clear about the causal roles of internal states throughout the learning process.

Note that pairs of individual states don't constitute nomic dependencies. Saying that an external condition F causes an internal state B just denotes a particular occurrence. The causal relationship, rather, is a dependency we capture by saying that things of type **F** are regularly followed by things of type **B**.[10] It is a relationship between types of objects, not between particular objects at particular times. Though individual occurrences don't constitute nomic dependencies, they may be particular instances of such a dependency (at a time). The lawful character of particular causal occurrences is grounded in a relationship between *types* of states. That is why we continue to use the word cause for individual occurrences. For the purposes of this paper, I will understand causal relationships to be strict, and nomological in character.[11]

2 Promotion and Learning

We are now in a position to understand how a type of internal state gets promoted to new executive capacities by changes to its causal properties through learning. These changes are found when ... individual learning is occurring, places where internal states *acquire* control duties or *change* their effect on motor output ...."[12] So a type of internal state gets harnessed to output by acquiring a new physical property, one which enables it to cause motions appropriate under the outside conditions it is indicating. The state type thereby gets recruited (promoted) for an executive role. This process is illustrated by considering how the type of internal indicator C gets its hand on the steering wheel:

> C is recruited as a cause of M because of what it indicates about F, the conditions on which the success of M depends. Learning of this sort is a way of shaping a structure's causal properties in accordance with its indicator properties. C is, so to speak, selected as a cause of M because of what it indicates about F.... learning of this sort must recruit indicators of F as causes of M.[13]

Internal indicators that are promoted are the same type of state before and after learning. Before learning, these states indicate conditions F, and after learning, they cause motions M (and hence comes to represent F). As Dretske says of a state that has been promoted:

> ... when it represents (say) F, it does so because it is a token of type B, and this type, in virtue of its indicator (informational) properties (its correlation with F) was recruited for control duties (as a cause, say, of movements of type M) because it was an indicator of F. By saying that it – this structure type – was recruited as a cause of M because of what it indicated about F, I simply mean that as learning progresses later tokens of B ... cause M ...[14]

So this state type B started as an indicator, and later acquired an executive capacity and began to cause motions. The model of an indicator state type being promoted is reinforced elsewhere:

> B is the type of physical condition whose correlation with condition (type) F makes tokens of B indicate (carry the information that) F (when they do so) and whose relationship with M (established through learning) makes tokens of B (when circumstances – motivational and otherwise – are right) cause M.[15]

The internal state type B has two important constitutive properties: the semantic property of indicating external conditions *and* the physical property of causing movements. This conjunction of properties of the internal state captures the executive nature of beliefs, and so I will refer to this conjunction as the *executive principle.* In addition, we have seen that the internal indicator state exhibiting these constitutive properties is of the same type before and after learning. I will call this notion the *promotion hypothesis.* We will return to evaluate the executive principle and the promotion hypothesis after examining a specific example of promotion.

3 The Car Axle

Consider the axle of a certain car. Suppose, for simplicity, that this car has only one gear working. By virtue of a direct connection to the car's wheels and connection to the engine via a clutch, the axle carries information. Due to these connections, the axle is a good indicator of the car's speed and of the number of rpm's of the car's engine. There is a nomic relationship between the vehicle's speed and the speed of rotation of the axle, and between the engine's rpm's and this rotation. These nomic relationships are between types of states: rotation and velocity on the one hand, and rotation and rpm's on the other. The spinning axle, therefore, is a state carrying information about vehicle speed and engine turnover.

Now, the types under which these indicator states fall can acquire *functions* of indicating.[16] Note that this is not the same as indication. Types of internal states which have a function of indicating have been selected to indicate what they do. For mechanical systems this requires the intervention of an intelligent agent, while for intelligent agents themselves, acquisition of function occurs through learning. Internal states which have a function of indicating are *executive*, that is, these states help control output in the system of which they are a part. In addition these states have the ability to misrepresent. Our axle can acquire the function to indicate speed by being hooked up via a flywheel and spring to a speedometer. This is the axle's analogue of learning. By being hooked up to a speedometer, the spinning axle is an internal state of a type that has acquired the function of indicating the vehicle's speed. Note that it is the spinning axle and not the speedometer which is playing the analogue role of a belief state here.[17] Internal executive states of a system cause outputs: belief states of an agent cause motions; spinning axle states of our automobile cause speedometer readings.

We have, then, that the spinning axle is of a type that has the function of indicating the car's speed. The type has this function due to a physical hookup with the speedometer. A spinning axle hooked up in this way therefore exemplifies two properties: the property of spinning, and the property of being hooked up to the speedometer. After this hookup, then, a spinning axle occurring causes a motion of the speedometer. Note that in virtue of the hookup, the spinning axle *represents* the vehicle's speed. In Dretske's notation, this would normally be written B(F), that is, the state B represents conditions F; B carries the content that F. The moving axle represents the car's velocity.

Note that B(F) is a relation of representation, rather than pure indication, which gives a fine-grainedness to the content of tokens B. The state of the axle spinning is now about the vehicle's speed rather than about the vehicle's speed *and* engine rpm's. As Dretske points out,

> The specificity of functions to particular properties, even when these properties are related in ways (e.g., by logical or nomological relations) that prevent one's being indicated without the other being indicated, is easy to illustrate with assigned functions, functions we give to instruments and detectors. . . . We can make something into a voltmeter (something having the function of indicating voltage differences) without thereby giving it the function of indicating the amount of current flowing even if, because of constant resistance, these two quantities covary in some lawful way.[18]

The spinning axle has been assigned the function of indicating car speed and not engine rpm's. It therefore represents conditions of type speed and not of type rpm.

Suppose we unhook the speedometer from the axle and hook up a tachometer. It is now clear what happens. By being hooked up to a tachometer, the axle acquires the function of indicating the engine's rate of turnover. The motions of the tachometer are a causal consequence of this hookup with the axle. A spinning axle hooked up in this way therefore exemplifies two properties: the property of spinning, and the property of being hooked up to the tachometer. After this hookup, then, a spinning axle occurring causes a motion of the tachometer. Similarly to what we noted for the speedometer, in virtue of the hookup, the spinning axle state now *represents* the motor's rpm's. Again using the above notation, this would be written B(G), that is, the state B represents conditions G; B carries the content that G. Thus the axle has been assigned the function of indicating engine rpm's and not car speed. It therefore represents conditions of type rpm and not of type speed.

Let's pause now, and take stock of the situation. First, we have seen from the previous section that pre-learning indicator states are of the same type as post-learning belief states. Second, we have seen two analogues of a learning situation. In both of these examples, types of states get functions of indicating. In the first case, spinning axle states are of a type that acquire the function of indicating conditions of vehicular speed, and in the second case, spinning axle states are of a type that acquire the function of indicating states of engine rpm. We have then, that the pre-learning spinning axle state is of the same type as the post-learning state which has the function of indicating vehicular speed. We also have that the pre-learning spinning axle state is of the same type as the post-learning state which has the function of indicating engine rpm. So the post- learning state indicating speed is of the same type as the state indicating rpm's.

But this conclusion cannot be correct. Compare instances of the two tokens along with their representational contents: the first is a representation that F, whereas the second is a representation that G. It is the nature of representations to have fine-grained contents. In this case the contents are not only fine-grained, but they also happen to differ. Different functions of indicating lead to different contents; it is this very fine-grainedness of functionality that leads to the fine- grained differences in contents.[19] Representational relations such as beliefs are different from informational relations. A belief that the engine is revving at 4000 rmp is different from a belief that the car is going 47 mph, simply by virtue of differences in content, even though the car may be going 47 mph at the same time it is revving at 4000 rpm and there may be a nomic relation connecting that speed with that rpm. If beliefs are to be type-individuated by their contents, then we have no choice but to say that beliefs that differ in content differ in type.

But that is not all. The conclusion is wrong for another reason: the two states cause different types of motions. We know that one spinning axle state causes a motion of the speedometer, and that another state causes a motion of the tachometer. In this way, the two output instruments will generally differ not only numerically but also structurally. For example, the speedometer may be an analogue mechanical pointer display, whereas the tachometer may be an digital electronic display. States of these output devices have radically different constituent physical properties, and hence are of different types. The two post-learning representational states (beliefs) cause different types of motions, and they do this because the causal relationships in effect are between different types of states.

Hence the two types of beliefs cause different types of motions. This is not surprising since it is by hooking an indicator state up to a type of movement that eliminates its indeterminacy of function; that is, it is this very causal connection that confers a function of indicating upon the type of state in question, thereby making states of that type into representations with a fine-grained content.[20] Thus beliefs that cause one type of motion are a different type from beliefs that cause another type of motion. If beliefs are to be type- individuated by their causal consequences, then we have no choice but to say that beliefs that differ in their consequences differ in type.

4 An Objection

One might object to the above by claiming that the sort of thing described in the car axle example just doesn't happen. One function of indicating isn't lost by a system when another is gained. And since there is no loss of function, the same type of indicator state will remain in effect even after promotion and learning. And what is accomplished by this is that the representational state that results is a single one with the function of indicating *both* conditions. Thus, the true analogue to belief in agents isn't the one laid out in the speedometer/tachometer example above. Rather, it is one where the speedometer connection is left intact when the tachometer connection is made. That is, the state to consider is not either connection in isolation, but rather the complex state of both devices being hooked simultaneously. This latter state now has the function to indicate both vehicular speed and engine rpm's.

A reply to this is that the car axle has just given an example of successive and different types of belief states arising from the same type of indicator state. This sort of thing can indeed happen, at least for automobiles. Of course I can disconnect the tachometer when I connect the speedometer. Tachometers may bore me. My axle therefore does lose the function to indicate speed when it acquires the function to indicate rpm's.

The same change in function of indication can occur in agents. Consider John Major's internal Maggie-Detector, which initially indicates that Thatcher is both a potential ally *and* a serious threat. After Major learns to be a smooth politician, his Maggie-Detector acquires the function of indicating that Maggie is a potential ally, and causes appropriate motions such as giving speeches supporting her. But after Maggie critizes Major on his performance as Prime Minister, Major has gone through a different learning situation. Now whenever his Maggie-Detector lights up, he has a *new* belief that Maggie is a serious threat, which produces appropriate motions such as giving speeches denouncing her views as outdated and irrelevant. This is a case of two distinct types of beliefs being promoted from the same indicator state, and in the process, one function of indicating is lost when another is gained. This results in different types of internal indicator states before and after learning.

The same thing happens for car axles. Different selection processes, or learning situations, were executed by the designer in order to 'recruit' the two different connections. I hooked up the speedometer because I required a representation of the speed. I hooked up the tachometer because I required a representation of engine turnover. I did not choose that particular tachometer connection because I wanted a representation of speed *and* rpm's. I wanted a representation of rpm's. Remember the fine-grainedness of representational content. This has its origin in the assignment of function to a type of indicator state. Representations of conditions of one type may be established independently of representations of conditions of another type, as was the case for our automobile. The second connection was added in order to obtain a particular representational content. It was not added in order to obtain a dual representation. After all, I can consider the content represented by one dial in isolation from the other. I don't have to consider both together just because there are two connections.

A second reply is that the executive nature of these representational states ensures that they make the system of which they are a part behave differently in different situations. The different states have different remote effects. In this way the movements of agents are akin to the readings on the car's two dials, the speedometer and tachometer: different movements (readings) issue from different beliefs (hookups with the axle). I learn that yucca plants are spiny from painful experience. I learn from my botany class that yuccas are succulents. Assume that there is a nomic relation between yucca plants being in front of me, and a certain yucca-style neural firing in my visual cortex when I fixate on one. Then the neural firing type indicates yuccas in front of me, in virtue of the regularity. Believing (or perceiving) spininess makes me behave in certain ways, and believing (or perceiving) succulence makes me behave in other, distinct, ways. This is so because we move for reasons, in order to achieve goals.[21] We behave differently with respect to objects depending on which of their properties best allows us to achieve those goals. Spininess has nothing to do with my digging up a yucca to achieve the goal to bring home a succulent plant. Succulence has nothing to do with running around the yucca to achieve my goal of avoiding getting impaled during the cross-country race. Because of this difference in output effects, beliefs with differing contents remain of distinct types even when they share the common origin of being *learned* from a single type of indicator state.

Finally, note that it is by having belief states hooked to motion that the types to which these states belong acquire their function of indicating. This resolves the indeterminacy of function indicator state types otherwise have.[22] Our axle originally indicates both **F**'s and **G**'s. According to the above prescription, by hooking up the speedometer we remove this indeterminacy. But then, if we accept that by hooking up the tachometer we give the axle the function to indicate **F**'s *and* **G**'s, then we give the axle back this very indeterminacy of function we set out to overcome in the first place by giving it representational abilities.

5 Learning Difficulties

The problems raised by the car axle example point to an inconsistency between the executive principle and the promotion hypothesis. Recall that the promotion hypothesis says that internal indicator states are the same type before and after learning. The executive principle says that internal state types indicate external conditions *and* cause certain types of motions. But causal relations between states are nomic relations: if states of one type cause states of another type, then they always do so. Before learning, internal indicator states do *no* cause motions of a given type. After learning and recruitment, the *same* type of state *does* cause motions of that type. Hence, internal indicator states are *not* of the same type before and after learning. One type does not cause motion; the other does. We thus have that internal indicator states both *are and are not of the same type* before and after learning. This conclusion can't be right. Hence, the executive principle and the promotion hypothesis are inconsistent.

This argument shows that the notion of promotion or recruitment is in trouble if the executive principle and the promotion hypothesis are both retained. Since internal (belief) states which indicate external conditions are taken to be of the same type both before and after learning, then whenever such a state occurs, a motion follows – even *before* learning. This brings into question what role recruitment could *ever* have, since internal indicators will always cause the appropriate motion under conditions F. What needs to be learned?

A central reason for these problems is that the constitutive properties of internal indicator states are real physical properties with causal consequences. Instances of these states presumably don't cause a motion before learning (they must be *recruited* to do this), therefore they don't have those crucial physical properties required for them to cause motions. Thus, before indicator states acquire causal efficacy they are of a different type than those which occur after learning, for the reason that the former lack the constitutive physical properties required for causing motions.

But now here is another problem. According to this model, after learning, indicator states are not only of a different *causal* type than before learning, but these states are no longer pure indicators: promotion has resulted in belief states. Beliefs are fallible, and so they can not be *strict* indicators since their contents may be false. And this means that they can no longer satisfy the causal relationship that has so far held for strict indicators, viz., that external conditions of a given type will cause internal indicator states of a certain type. After all, it can no longer be true for representational states, as opposed to informational states, that such a law holds for them. My long lost brother now stands in front of me, but I don't believe my eyes. This is the nature of belief. Hence internal states will also differ in type with respect to content: before learning, these states carried informational content; after learning, they carry fallible content. These dilemmas show there are problems with the theory of content we have been considering. It appears that learning cannot promote types of states from not being causes into being causes of motion, for, as a consequence of learning, internal states begin to play a different causal role – and that means either that one type of state disappears when another appears, or that a new type is created which is different from the first type. Perhaps, then, what learning does is replace one type of indicator with another type of indicator. But this isn't quite right either, for the latter type is a representational type, not an indicator type, according to the theory.

6 Innocent and Efficacious Properties

It could be replied that there is a level at which the states do not change type after promotion. Rachel has a can of red playdough. When she takes it out of its container, it is cylindrical. After crafting by the toddler it becomes a (nearly) perfect sphere. Throughout, however, it has remained red. These shapes, which are physical properties, have different causal consequences: one type (cylindrical) can roll only along one axis, while the other type (spherical) rolls along any axis. They were also caused by different causal (nomic) processes: the cylindrical shape was caused by a container, while the spherical shape was caused by pressing between cupped hands.

We can think of Rachel as taking the red cylinder and *promoting* or *selecting* it to a red sphere. Being cylindrical and being spherical are two different properties that enter into different causal relations. Before promotion the red cylinder only rolled along one axis. After promotion the red sphere could roll along any axis, and could be used in new ways; it had a new property. But it was the red object that was promoted. The object is of type RED throughout its acquiring different physical properties. Surely this is the sense in which Dretske means an object remains the same type before and after promotion and learning.

Let me call this the *innocent* intrerpretation of type identification, and call a property such as RED an *innocent property*. I call RED innocent because this property does not enter into the causal relations which signify instances of promotion. RED isn't caused by a container or by a hand as cylindricality and sphericity are, and RED doesn't enter into the degrees of freedom for the motions of cylinders and spheres. Let me call *efficacious properties* those properties that can enter into causal relations – in this example, properties such as shapes. Likewise, types will be innocent or efficacious, depending on the properties which determine the types. Therefore by holding that internal states do not change types upon promotion, Dretske must be referring to innocent properties and types in his theory of learning.

But this theory of learning does not choose innocent properties and types for the internal states promoted during learning. It explicitly chooses properties which indicate, and therefore are caused by, outside conditions, and other executive properties that *cause* motions. Recall that, before learning an internal state type **B** indicates (and hence is caused by) external conditions of type **F**, but doesn't cause motions of type **M**; and that after learning, the same type **B** represents **F**, and now causes motions **M**. Properties such as **B** are not innocent properties; they are efficacious properties: properties that enter into causal relations. And again, since **B** has different contents and different causal consequences before and after learning, it cannot remain of the same type before and after learning. Efficacious properties are individuated by causal relations; hence states with efficacious properties that enter into different nomic relations are of different types. Innocent properties do not figure in determining the types of states promoted during learning.

7 Conclusions

We learned from the car axle example that different types of representational states may indeed issue from a single type of indicator state, and these new states are typed according to their various causal roles. The conclusion is that these fine-grained beliefs are distinct in content and causal role from the original internal indicator state. We have also seen that Dretske's theory types internal states by their efficacious properties, and not their innocent properties. Thus the promotion hypothesis must be rejected, since we have seen that indicator states must indeed change type as a consequence of promotion and learning.

A more general conclusion, however, is that mental states need to be typed strictly according to their constitutive properties. Goldman has made this claim for acts, as has Kim for events. Both Goldman and Kim are strict about types in this way because without being careful about constitutive properties, their acts/events will have untoward causal consequences. It is important for Goldman to distinguish between the acts John's pulling the trigger and John's killing Smith because the two acts have different causal consequences: John's pulling the trigger causes the gun to fire, whereas John's killing Smith does not.[23] Kim's strict typing of events by their constitutive properties is also designed to avoid such difficulties.[24]

The moral for the theory of content we have been considering is the same: one needs to be careful about the typing of mental states. *Causation* is the crucial link throughout the entire analysis: it underlies the pickup and utilization of information, and is responsible for ensuing behavior. That is why the concept of promotion flaws Dretske's theory of learning by not adequately distinguishing the causal roles of indicator and representational states.

There are three related lessons to be learned from this analysis. First, pre-learning indicator states *differ in type* from post- learning belief states which succeed them. Second, these two types of states differ both in virtue of their contents, which they acquire due to causal relations with outside conditions, and in virtue of their abilities to cause further, succeeding, states (movements). Thus it is causal relationships, *not instances of promotion*, that are the final arbiters when determining types of mental states. By examining the causal relationships that mental states enter into with respect to outside conditions and subsequent movements, we may determine the types of these states. Third, because promotion cannot work in the way proposed, Dretske's theory of representation and learning is weakened. But I believe there are modifications available to restore the theory of representational content.

I see two compatible approaches towards formulating a concept of promotion which can work within Dretske's theory of representation. The first approach is to hold that it is concrete objects, rather than *types* of states, which get promoted to new causal roles.[25] These concrete objects will reside inside the head, and must exemplify the very physical *properties* which determine their mental type (indicator vs. representational) and their causal role. Thus the concrete object in question remains the same, and so may be tracked through instances of promotion, but will acquire different properties over time. This approach accepts that the type of state carrying representational content differs from the original indicator state type.

The second approach is to hold that the internal state type remains the same, while, through learning, a connecting state is formed or installed in such a way that when the indicator state is activated, then the connecting state causes an output motion to proceed as required by the theory. The problematic burden of causing motion that the original theory required of the indicator state is therefore taken on by the new, connecting state. This is I think the most promising avenue, in part because it has (some) biological plausibility. Let me explain. Suppose we accept, by way of an example, that we have some sort of chair-indicator which fires in our brain when we visually fixate on a chair. The causal process might look something like this: we look at a chair, an upside down image of it is projected onto the retina, the retinal cells fire in a pattern characteristic of chairs, and this causes, by way of the optic nerve, the Lateral Geniculate Nucleus, and other connections, a characteristic pattern of activation (chair indicator B) in certain neurons in the visual cortex. So much for the hardwiring we normally inherit at birth. Assume this process occurs both for infants, who don't yet have beliefs about chairs, and for adults, who do.

The biological plausibility arises through neural changes which are a direct effect of learning, such as Long Term Potentiation, or LTP. LTP appears to be a chemical process which allows changes to our neural substrate as a consequence of learning.26 In particular, learning can induce, via mechanisms of LTP, connections between sensory neurons (which indicate outside conditions) and motor neurons (which cause motions).27 View learning then as installing a new matrix of connections, call it W, between sensory and motor neurons.28 W becomes a permanent part of the structure of our brain. Now when our internal indicator B goes off, W is in place, and the two states together, B and W, cause a motion M. Notice that the causal burden has been taken off of B, allowing it to be promoted in the sense Dretske's theory would like. Before learning B didn't cause anything. Learning then installs W. B and W together now cause M. B hasn't obtained a new causal property because if you take away the separate state W after learning (chemically or surgically alter the synaptic connections) then B alone won't cause anything. But it will continue to be a chair indicator.

Here, then, is a way that a type of state can be promoted into a representational state. The difference is that learning installs a new state W to enable the promotion of B. B still becomes executive, since after learning, when B occurs, M will ensue. B also has the capacity for misrepresentation that Dretske's theory requires: if B fires randomly, or by accident after learning, it now will cause motions M, even when F is not around. This is a form of misrepresentation that is not possible for a pure indicator, because indicators cannot cause inappropriate motions when they indicate, as they are causally inert.

Both of these approaches, I believe, show some promise in accounting for the changes in content and causal role that Dretske's theory requires for representational states. The latter one, however, can maintain the indicator and representational properties of a single type that is promoted, while avoiding the difficulties of giving that same type a new causal role.

NOTES

Work on this paper was made possible by a McDonnell-Pew Fellowship from Oxford University. I would like to thank John Perry for critical comments and sage guidance on earlier versions of this paper, and Martin Davies for a particularly helpful piece of advice.

[1]See, for example, Dretske (1986), (1988a), (1988b), (1989), (1990a), (1990b), and (1991).

[2]Dretske (1991, pp. 214-216) is explicit that it is types of states that get promoted. Not individual states. Also note that promotion, recruitment and selection of types are equivalent denotations in his analysis.

[3]Dretske (1988, p. 101).

[4]Dretske (1981), (1988), and Skokowski (1994).

[5]This simplification follows the development in ch. 4 of Dretske (1988).

[6]Dretske (1991, pp. 214-216) is explicit that it is types of states that get promoted, and that type stays unchanged as a consequence of promotion. Also see Section 2 immediately below.

[7]Dretske (1991), pp. 214-216.

[8]Kim's (1973) gives a clear discussion of events and causal relations between types of events.

[9]This usage follows Dretske (1988), (1991), Kim (1973), (1991), and Goldman (1970).

[10]Causal correlations are between types (properties), not tokens. See Dretske (1991, pp. 214-215) and Kim(1973)

[11]Probabilistic accounts of action are definitely interesting, but they do not figure in Dretske's analysis. Hence, I will not consider them in this paper.

[12]See Dretske (1988, p. 95).

[13]See Dretske (1988, p. 101).

[14]See Dretske (1991, p. 215). I have substituted my notation of types (B and F) for his type notation.

[15]See Dretske (1991, p. 214, 215). I have substituted my notation of types (B and F) for his type notation.

[16]Dretske (1988).

[17]This distinction is made explicit in Dretske (1988, p. 105), when indicator functions of thermostats are discussed. The movement of the bi-metallic strip plays a "purely cognitive" role within the thermostat. As such it is an internal state which causes outputs. The subsequent states it causes, such as the shutting off of the furnace, or the movement of the temperature readout dial, are outputs, or motions, resulting from this 'cognitive' state, and are not to be confused with the cognitive state itself. In my example, the axle plays the "purely cognitive" role of an internal state.

[18]See Dretske (1988, p. 76).

[19]See Dretske (1988, p. 76, 77).
[20]See Dretske (1988, p. 70 and Chapter 4).
[21]See Davidson (1980), and Dretske (1988) and (1989).
[22]Ibid.
[23]See Goldman (1970, p. 2).
[24]Kim (1970, p. 227).
[25]For automobiles, the car axle may serve as the internal concrete object, whereas for agents, a bundle of neurons may do the trick.
[26]Thompson (1986), and Cotman and Lynch (1989).
[27]Greenough, W., Larson, J., and Withers, G. (1985).
[28]Skokowski (1992).

8 REFERENCES

Armstrong, D.M. 1978: A Theory of Universals. Cambridge: Cambridge University Press.

Cotman, C. and Lynch, G. 1989: The Neurobiology of Learning and Memory, Cognition, Vol 33.

Davidson, D. 1980, "Actions, Reasons and Causes", in Donald Davidson, Essays on Actions and Events, New York: Oxford University Press.

Dretske, F. 1981: Knowledge and the Flow of Information. Cambridge, MA: MIT Press.

Dretske, F. 1986: "Misrepresentation", in Bogdan, R. (ed.), Belief. Oxford: Oxford University Press.

Dretske, F. 1988a: "The Explanatory Role of Content". Contents of Thought: Proceedings of the 1985 Oberlin Colloquium in Philosophy. Tucson, AZ: University of Arizona Press.

Dretske, F. 1988b: Explaining Behavior. Cambridge, MA: MIT Press.

Dretske, F. 1989: "Reasons and Causes", in Tomberlin, J. (ed.), Philosophical Perspectives, vol. 3, Philosophy of Mind and Action Theory. Atascadero, CA: Ridgeview Publishing.

Dretske, F. 1990a: "Does Meaning Matter", in Villanueva, R. (ed.), Information, Semantics and Epistemology. Oxford: Basil Blackwell.

Dretske, F. 1990b: "Putting Information to Work", in Hanson, R. (ed.), Information, Language, and Cognition. Vancouver, B.C.: University of British Columbia Press.

Dretske, F. 1991: "Replies to Critics", in McLaughlin, B. (ed.), Dretske and his Critics. Cambridge, MA: Basil Blackwell.

Goldman, A. 1970: A Theory of Human Action. Englewood Cliffs, NJ: Prentice-Hall.

Greenough, W., Larson, J., and Withers, G. 1986: Effects of Unilateral and Bilateral Training in a Reaching Task on Dendritic Branching of Neurons in the Rat Motor-Sensory Forelimb Cortex, Behavioral Neural Biology, Vol. 44, pp. 301-314.

Kim, J. 1973: "Causation, Nomic Subsumption, and the Concept of Event". Journal of Philosophy, 70, No. 8, pp. 217-236.

Skokowski, P. 1994: "Can Computers Carry Content Inexplicitly?". Minds and Machines, 4, No. 3, pp. 333-344.

Skokowski, P. 1992: From Neural Networks to Human Agents. Ph.D. Dissertation, Stanford University.

Thompson, R.F. 1986, The Neurobiology of Learning and Memory, Science, Vol. 233, p. 941-947.

Topology via Constructive Logic

Steven Vickers

Department of Computing
Imperial College of Science, Technology and Medicine

Abstract By working constructively in the sense of geometric logic, topology can be hidden. This applies also to toposes as generalized topological spaces.

1 Introduction

One aim of my book *Topology via Logic* [6] was to describe the use of toplogy in the denotational semantics of computer programming languages, explaining the topology through a logic of observations that describes a computer program by "what you can observe just by using it". Surprisingly, however, a mathematical structure introduced there, the so-called *topological systems*, found parallels in the work of situation theory, suggesting that the observational analysis has wider applicability. I say this merely to justify the presence of this article in this volume, for its content is more technical.

Specifically, I wish to show how, once one has accepted the desirability of topology and continuity, its use can be *simplified* by working within the constraints of constructive mathematics. Thus I am trying to sell constructivity not as a piece of dogma but for its practical usefulness.

There are various flavours or schools of constructivism, and the one I shall describe is the "geometric" (the name derives from historical roots in algebraic geometry rather than from any geometry evident in its use). Geometric logic is — essentially — described in [2] and [5] (amongst others), but the reader should beware of the terminology. Geometric logic is generally understood to include infinitary disjunctions, but Mac Lane and Moerdijk restrict their definition to the fragment in which all disjunctions are finitary. This is usually called *coherent*. Note that the classical completeness result (Corollary X.7.2 in Mac Lane and Moerdijk, an application of Deligne's theorem) holds only for coherent theories.

Logic, Language, and Computation, Vol. II, edited by Lawrence S. Moss, Jonathan Ginzburg, and Maarten de Rijke.

I intend to be brief, so I shall assume that the reader already has at least some aquaintance with the first few chapters of [6].

2 Observational Logic

Though it is easy to think of classical logic as a universal language of statements, in any given context it is reasonable to ask whether all its connectives are equally meaningful. Specifically, it is argued in [6] that if formulae represent (finite) observations, then the only reasonable connectives are conjunction and disjunction (and **true** and **false**); but that infinitary disjunctions are also reasonable.

The corresponding logic is (propositional) *geometric* logic. However, the definition of geometric theories is slightly surprising, for the extralogical axioms of a theory are more general in form than just formulae.

A propositional geometric theory is defined by —

- a set of propositional sysmbols
- a set of axioms of the form $\phi \vdash \psi$, where ϕ and ψ are formulae built up from the primitive symbols using the geometric connectives $\wedge$, **true**, $\bigvee$ and **false** (the big $\bigvee$ is intended to indicate arbitrary disjunctions, possibly infinitary).

(Note that negation can be expressed only to a limited extent — negated *formulae* appear as *axioms* $\phi \vdash$ **false**.) The observational intuition is that formulae represent observations, while axioms — how observations relate to each other — represent scientific hypotheses or background assumptions.

Although geometric theories are not mentioned as such in [6], it is evident that they are equivalent to presentations of frames by generators and relations: the propositional symbols are the generators, and the axioms the relations. The frame $\mathrm{Fr}\langle T\rangle$ corresponding to a theory T should be thought of as the "Lindenbaum algebra" for T — the algebra of formulae modulo equivalence.

Let us write $\Omega[T]$ for this frame, so that the corresponding locale is written [T]. What are its points? They are the frame homomorphisms from $\mathrm{Fr}\langle T\rangle$ to Ω (the frame of truth values — classically, $\Omega = \{\mathbf{false}, \mathbf{true}\}$). The universal property of "presenting by generators and relations" says that such a homomorphism is equivalent to a function assigning truth values to the propositional symbols of T, in such a way that the axioms are respected; but that is exactly a model of T: the points of $[T]$ are equivalent to models of T. We shall try to develop the idea that a locale is "the space of models" for a geometric theory.

3 An example

An example given in [6] is that of bitstreams — finite (unterminated) or infinite sequences of zeros and ones. A first observational theory Th_1 takes propositional symbols of the form $[s_n = x]$ where n is an element of the set $\mathbf{N}$ of natural numbers and x is an element of $2 = \{0, 1\}$. (This is a symbol *schema*, describing an $(\mathbf{N} \times 2)$-indexed family of symbols.) The axioms are given by schemas

$[s_n = 0] \wedge [s_n = 1] \vdash \mathbf{false}$

$[s_{n+1} = 0] \vee [s_{n+1} = 1] \vdash [s_n = 0] \vee [s_n = 1]$

Let us immediately analyse the models of this theory. A model interprets each propositional symbol as a truth value, and hence corresponds to a set

$s = \{(n, x) \in \mathbf{N} \times 2 : [s_n = x] \text{ is interpreted as } \mathbf{true}\}$

In addition, the axioms must be respected: so $(n, 0)$ and $(n, 1)$ cannot both be in s — this says just that s is a partial function from $\mathbf{N}$ to 2. Moreover, by the other axiom the domain of definition of s is an initial segment of $\mathbf{N}$ (if it contains n then it also contains all natural numbers less than n). So the points of $[Th_1]$ are equivalent to the partial functions from $\mathbf{N}$ to 2, with domain of definition an initial segment.

An equivalent formulation Th_2 was also given, with propositional symbols **starts** l for l in 2^* (i.e. l a *finite* sequence of elements of 2) and axioms

starts $l \vdash$ **starts** m if $m \sqsubseteq l$ (i.e. if m is a prefix of l)

starts $l \wedge$ **starts** $m \vdash$ **false** if neither l nor m prefixes the other

By a similar analysis, a model of this is equivalent to an *ideal* of 2^*, a subset of 2^* that is lower closed (under the prefix ordering $\sqsubseteq$) and in which any finite subset $S \subseteq_{\text{fin}} I$ has an upper bound in I. So the points of $[Th_2]$ are equivalent to the ideals of 2^*.

It was left as an exercise in [6] to show that the two theories are equivalent — they have isomorphic Lindenbaum frames; we shall look at part of this from a slightly different perspective.

Let us define a transformation F, transforming models of Th_2 (ideals) to models of Th_1 (partial functions on $\mathbf{N}$):

$F(I) = \{(n, x) \in \mathbf{N} \times 2 : \exists l \in 2^*.(\#l = n \wedge l{+}{+}[x] \in I)\}$

($\#$ is the length function on finite lists, $++$ is concatenation, $[-]$ constructs singleton lists. We are assuming that the natural numbers n start with 0.) It is easy enough to see that $F(I)$ is indeed a partial function whose domain of definition is an initial segment.

These ideals and partial functions were only convenient representations of more strictly defined models of propositional theories; let us examine how the construction works on these.

An ideal I corresponds to a 2^*-indexed family of truth values $[\![\mathbf{starts}\ l]\!]$ (i.e. the truth values of the formulae $l \in I$). In the corresponding model of Th_1, $[s_n = x]$ gets the value **true** iff $(n, x) \in F(I)$, and it follows that each $[s_n = x]$ gets the truth value

$\bigvee\{[\![\mathbf{starts}\ (l{+}{+}[x])]\!] : \#l = n\}$

This is a geometric combination of the given truth values, so we can say (at least at the propositional level) that we have a gometric construction of models of Th_1 out of models for Th_2.

Now let us look at something closer to [6], namely the inverse images under F. A proposition $[s_n = x]$ can be viewed as a collection of models of Th_1, namely those for which $[s_n = x]$ is interpreted as **true**. Consider its inverse image under F:

$I \in F^{-1}([s_n = x]) - F(I) \in [s_n = x] - (n, x) \in F(I)$

$-$ for some $l \in 2^*, \#l = n$ and $l{+}{+}[x] \in I$

$- \ I \in \bigvee\{\mathbf{starts}\ (l{+}{+}[x]) : \#l = n\}$

We thus see essentially the same formula used to calculate both inverse images of propositions and direct images of models.

Once we have constructed the inverse images of the primitive formulae $[s_n = x]$, we know those of the more general ones — for inverse image preserves unions and intersections. We find that we get (according to the techniques of [6]) a frame homomorphism from $\Omega[Th_1]$ to $\Omega[Th_2]$, in other words a continuous map from the locale $[Th_2]$ to $[Th_1]$. What this suggests is a connection between, on the one hand, continuous maps between locales (i.e. — by the usual definition — frame homomorphisms going backwards), and, on the other, geometric transformations of models. This is quite general.

Theorem 3.1 *Let T and U be two propositional geometric theories. Then the following are equivalent:*

1. *frame homomorphisms from $\Omega[U]$ to $\Omega[T]$*

2. *geometric transformations of models of T into models of U*

Proof (sketch) The key is that geometric constructions do not rely on classical logic (without negation, excluded middle is not even expressible, let alone assumed), and the frame $\Omega[T]$ can be viewed as a non-classical algebra of truth values. In such an algebra, we can seek non-classical models of propositional geometric theories. In particular, the symbols of T have their obvious interpretation in $\Omega[T]$, and this interpretation respects the axioms of T (it is forced to by the very construction of $\Omega[T]$): this gives a "generic" model of T in $\Omega[T]$, and everything else in $\Omega[T]$ is constructed geometrically from it. In these terms, a frame homomorphism from $\Omega[U]$ to $\Omega[T]$ is just a model of U in $\Omega[T]$, and so a model of U constructed geometrically from the generic model of T. But being a generic model means that it has no properties whatsoever other than those that follow from being a model of T. It follows that any construction on the generic model can be specialized to any specific model. Hence a geometric construction of a model of U from the generic model of T is equivalent to a geometric transformation of arbitrary models of T into models of U. □

Slogan: continuity = geometricity

4 Predicate geometric theories

The models of Th_1 and Th_2 were most naturally expressed as models of *predicate* theories (many-sorted, first order):

For Th_1, we use a binary predicate $s(n,x)$ ($n : \mathbf{N}$, $x : 2$) with axioms —
$s(n,x) \wedge s(n,y) \vdash x = y$
$s(n+1,x) \vdash \exists y.s(n,y)$

For Th_2, we use a unary predicate $starts(l)$ ($l : 2^*$) with axioms —
$starts(l) \wedge m \sqsubseteq l \vdash starts(m)$
$\mathbf{true} \vdash starts(\epsilon)$ (ϵ here is the empty list)
$starts(l) \wedge starts(m) \vdash \exists n.(starts(n) \wedge l \sqsubseteq n \wedge m \sqsubseteq n)$
(We have directly formulated the ideal condition.)

The geometric theories as defined were propositional, but can the ideas be extended to predicate theories? In fact, predicate geometric theories are well known. The connectives for formulae include not only $\wedge$ and $\bigvee$, but also $=$ and $\exists$; then a predicate geometric theory comprises —

- sorts
- function and predicate symbols, each with declared arity (number and sorts of arguments and result)
- axioms of the form $\phi \vdash_S \psi$ where ϕ and ψ are formulae, constructed from the symbols using the geometric connectives, and whose free variables are all taken from the finite set S (see [4], p.245).

This looks insufficient for our predicate theories for Th_1 and Th_2 — there is much that appears extralogical, such as $\mathbf{N}$, $+$, 2^*, #, ++, etc. However, these can all be characterized uniquely up to isomorphism by geometric theory, so the theories as given can be augmented by extra vocabulary and axioms to define these symbols. For instance, for $\mathbf{N}$ and $+$, use $0 : \mathbf{N}, s : \mathbf{N} \to \mathbf{N}$ with axioms
$s(x) = 0 \vdash_{\{x\}} \mathbf{false}$
$s(x) = s(y) \vdash_{\{x,y\}} x = y$

$\mathbf{true} \vdash_{\{x\}} \bigvee_n x = s^n(0)$

(This last one has the air of cheating — it presumes an external $\mathbf{N}$ to index the infinite disjunction. However, there has to be some kind of trick at this point, for Gödel's theorem tells us that $\mathbf{N}$ can't be characterized in finitary first order logic. We shall return to this point later.) Then $+ : \mathbf{N} \times \mathbf{N} \to \mathbf{N}$ is characterized by

$\mathbf{true} \vdash_{\{y\}} 0 + y = y$

$\mathbf{true} \vdash_{\{x,y\}} s(x) + y = s(x + y)$

The effect is to give a collection of geometric constructions that can be used within the theories. These include —

- Cartesian product
- Disjoint union
- Equalizers and coequalizers (quotients)
- Free algebra constructions (e.g. $\mathbf{N}$ and list types X^*)
- Recursively defined functions
- Finite powersets $\mathcal{F}X$ (isomorphic to free semilattices)
- Universal quantification bounded over finite sets: $\forall x \in S.P(x)$, where $S : \mathcal{F}X$

However, they do not include exponentiation X^Y or full power sets $\mathcal{P}X$ (the logic is weak second order). Technically, the "geometric constructions" are those that are preserved by the inverse image functors of geometric morphisms between toposes.

It now turns out that the transformation F of models (Th_2 to Th_1) of the *predicate* theories can be expressed geometrically. So also can the corresponding inverse transformation G from Th_1 to Th_2, and the isomorphisms $s \cong F(G(s))$, $I \cong G(F(I))$ needed to show the theories equivalent.

Hence: we argued that continuity was geometricity on models for propositional geometric theories; but in practice it is more convenient to work with equivalent predicate theories.

Of course, there is a non-trivial technical claim here: that the propositional and predicate notions of "geometric transformation" agree. This comes out of the machinery of sheaf theory.

5 Observational intuition

Expanding on [6]'s observational intuition for the propositional geometric logic, one can also [7] give an observational account of the predicate logic. The idea is to describe a set not as a fully comprehended collection of elements, but as instructions for dealing with such elements as you might encounter:

1. how to know when you've "apprehended" an element of the set;
2. how to know when you've observed two apprehended elements to be equal.

(cf. Bishop's [see 1] definition of a set as comprising a stock of representations of elements, and a defined equality relation on them.)

Though this is informal, the intuition fits well with the constructions listed as geometric. For instance for a free algebra, you know how to recognize terms and how to check proofs of equality between terms. (Notice how for algebraic theories with undecidable word problem, *in*equality between terms is not algorithmically checkable.)

In practice, these intuitions provide a good bench mark for testing the geometric validity of arguments. For example, we claimed that the finite powerset construction $\mathcal{F}$ was geometric. If X is described observationally as above, then so is $\mathcal{F}X$: a finite subset of X is apprehended by apprehending all its elements and listing them (though because *in*equality is not necessarily observable, you can't guarantee that all the elements of the list are distinct). To observe that $\{x_1, \cdots, x_m\} = \{y_1, \cdots, y_n\}$, you observe that each x_i is equal to some y_j and vice versa.

Suppose now that $S = \{x_1, \cdots, x_m\}$ is a finite set, and $\phi(x)$ is an observable property. Is $\{x \in S : \phi(x)\}$ finite? Not in general! To list *all* its elements, you'd have to know that the unlisted elements don't have property ϕ, and the problem is that $\neg\phi$ is not necessarily observable. It turns out that this and similar unexpected behaviour is already known to topos theorists; the observational account gives a rough and ready way of anticipating it.

6 Generalized topological spaces (toposes)

The predicate versions of Th_1 and Th_2 were equivalent to propositional theories. (Technically, this follows from the fact that all the sorts — $\mathbf{N}$, 2^*, etc. — were geometrically derivable out of nothing: "propositional" means no *essentially new* sorts.) However, a truly predicate theory can also be thought of in this spatial way. A propositional theory was thought of as describing a "locale", its "space" of models (both classical and non-classical); but technically it was represented as a frame, the topology. A continuous map between locales is really a geometric transformation of models into models, technically representable in reverse by the inverse image function, a frame homomorphism.

Similary, a predicate theory can be thought of as describing a "generalized space" of its models, and this is a *topos* in the sense of Grothendieck's dictum, "a topos is a generalized topological space". However, the technical definition, the "generalized topology", is more complicated. The Lindenbaum frame of propositions (formulae without free variables) is no longer adequate for predicate theories, and has to be extended to a "Lindenbaum category of sets", the category you see constructed in — for instance — [5] as the classifying topos of the theory. (This is "topos as generalized category of sets". Note that this "generalized topology" is different from the "Grothendieck topology" that is used at a certain stage in the construction. The category and Grothendieck topology that comprise a "site" are more analogous to the generators and relations of a presentation, something that can be seen more clearly in Johnstone's [3] sites for frames — the Grothendieck topology is the analogue of the coverage.) The "continuous maps" — the geometric transformations of models — now appear as geometric morphisms between toposes.

Hence our reinterpretation of continuity as geometricity has also cast light on the notion of topos as generalized space, a notion of which it is easy to lose sight in the standard accounts of toposes.

7 Arithmetic Universes

What follows is more speculative, though existing results lend support to the broad argument suggested.

A crucial feature of geometric logic is the arbitrary disjunctions, that is to say disjunctions of arbitrary sets of formulae. This was used to justify inductive and recursive constructions as geometric, and suggest a "geometric mathematics" that is algorithmic in flavour.

However, there is a gap here. The arbitrary set-indexed disjunctions of geometric logic encompass far more than the recursively indexed ones, and in fact the extent of geometric tranformations (as continuous maps) depends on your underlying idea of what sets are — for this determines what disjunctions you can form. Thus geometric logic is not absolute in itself, but relative to the chosen set theory. In fact, this leads to certain anomalies in the observational interpretation.

A simple one raised by Mike Smyth (and mentioned in [6]) concerns the discrete topology on the natural numbers **N**. For sure, every singleton $\{n\}$ represents a finitely observable property of natural numbers, so any disjunction of singletons — i.e. any subset of **N** — should also be finitely observable. Consider then, for an algorithm A,

$\bigvee\{\{n\} : A(n) \text{ does not terminate}\}$

Can this really be "finitely observable"? That would seem to imply a solution to the halting problem, which of course is impossible. The catch is that we have used classical set theory to comprehend the disjuncts, and this has smuggled in inobservable features. Thus geometric logic based on classical set theory does not exactly capture the observational ideas.

A likely-looking way out is to restrict the infinities to effective ones, by taking a primitive collection of "geometric constructions" (such as those listed in Section 4) as directly defining our notion of set theory. The logical disjunctions come out of set-theoretic disjoint unions (taking images to obtain non-disjoint unions). These would be of only finitely many sets (properly, geometric logic countenances disjoint unions of infinitely many sets), but this would be partially compensated for by the inductive constructions. It is conjectured that this can be precisely formalized in category theory, by using Joyal's *arithmetic universes* [unfortunately unpublished]: that the categorical structure postulated in an arithmetic universe models the "effective geometric constructions". It is hoped that by thus restricting the notion of geometric construction (and hence the corresponding notion of continuous map), a fragment of topology can be found that genuinely matches the observational ideas.

To return to the "cheating" characterization of **N** given in Section 4, **N** would be characterized not logically, with an externally indexed countable disjunction, but categorically, as a free algebra with constant 0 and unary operator s.

8 Conclusions

By keeping one's mathematics constructive, one can make a lot of topology implicit: if the points are described as models of a geometric theory, then the topology is defined implicitly, and if transformations are defined geometrically then continuity is automatic. This works not only for ordinary topology, but also for Grothendieck's "generalized topological spaces" (toposes), generalizing from propositional to predicate theories.

On a practical level, constructive reasoning can thus can lighten the burden of topological discussion; at a deeper level it is hoped that the approach can bring a true reconcilliation between topology and effective mathematics.

Further exposition of these ideas can be found in [8] and [9].

9 Bibliography

Some of my own papers are also available electronically — see Web page `http://theory.doc.ic.ac.uk:80/people/Vickers/`

1. E. Bishop and D. Bridges, *Constructive Analysis* (Springer Verlag, 1985).

2. P.T. Johnstone, *Topos Theory* (Academic Press, London, 1977).

3. P.T. Johnstone, *Stone Spaces* (Cambridge University Press, 1982).

4. J. Lambek and P.J. Scott, *Introduction to Higher Order Categorical Logic* (Cambridge University Press, 1986).

5. S. Mac Lane and I. Moerdijk, *Sheaves in Geometry and Logic* (Springer Verlag, 1992).

6. S.J. Vickers, *Topology via Logic* (Cambridge University Press, 1988).

7. S.J. Vickers, "Geometric Theories and Databases", in Fourman, Johnstone and Pitts (eds) *Applications of Categories in Computer Science* (Cambridge University Press, 1992), 288–314.

8. S.J. Vickers, "Toposes pour les Nuls", Semantics Society Newsletter 4 (1995). Also available as Imperial College Research Report DoC 96/4 (1996).

9. S.J. Vickers, "Toposes pour les Vraiment Nuls", in Edalat, Jourdan and McCusker (eds) *Advances in Theory and Formal Methods of Computing 1996* (Imperial College Press, 1996), 1–12.

Remarks on the Epistemic Rôle of Discourse Referents

Thomas Ede Zimmermann

Institut für Maschinelle Sprachverarbeitung
Universität Stuttgart

In this paper I will be concerned with the notion of information underlying so-called *dynamic* approaches to natural language semantics[1], according to which sentence (and discourse) meaning is described in terms of information states containing discourse referents, thus being of a finer structure than the propositions of possible worlds semantics. A comparison of standard (static) theories of direct reference will reveal a conceptual gap in the core part of the dynamic enterprise: the specific informational value of discourse referents is left unexplained. Various alternative ways of filling this gap by giving discourse referents an epistemic interpretation will then be discussed and evaluated. Most of them have a consequence which certain formulations of dynamic semantics explicitly tried to avoid: *representationalism*, i.e. the doctrine that information states are more fine-grained than their informational contents.

1 Information States and Direct Reference

According to *static* semantic theories[2], the meaning of a sentence (containing no context-dependent expressions) can be identified with its content, the proposition it expresses, which can be thought of as (characterizing) a *neutral information state*, i.e. a set of possible situations, or indices (whose internal structure must be independent of, or prior to, the structure of the

Logic, Language, and Computation, Vol. II, edited by Lawrence S. Moss, Jonathan Ginzburg, and Maarten de Rijke.

[1]Although I have the feeling that the importance of *dynamics* for these theories is generally somewhat overestimated, I am nonetheless following common practice and using it as a cover-term for the kind of treatment of non-quantificational noun phrases as developed by Kamp [15], Heim [13], and others. Terminological details aside, whatever I say below is intended to apply to all versions of dynamic semantics alike.

[2]By this I mean anything along the lines of Montague [22].

sentences describing them[3]). A sentence determines the information state I that is consistent with precisely the indices it describes; a neutral state I may, e.g., correspond to the information expressed by:

(1) **The butler is the murderer.**

by containing precisely those worlds w in which the butler is identical with the murderer:

$$I = \{w \mid (\exists x)(\exists y)\ [x = \textit{the butler in } w \ \&\, y = \textit{the murderer in } w \,\&\, x = y]\}$$

The butler not necessarily being the murderer, the denotations of the two noun phrases in (1) do not coincide at all possible indices. In particular, a person x may be in a (i) consistent or even (i') truthful information state I which is (ii) compatible with the butler's not being the murderer: (i) I is not empty, or even (i') contains the situation x is in while being in state I, and (ii) I also contains at least one index at which the butler is not the murderer.

In *dynamic* semantics, sentence (and discourse) meanings are certain transformations of *situated information states*, which are themselves world-dependent relations between (existing) individuals; a situated state R may, for example, correspond to the information expressed by (1) by relating two individuals in precisely those worlds w in which the first is the butler, the second is the murderer, and the two are identical:

$$R_w = \{\langle x, y\rangle \mid (\exists x)(\exists y)\ [x = \textit{the butler in } w \ \&\, y = \textit{the murderer in } w \,\&\, x = y]\}$$

The *interpretation* of a sentence transforms an input state R into a more specific state S: all tuples in S_w expand elements in R_w, so that, in general, not all of the original possibilities are consistent with the update state created by the sentence. For instance, the transformation induced by an identity statement like (1) may expand the members of a given R_w by the common denotation of the equated terms in w, and will return an empty S_w if the denotations do not coincide. Being referential (= non-quantificational) noun phrases, either of the terms in (1) increase the length (= arity) of information states. But they do so in different ways, adding different individuals in those possible worlds where their denotations do not coincide. In particular, a person may be in a (i) consistent or even (i') truthful situated information state R which (ii) can be consistently updated

[3]This is to avoid circularity in the notion of informational content. I will not elaborate this point here; see Lewis [21], ch. 3, for further discussion.

with the butler's not being the murderer: (i) R or even (i') its extension is not empty, and (ii) there is at least one world w in which the extension of the update state is satisfied (= non-empty), which is precisely the case if R_w is satisfied and **the butler** and **the murderer** introduce different individuals in w.

Both of the above approaches account for Frege's Puzzle[4], although they do so in slightly different ways. Whereas the static account models the informativeness of (1) by the non-universality of a corresponding set of indices, the dynamic account imposes further structure on that set. Thus, as far as this particular epistemological problem is concerned, the static account is more parsimonious than its dynamic rival. However, what may appear like a dynamic overkill, viz. the introduction of *discourse referents*, is well motivated in terms of discourse anaphora and donkey sentences – or so I will assume[5].

According to a well-known extension of the static point of view[6], context-dependent items (i.e. demonstratives and other indexicals) are *directly referential*, which means that their reference is independent of the particular situation: they are interpreted by the individuals they denote, as determined by speakers' intentions and actions. Hence their denotation varies with, and only with, the context (= utterance situation), and so does the content of an expression containing a demonstrative (or any other directly referential expressions). Consequently, a very slow utterance of

(2) **He** [pointing to the butler in the morning] **is identical with him** [pointing to the butler in the evening].

characterizes a trivial information state, i.e. one that does not differentiate between indices. But since (2) can be of *informational value* to a hearer, Frege's Puzzle recurs in a demonstrative version.

Speakers' information states thus cannot be the neutral information states of static semantics; they are, rather, *perspectival*: like meaning, which can be construed as a function assigning contents to contexts, informational value requires abstraction from context. Indeed, the informational value of a sentence can be taken to be the *diagonal* of its meaning, i.e. the set of contexts that are themselves members of the proposition the sentence expresses in that context; and perspectival information states are then sets of contexts. Consequently, information growth (or any other epistemic change) is a relation between context sets[7].

---

[4]See Frege [7]. I am assuming that (1) does not contain any context-dependent elements so that it is indeed a version of Frege's Puzzle.

[5]See especially Heim [14] for the kind of motivation I have in mind.

[6]See, e.g., Kaplan [19].

[7]See Haas-Spohn [11], p. 236ff., for a fuller story of context change along these lines. The modelling of (perspectival) information by diagonalization goes back to Stal-

For later reference, it will be useful to represent static sentence meanings by λ-terms of the form:

(3) $\lambda c\ \lambda i\ R(c,i)$,

where c runs over contexts and $\lambda i R(c,i)$ represents a proposition. The diagonalization of the above meaning can then be represented as: $\lambda c R(c,c)$.

According to a recent extension of the dynamic point of view[8], demonstratives denote individuals, as determined by speakers' intentions and actions: their interpretation varies with, and only with, certain contextual parameters, and so does the interpretation of any sentence containing a demonstrative. Consequently, a very slow utterance of (2) above will transform a given situated information state in a trivial way, whenever the contextual parameters are as described in the brackets; and again we get a demonstrative version of Frege's Puzzle.

Speakers' information states thus cannot be the situated information states of dynamic semantics; they are, rather, *global*: like meaning, which can be construed as a function assigning to any [sufficiently long] list of contextual parameters an interpretation of the sentence, informational value requires abstraction from context. Indeed, the informational value of a sentence is taken to *be* its meaning; and global information states are the static counterparts of meanings, i.e. functions from context(ual parameters) to situated information states. Consequently, information growth (or any other epistemic change) is a relation between global information states.

For later reference, it will be useful to represent global information states by λ-terms of the form:

(4) $\lambda\langle o_1,\ldots,o_n\rangle\lambda\langle s_1,\ldots,s_m,w\rangle R(o_1,\ldots,o_n,s_1,\ldots,s_m,w)$,

where the o_j represent the relevant contextual parameters (providing *objects*), and $\lambda\langle\vec{s},w\rangle R(\vec{o},\vec{s},w)$ denotes a relation holding between the *discourse referents* [or *subjects*] $s_1,\ldots,s_m$ in a world w.[9]

Obviously, the static and the dynamic account of context-dependence have certain features in common: both invoke a two-step interpretation process in that the meanings of demonstratives are determined on a separate level; both interpret demonstratives directly, by their referents; and on

naker [27]; Kaplan [19] only had meaning (his *characters*). Whether the diagonals are again neutral information states depends on whether contexts are (special kinds of) possible situations; see Lewis [20] for further discussion.

[8]The apparatus developed in Dekker [3] is actually not aimed at a theory of meaning but should rather be understood as a theory of perspectival information, or diagonalized meaning. In particular, subjectivization and diagonalization are not as parallel as I say. I apologize for this (initially unintended) misconstrual – but then even Paul Dekker (p.c.) agrees it is a productive and a fruitful one.

[9]I use $\lambda\langle x_1,\ldots,x_n\rangle\phi$ as short for '$\lambda X(\exists x_1)\cdots(\exists x_n)[X=\langle x_1,\ldots,x_n\rangle\&\phi]$'.

both accounts, speakers' epistemic perspectives are modelled by abstraction from context(ual parameters). On the other hand, there are certain equally obvious differences, including one in granularity: neutral information states contain nothing but the situations they characterize, whereas situated information states contain both the worlds characterized and individuals living in these worlds. Of course, this particular difference has nothing to do with demonstratives, but the two approaches also differ in their construal of contexts. Statically, direct reference may depend on arbitrary features of the context. Dynamically, demonstrative reference depends on a (sentence-dependent) list of contextual parameters. And though it may seem accidental that the dynamic account does not cover more cases of context-dependence that would call for more contextual structuring, the obvious formal parallel between contextual parameters, i.e. objects, and discourse referents plays an important rôle in epistemic dynamics, to which I will now turn.

One way in which objects can be put to use in dynamic interpretation is to account for their anaphoric behaviour: like all referential noun phrases, demonstratives can be antecedents to discourse-anaphoric pronouns; but unlike indefinites and definite descriptions, they keep their referring function in donkey constructions. Here are two simple examples to illustrate these points:

(5) **A farmer grows this** [pointing to rice]**. He eats it with curry.**

(6) **If a farmer grows this** [pointing to rice]**, he eats it with curry.**

In order to capture this behaviour of demonstratives, one only has to make sure that they are accessible to later pronouns, while at the same time they do not get bound by any dynamic operators. I will not develop this rather obvious point here[10], but it should be clear that discourse accessibility involves some operation that turns an object into a subject or, equivalently, a global information state involving a certain object into one that contains it as a subject. More specifically, for suitable j, a j-*leap*, turns state (4) into:

(7) $$\lambda\langle o_1, \ldots, o_j, \ldots, o_n\rangle \lambda\langle o_j, s_1, \ldots, s_m, w\rangle R(o_1, \ldots, o_n, s_1, \ldots, s_m, w)$$

Note that a j-leap adds an object as a subject without changing its original status as an object. But combining it with a j-*drop* we obtain from (7):

(8) $$\lambda\langle o_1, \ldots, o_{j-1}, o_{j+1}, \ldots, o_n\rangle \lambda\langle o_j, s_1, \ldots, s_m, w\rangle R(o_1, \ldots, o_n, \vec{s}, w)$$

[10] See Zimmermann [30], p. 217 f., for more on this.

This combination of j-leap and j-drop is called *j-subjectification*[11], and it is of some theoretical interest. For there is a close connection between it and diagonalization. To see this, consider *total* subjectification, i.e. the consecutive application of N_σ-subjectification to a given state σ (where N_σ is the number of σ's objects), until it is no longer applicable. When applied to (4), total subjectification will yield

(9) $\lambda\langle o_1, \ldots, o_n, s_1, \ldots, s_m, w\rangle R(o_1, \ldots, o_n, s_1, \ldots, s_m, w)$,

thus turning a (binary) global state into a (unary) situated state with essentially the same content – or so it seems. In particular, the contextual contributions of information, the objects, now end up as subjects, i.e. as parts of a context-insensitive information state. And, from a formal point of view, something similar happens when meanings undergo diagonalization, where contextual dependence is merged into undifferentiated dependency. But there is also a difference: by turning objects into subjects, total subjectification preserves the fine-grainedness of contexts. However, apart from its obvious effect on the dynamics of pronominalization, this difference is illusory: contexts may have a rich internal structure, in which case the static meaning (3) can be rewritten as

(10) $\lambda\langle o_1, \ldots, o_n\rangle \lambda i R(o_1, \ldots, o_n, i)$,

which diagonalization will turn into

(11) $\lambda\langle o_1, \ldots, o_n\rangle \lambda i R(o_1, \ldots, o_n, \langle o_1, \ldots, o_n\rangle)$,

It may appear odd that a list of contextual parameters occupies the index's R-position, but the type-mismatch easily dissolves under the assumption that indices, like contexts, can be split up into various parameters (world, time, place, ...) each of which uniquely corresponds to a contextual parameter (though not necessarily *vice versa*): i itself could then be represented as a tuple of the form $\langle i_1, \ldots, i_k\rangle$, where $k \leq n$ and diagonalization results in replacing each i_j by the corresponding o_j[12]. Hence from an abstract and formal point of view, diagonalization and total subjectification do appear to be essentially the same operation. The deep difference emerges when we turn to the original motivation behind diagonalization.

Imagine Herman making a phone call to his friend Paul with a breathy voice, thereby concealing his identity. In this situation Paul could express his conviction that *whoever is being talked to has a funny voice* by uttering:

(12) **You have a funny voice.**

[11] The term is due to Dekker [3], where it is introduced as a primitive operation.
[12] See Zimmermann [29], p. 174ff., for more on this.

Given that **you** assigns the hearer to any context, the italicized proposition turns out to be the diagonal of the meaning of (12). Hence the diagonal puts the information contained in the meaning of (12) into the speaker's perspective: even if Paul does not know that it is Herman he is talking to in that context, he can address him as **you** because the meaning of **you** will pick out whichever person he is talking to in that context. So, under diagonalization, the descriptive content of a pronoun like **you** is on a par with that of a non-rigid predicate like **have a funny voice**: both extensions depend on one and the same parameter.[13]

Of course, the descriptive content of demonstratives and other context-dependent expressions is also present in the meaning; and this is no surprise, for diagonalization is an operation on meaning and thus completely determined by the latter. But although it is not a one-one mapping, it does seem to capture all informationally relevant aspects of meaning, and nothing else[14]. That is what makes it so interesting.

Let us now return to the dynamic treatment of demonstratives. Since static meanings correspond to global information states, one would expect the latter to be able to capture the descriptive content of a demonstrative. This is indeed the case: if we formalize (2) as (1), the sentence is accepted in a global information state σ if that state is only compatible with contexts in which the referent of d_1 (= the individual pointed to first) is identical with the referent of d_2 (= the individual pointed to next), whoever this individual may be:

(2') $(d_1 = d_2)$

But now something unexpected happens: if we apply (total) subjectification to σ, the demonstratives in (2) lose their descriptive content. For the resulting state σ^* accepts the same (closed) formulae as σ:

$$\sigma^* = \{\langle w, d, d\rangle \mid w \in \sigma, d \in D_w\}$$

Of course, the result of total subjectification would have looked different had the initial state contained additional information about the objects denoted by the demonstratives. But even then the very information that these are the objects pointed at would have been lost. How could that happen, given the close analogy between subjectification and diagonalization?

One may argue that the information that d_1 was the first object pointed to was never there in the first place: this connection was only made by our loose and informal understanding of contextual parameters, but the formal

[13] I deliberately ignore the fact that, for reasons given in Lewis [20], this parameter should be thought of as a context rather than an index. All that matters here is the fact that diagonalization, like total subjectification, leaves only one parameter.

[14] See Zimmermann [29], p. 181f., and Haas-Spohn [12], p. 70f., for further discussion.

framework can do without this. Fair enough, but if so we are left without any understanding of contextual parameters and hence without any idea as to how dynamic semantics is supposed to account for epistemic change. This is certainly an unwelcome result.

Let us instead take for granted that, *by virtue of our meta-theoretical understanding*, each contextual parameter j comes with the descriptive content of being the jth individual pointed to. It is then quite obvious why subjectification loses this content[15]: the object filling the jth parameter is turned into the first discourse referent; but no meta-theoretical understanding tells us that the first subject slot is occupied by the jth individual (previously?) pointed to. And, of course, no meta-theoretical understanding about subject-rôles could or should tell us that: for one thing, any object may become the first discourse referent by subjectification, so that j would not be unique; and, like any subject, the first one need not be obtained by subjectification and, indeed, never have been pointed to.

If the task were to fix subjectification in a way which gets it closer to diagonalization, there would be various ways to go about. I prefer to leave this to the reader. I guess the resulting operation should have more intuitive appeal than the original one, and that it could be defined so as to generalize from demonstratives to other kinds of directly referential expressions. My present interest lies elsewhere, viz. in the very asymmetry between subjects and objects which is, I claim, responsible for the failure of subjectification as an *analogon* to diagonalization. For not only do the subjects not have the intuitive content associated with objects; they do not seem to have any intuitive content whatsoever.

Surely this is a problem; for we have just seen that depriving the objects of their intuitive content leaves us with an uninterpreted theory of epistemic change. But then the same will happen if we cannot give content to the subjects or discourse referents – and I will argue that this is indeed difficult. This problem is fundamental to any dynamic approach to meaning and information. So let us face it.

2 Information States and Discourse Referents

In order to isolate the particular informational contribution that subjects are supposed to make, let us consider a minimal pair of information states differing only in their respective discourse referents. The scenario is a somewhat artificial *gedankenexperiment*, but I hope it brings out clearly what I take to be missing from the dynamic account of information:

[15]This information drop was Dekker's original motivation behind the operation.

The Dutch Twins This is the story of J and M, a pair of twins who prefer to stay anonymous. J and M are very fond of each other, and they are also very concerned about each other's knowledge: whenever one of them learns something, he immediately informs his brother, without omitting the least detail, let alone deceiving him. Since they have been together for longer than anyone can remember, they have ended up in (qualitatively) identical information states, which they certainly do not regret. Right now they are standing at the window of their common office overlooking the local green, which they are used to refer to as **the park**, when suddenly J makes a discovery and, as always, immediately informs M by uttering:

(13) **A man is walking through the park.**

At the same time M, who is in the same position as J[16], perceives the same disturbing scene and, consequently, immediately tries to update his brother's information state. But he prefers to achieve this goal by uttering:

(14) **It is not so that no man is walking through the park.**

In spite of being occupied with producing a sentence, both J and M are able to understand their respective brother's utterance and react with an immediate update.

The terms *update* and *information state* appearing in this story ought to be taken in a pre-theoretic sense. Now, according to dynamic epistemic semantics, M and J would have to end up in distinct (global) information states, σ_M and σ_J: by accepting (13), M has increased the number of discourse referents by 1, but J's subjects remain the same. And since, by assumption, M and J had started from the same state σ, it follows that $\sigma_M \neq \sigma_J$. But neither J nor M learns anything from his brother's utterance – or so it seems. We thus have a genuine puzzle, which I will call the *intentional puzzle.*

One thing a general theoretical account of the epistemic rôle of discourse referents would have to do, then, is precisely to resolve the intentional puzzle by explaining how J and M can be in distinct information states, even though it looks like they are not. Now, although the number and variety of solutions to the intentional puzzle are, in principle, open, the following list exhausts the kinds of accounts I have actually come across (or made up, as the case may be):

Solutions to the intentional puzzle

[16]Of course, M cannot be in the *very* same position as J. But their positions can be close enough to give them [qualitatively] identical perceptual input.

1. *Zero solution* A discourse referent represents the information obtained by existentially quantifying it away.

2. *Meta-discourse solution* A discourse referent represents the information that a certain kind of noun phrase (e.g., an indefinite) has been used.

3. *Dynamic solution* A discourse referent can distinguish between information states which themselves can only be individuated in terms of their update behaviour.

4. *De re solution* A discourse referent represents the object that the content of its information state is about.

5. *Auto-epistemic solution* A discourse referent reflects the belief that the information obtained by quantifying it away causally depends on a specific individual.

6. *De origine solution* A discourse referent represents a source of the informational content of its information state.

I am going to discuss these six approaches in turn; in doing so, I hope that some of the above formulations that may appear vague or cryptic will become clear enough to reveal what kind of solution is referred to. It will turn out that none of them are entirely satisfying, though I think that I can present some evidence of favour of the last one. But let us start at the beginning:

ad 1.
The zero solution[17] tells us, in effect, that we should not take dynamic metaphors too literally. After all, whatever M knows about men walking in the park at the time of J's utterance of (13) is also known to J. And what else could the additional discourse referent be about that distinguishes σ_M from σ_J? So even though σ_M and σ_J are distinct epistemic states, they coincide in their content, they are *informationally equivalent*. How can that be?

As far as I can see, any mismatch between epistemic states and information states directly leads to some form of *representationalism*[18]: the state σ may be said to *represent* the information it contains; and to *be in* state σ then amounts to somehow having access to this representation (in a sense to be explained by the representationalist). In particular, two distinct states

[17]Although I know of no explicit source, the zero solution seems to be popular among semanticists who favour a static approach to information states.

[18]One may even *define* representationalism as this kind of mismatch; but see Dekker [2] for other possible definitions.

σ and σ' may represent the same information, which will be the case if their *existential closures* $\mathsf{Cl}_\exists(\sigma)$ and $\mathsf{Cl}_\exists(\sigma')$ coincide, where:

$$\begin{aligned} & \mathsf{Cl}_\exists(\lambda\langle o_1,\ldots,o_j,\ldots,o_n\rangle\lambda\langle s_1,\ldots,s_m,w\rangle R(o_1,\ldots,o_n,s_1,\ldots,s_m,w)) \\ = \; & \lambda\langle o_1,\ldots,o_j,\ldots,o_n\rangle\lambda w(\exists s_1)\ldots(\exists s_m)R(o_1,\ldots,o_n,s_1,\ldots,s_m,w)) \end{aligned}$$

According to this view, (situated) information states are essentially structured propositions consisting of a matrix and a block of existential quantifiers represented by the subjects[19]. Hence, following the zero solution, even compositional reformulations of DRT have not managed to escape representationalism with its language-dependent notion of information states.[20]

ad 2.
Unlike its predecessor, the meta-discourse solution[21] takes the prediction that σ_M, σ_J, and σ are not one and the same information state at face value. After all, in interpreting their respective brother's utterance, J and M do learn something; for instance, J learns that M has just made an utterance of (13), M learns that J tried to convince him that (14) is true, etc. Indeed, J and M seem to learn different things from their respective brother's utterance; and maybe dynamic semantics can be understood to capture at least some of these subtle differences, rather than idealizing them away. In particular, the fact that J used an indefinite to inform his brother about the latest developments in the park is recorded in M's, but not in J's, updated information state by the presence of an additional discourse referent.

As it stands, the meta-discourse solution is rather vague. Among other things, (a) the relation between the subject and the noun phrase by which it is introduced is in need of further specification; and (b) the question who

[19]The difference between situated and global information states is irrelevant to my present concerns of understanding the rôle of discourse referents. Using the format (but not the notation) of Cresswell & von Stechow [1], the structured proposition corresponding to the situated information state

$$\lambda\langle s_1,\ldots,s_m,w\rangle R(\vec{o},s_1,\ldots,s_m,w)$$

can be defined by

$$\begin{aligned} \langle\lambda\langle Q_1,\ldots,Q_m\rangle\lambda w\; Q_1(\lambda s_1\ldots Q_m(\lambda s_m R(\vec{o},s_1,\ldots,s_m,w)\ldots)), \\ \lambda P(\exists s_1)P(s_1),\ldots,\lambda P(\exists s_m)P(s_m)\rangle \end{aligned}$$

The connection between structured propositions and dynamic semantics was poitned out to me by Rainer Bäuerle a long time ago.

[20]See footnote 3 above.

[21]Cf. Stalnaker [28] for a recent formulation; the present section is largely inspired by this paper and by discussions with Robert van Rooy, who convinced me that the meta-discourse solution is more adequate than I first thought.

used that noun phrase might turn out to be relevant too. (a) is among the classical objectives of dynamic semantics, whereas (b) has only recently received attention.[22] But however these and similar details are filled in, at least three problems remain:

(c) a meta-discourse version of the intentional puzzle;

(d) the semantics of (generalized) donkey sentences;

(e) a translation argument.

In order to obtain (c), one only need to compare M's new information state σ_M to that state σ^* that results from the old σ by dropping those worlds[23] in which J has not uttered a noun phrase of the relevant kind; then σ_M and σ^* are informationally equivalent, even though they are distinct information states. This variant of the intentional puzzle should not come as a surprise: the meta-discourse solution differs from the zero solution only in ascribing certain descriptive contents to discourse referents – but these *subject contents* can still be construed as restricting the discourse referents' quantificational domains without affecting their status as existential quantifiers. We thus see that solving the intentional puzzle is not just a matter of supplying appropriate descriptive contents for discourse referents.[24]

As far as I can see, descriptive contents only help solving the intentional puzzle if they are used to restrict the class of underlying propositions – thereby slightly changing the dynamic framework. More specifically, assuming that subjects are always associated with descriptive contents, let us call a situated information state σ *clean* if there is no subject content Δ such that $\mathsf{Cl}_\exists(\sigma) \subseteq \mathsf{Cl}_\exists(\Delta)$. The idea is that a clean state only carries information about the subject matter of the conversation, not about how its subjects are given. We could then map each clean $\sigma = \lambda\langle s_1, \ldots, s_m, w\rangle R(s_1, \ldots, s_m, w)$ onto a dirty state

$$\sigma^\# = \lambda\langle s_1, \ldots, s_m, w\rangle[R(\sigma_1, \ldots, \sigma_m, w)\&\Delta_1(s_1, w)\& \ldots \&\Delta_m(s_m, w)],$$

[22]See Groenendijk *et al.* [9]. As to (a), it is tempting to relate the subject to the descriptive content of the (referential) NP. However, as Groenendijk *et al.* [10], p. 237f, have pointed out, this might not be the whole story.

[23]The result of *dropping* a world w^* *from* a global information state

$$\lambda\langle o_1, \ldots, o_n\rangle\lambda\langle s_1, \ldots, s_m, w\rangle R(o_1, \ldots, o_n, s_1, \ldots, s_m, w)$$

is

$$\lambda\langle o_1, \ldots, o_n\rangle\lambda\langle s_1, \ldots, s_m, w\rangle[R(o_1, \ldots, o_n, s_1, \ldots, s_m, w)\&w \neq w^*].$$

[24]Analogous remarks apply to any other account of discourse referents in terms of descriptive content. For instance, if each subject were equipped with a rank ρ in a contextually given *salience order*, then σ_M would contain the same information as that state which results from σ by dropping all the worlds in which no man of rank ρ is walking through the park.

where $\Delta_1, \ldots, \Delta_m$ are the respective subject contents of $\sigma_1, \ldots, \sigma_m$. Hence $\sigma^{\#}$ can be thought of as the full information carried by σ in a given situation. If we now confine semantic interpretation to clean information states, which are in turn (e.g., pragmatically) interpreted by their dirty $^{\#}$-images, the intentional puzzle will have lost much of its bite: σ_M is a clean information state not involving any discourse information (and thus merely a rough approximation of M's actual state), whereas $\sigma^* = \sigma_M^{\#}$ is dirty (and less idealized).

The same technique also proves to be fruitful to solve the problem (d) of analyzing donkey sentences – and particularly those involving generalized quantifiers (like **If a farmer owns a donkey he usually beats it**): these sentences seem to quantify over the relations expressed by the two clauses they connect, whereas the meta-discourse solution says that, strictly speaking, these clauses do not express any relations. However, if we distinguish between the semantic contribution of a sentence, which does involve a relation, and its conversational meaning, which does not, the construction can be interpreted along familiar, dynamic lines.

Finally, there is the translation argument (e). If σ_M contains information about the use of the English NP **a man**, then (13) and its Dutch counterpart (15) would have to differ in their meanings, whereas (14) and (16) would not; for (13) and (15) would contain information about English and Dutch, respectively, whereas (14) and (16) are entirely about parks and persons.

(15) **Er loopt een man door het park.**

(16) **Het is niet zo dat er geen man door het park loopt.**

I find this conclusion hard to swallow, and I am not sure how the meta-discourse solution could escape it. Maybe discourse referents do not have to make reference to any specific language such as English or Dutch, as long as the information they contain is enough to single out the kind of NP that can serve as an antecedent in discourse anaphora. But then I do not know of any such non-circular characterization excluding (hypothetical) NPs that look like English indefinites but are interpreted as true existentials unable to induce discourse referents. Another line would be to endorse the old structuralist dogma that all meaning is language-dependent; the purported synonymy of (13) and (15) could then be regarded as an instance of functional equivalence.[25]

[25]Haas-Spohn [12], p. 146, has recently defended that dogma, and also pointed out that translation frequently relies on functional equivalence rather than synonymy (*ibid.*, p. 110, fn. 17).

ad 3.
According to the dynamic solution, discourse referents are needed in distinguishing update potentials of information states. The idea is that, coreferentiality provided, (13) puts a hearer in a position (i.e. information state, on the dynamic solution) of processing further information like (17): although σ_M and σ_J may, in some sense, agree in informational content, they must be distinguished in terms of update potential; once being in different states, J and M differ in what they are prepared to learn.

(17) **He is wearing blue suede shoes.**

The underlying intuition is that (13) and (17) are understood as being *about* the same individual; moreover, the relevant concept of aboutness is said to be irreducible to any static notion of content and must therefore be reconstructed in terms of shared subjects. Let me briefly discuss a typical example[26] purported to show that discourse referents are really necessary for a full understanding of epistemic dynamics.

A Philosopher's Nightmare Sitting in an Amsterdam street café, German philosopher W sees J running past him, on his way to his brother. Although W recognizes him as one of the Dutch twins whom he has seen at two conferences, he does not recognize him as J, i.e. he could not tell which one of them he is watching. Indeed, W lacks any criteria of telling them apart and would thus not even be in a position to wonder which of the twins he is watching; all he knows is that they are two and that they share any property he ascribes to either of them. Now, while J is walking by, W notices two details about him: the first is that he is smoking a cigarette, which leads W to the conclusion that he must be a smoker; the other detail is J's fashionable glasses which W takes to be an indication of short-sightedness. W wants to take in these two pieces of information but does not quite know how to do that: since he did not have any means of identifying J in his pre-café information state, he now lacks any way of attributing these new properties *to him*. He tries out various approaches to updating, but none of them seem to be adequate: the proposition that any Dutch twin he has met at a conference is a short-sighted smoker is certainly too strong to be supported by his observations, whereas adding the information that one of the Dutch twins he has met at a conference is a short-sighted smoker is more than he actually observed: the fact that J is one of the people he met at a conference is not anything he has just learnt. Totally frustrated, W pays his bill and walks back to his hotel unupdated.

Given the dynamic approach to information states, it would seem that W could be helped: his information state σ_a at the beginning of the

[26]This is a variation of an example from Spohn [26], who defends the dynamic solution, the *locus classicus* of which is Perry [23].

above episode would have contained two (qualitatively) indistinguishable discourse referents x and y and all W would have had to do in the coffeeshop is to add the information that x is a short-sighted smoker, or that y is, or that either x or y is. In other words, rather than being unable to find an adequate update, W would even have had a choice between three such updates. This is why we should accept the dynamic view of information states.

The argument does not convince me. In fact, I think that closer inspection of σ_a would reveal that the incoming information can be described in propositional terms. For in order to be able to recognize J as one of the Dutch twins he had met at a conference or two, in σ_a W would have to ascribe to each of them certain observable properties P which he ascribes only to them and which he also takes to identify them. In other words, σ_a would have to imply something of the form:

(18) $$(\forall x)[P(x) \leftrightarrow DT(x)] ,$$

where DT is short for 'being one of the Dutch twins W has met before'[27]. But then, for some such P, the proposition defined in (19) would adequately summarize W's observation:

(19) $$(\exists x)[P(x) \& SSS(x)] ,$$

where SSS is short for 'being a short-sighted smoker'. Indeed given (19), which is unique up to proper identification of P (which, again, can be motivated in terms of W's recognition capabilities), W is saved from a Buridanian dilemma which he would have faced on the dynamic account: he seems to lack any grounds for preferring one of three equally adequate update possibilities, which should frustrate him just as much as the alleged inadequacy of static information states.[28]

The same kind of reasoning applies to a knowledge revision variant of the above argument: if W suddenly remembers that he had once seen both

[27]Following Lewis [20], DT would have to be defined in terms of the speaker's perspective, so that the resulting information state would be a property rather than a proposition; the proper paraphrase would therefore be: 'being one of the Dutch twins I have met before'. Again, I take the liberty of ignoring this complication because I think it is independent of the present discussion.

[28]One may object that one of the three updating possibilities is to be preferred. Given the symmetry of the case, this would have to be the disjunctive update, which is indeed logically weaker than its alternatives and appears to be the more careful and thus rational choice. But, as long as there are no *additional (presumably causal) links* between the Dutch twins and W's corresponding discourse referents, neither of the specific updates could go wrong without the disjunction failing at the same time: the specific update would simply be about whichever of the twins happens to verify the disjunction. On the other hand, the externalist link hypothesis would have to be justified on independent grounds.

the Dutch twins when one – and only one – of them was wearing spectacles, σ_a would have to be revised so as to imply a specific qualitative difference between the two *DT*s and, a fortiori, between the two *P*s. The updating formula (19) would then still be the same – which is intuitively correct because the revision (caused by memory) should not have any effect on the perceptual content. But even though these specific examples can somehow be dealt with, one may still wonder whether some kind of continued identity example could not be constructed. Instead of giving a general impossibility proof to that effect – which I cannot – let me tell yet another story designed to show how difficult it is to come up with a plausible defense of the dynamic solution to the intentional puzzle:

Etypes This is the story of an extraterrestrial species, the Etypes, who live in Etypeland. Etypes look a lot like human beings and they have also developed a rather human culture (cathedrals, video-games, etc.) – but the languages they speak are extremely different from and, in a way, inferior to ours: they are static. This is no accident: their genetic design prevents Etypes to ever be in information states involving subjects. So, e.g., an Etype never says anything as complex as (13), though they are quite able to say things like (14). Or rather, they could utter (13) but then what they would mean is the proposition consisting of all worlds in which there is a man walking in the park. Needless to say, the Etypes' only, extremely primitive mode of updating is by taking intersections. Still, it turns out that communication among Etypes is pretty successful (see above) – and this in spite of the fact that they never continue a sentence like (13) by (17) - or when they do, it is just short for:

(20) **One of the men walking through the park is wearing blue suede shoes.**

Finally, there are no donkeys in Etypeland. [29]

It seems that Etypes do not miss any particular piece of information (about the non-linguistic world) – yet according to the dynamic solution, the epistemic states we human beings are in most of the time would be inaccessible to them. It also seems to me that Etypes do not principally lack any learning strategies (= possible update histories) – yet according to the dynamic solution the strategies involving identification of discourse referents are unavailable for them. Of course, my impressions are still motivated by the above propositional paraphrases, but it is hard for me to imagine how the general paraphrasing strategy illustrated by (20) could ever fail.[30]

[29] See Heim [14] for the reasons why this is so (and for differences between human and Etypical discourse anaphora).

[30] In fact, van der Does [4] has given a systematic account of this strategy.

ad 4.

According to the *de re* solution, information states represent the internal parts of an external representation of knowledge, along the lines of standard *de re* construals[31]: knowledge – or more generally: any attitude – *about* an individual is a relation between a self S, the individual X and a property P ascribed to that individual. Roughly, the relation holds if X objectively bears a special kind of relation R to S and at the same time knows (believes, ...) that whoever he or she (S) is R-related to has P. As a case in point, J's special way of informing his brother about the goings-on in the park might put the latter into an acquaintance relation with the man observed by J, whereas – or so the *de re* solution would have it – M's report does not establish such a relation.

The *de re* solution is obviously wrong. For, quite apart from possible worries about the difference it makes between the two kinds of reports, it only works under especially felicitous circumstances. To see this, one need only imagine that, in the Dutch Twins scenario, both J and M had fallen victims to a delusion: nobody is walking through the park, the Amsterdam smog created a *fata morgana*. In that case, updating would have proceeded in the very same fashion, but J's report would not have been about any particular person.

ad 5.

According to the auto-epistemic solution[32], the idea behind the *de re* solution can be saved by internalizing aboutness. Deluded or not, what makes the twins' ideas about men walking through the park so specific, is that they believe them to have been caused by a specific individual: having swallowed J's report, M believes there to be some x that caused J to believe the propositional content of his report and, consequently, believes his own information state to causally depend on x.

My objection against the auto-epistemic solution is even less compelling than those against the first three solutions. Discourse referents, so it goes, are primarily a means of describing anaphoric relations and therefore ought to be of use in accounting for cases of intentional identity as in[33]:

(21) **Hob thinks a witch blighted Bob's mare, and Nob thinks she killed Cob's sow.**

[31] Cf. Kaplan [18].

[32] I have not been able to find an explicit statement of the auto-epistemic understanding of discourse referents. Kamp [16] comes close to it – but only in connection with direct reference.

[33] Cf. Geach [8]; I have followed Edelberg [6], p. 561, and slightly simplified the original example.

On the relevant understanding (not necessarily: reading), (21) does not imply the existence of (real) witches and the underlined pronoun is anaphoric to the underlined indefinite. One may therefore wonder whether it is possible to interpret this anaphoric relationship by means of identifying discourse referents across Hob's and Nob's information states σ_H and σ_N. On the auto-epistemic approach, this means that σ_H would have to imply that Hob is causally related to a witch represented by a discourse referent d, σ_N would have to imply that Nob is causally related to a witch represented by a discourse referent d', and d and d' would somehow have to stand in a suitable relation R. But, then, what would R be?

I can only speculate, but it seems to me that the auto-epistemic solution does not provide the right ingredients for defining such R. E.g., dRd' cannot simply mean that d and d' are informationally equivalent in the sense that, roughly, $\lambda d \mathsf{Cl}_\exists(\sigma_H(d)) = \lambda d' \mathsf{Cl}_\exists(\sigma_N(d'))$; for if they were, then so would σ_H and σ_N, which, of course, is not necessary for (21) to be true. But then one might still hope that informational equivalence can be weakened to a relation that holds between discourse referents from (possibly) distinct information states if they somehow play analogous rôles. Again I do not think that this is correct, but this time because it is too weak. To see this, consider the following story[34], originally designed to show that (21) does not require σ_N to relate d' to (a representation of) σ_N:

Edelberg's Astronomers Two teams of astronomers have independently been investigating the peculiar motion of superclusters of galaxies – that is, the motion of *clusters* of clusters of galaxies, over and above that due to the Hubble expansion of the universe. Neither team knows about the work of the other, but both independently and correctly ascertain the peculiar motions of the Hydra-Centaurus supercluster, of the Local Supercluster, and of our own Local Group. Both teams attempt to explain the vectors of the peculiar motions in the same way: by postulating an "overdensity" of galaxies at roughly twice the distance between the Hydra-Centaurus supercluster and our own galaxy. The idea is that an enormous collection of galaxies, "a distant concentration of mass that appears to be larger than any proposed by existing cosmologies," lies beyond the Hydra-Centaurus supercluster, drawing it, as well as our own Local Supercluster of galaxies, toward it. The American team calls the structure "The Great Attractor", the Soviet team calls it "The Overdensity" (in Russian). Due only to certain differences in instrumentation and atmospheric conditions at the times and locations of observations, the two teams conjecture the structure to

[34]The following passage is from Edelberg [6], p. 574f, where it is marked as a "corruption of Dressler [5]"; I have only added a heading and corrected an obvious mistake ('distance' instead of 'difference').

be at "slightly" different distances. The Americans say it is twice the distance of the Hydra-Centaurus supercluster; the Soviets say it is at 2.1 time the distance. In reality, let us suppose, the Great Attractor does not exist at all: the peculiar motions of the various superclusters are each caused by independent factors.

According to this story, the following variant of (21) is true:

(22) **The American team believes that an immense overdensity of galaxies at a certain distance causes a peculiar motion of superclusters, but the Soviet team thinks it is slightly further away.**

The relation between the underlined expressions seems to be analogous to that in (21); yet I think that the relation between the corresponding discourse referents would have to be closer than just one of "rough similarity of explanatory rôle"[35]. This becomes clear if we consider a Twin Earth variant of the above story[36], according to which the two teams live in two distinct regions of the universe that happen to be qualitatively identical. In that case (22) is clearly false. I think what is going on here is this: in the original story, (22) is true because the Americans and the Soviets – had they been right – would have had beliefs about the same phenomenon; on the Twin Earth version, (22) is false because the two teams – had they both been right – would have had beliefs about distinct phenomena. And I do not see how this could be captured by any internalist account of specificity of discourse referents: there appears to be more to discourse referents than content.

ad 6.

According to the *de origine* solution, the *res* in the *de re* solution must be substituted with their sources. The idea is to relate each discourse referent to the (external) event of entering the information state in which it occurs: whether or not anything matches the descriptive content of the discourse referent, its very presence is due to an act of belief acqusition that is itself believed to be causally related to an object that does fit the description. Thus if the Dutch twins had been subject to a *fata morgana*, M would still believe his being informed by J to be causally related to a man in the park – as he does in the above scenario. In both cases, an event of being (mis-) informed by J is the relevant source of M's belief to which he stands in a *de re* relation. Similarly, in the case of Edelberg's Astronomers the relevant source would be the peculiar motion of superclusters – and it is the same

[35]The term, which I take to be referring to an internal property of information states is due to Edelberg [6], p. 576, where it is intended to characterize a sufficient condition for intentional identity.

[36]The method of Twin Earth *gedankenexperiments* is, of course, due to Putnam [24].

event for both groups of scientists, but it would not be on the Twin Earth version.[37]

The relation between the informational source and the epistemic state is thus one of *de re* belief; and it can be construed in a non-representationalist way. This is particularly important because, in a sense, the *de origine* account is but a variant of the meta-discourse account. But there is at least one difference: the former, but not the latter, establishes a language-independent link between discourse and information states. The importance of this point emerges in the puzzle of Etypical languages. On the meta-discourse account, Etypes simply speak a language without (specific) indefinites; they are thus strange but nevertheless conceivable creatures. However, on the *de origine* account, the Etypes' information states not involving any subjects, means that their epistemic states are not causally related to their sources in the same way as ours are; i.e. that they are never brought about by communication, perception, or whatever source; and this assumption is clearly at odds with the rest of the story. So, according to this approach the whole scenario turns out to be incoherent. I take this to speak in favour of the *de origine* account.

However sketchy it may be, the *de origine* solution clearly involves a rather radical departure from established theories of (direct) reference. For it reduces knowledge about specific objects to knowledge of informative events. One may hope that this feature could be avoided at the price of an asymmetry between correct and erroneous specific beliefs: events only seem to be essential in accounting for the latter; the former could still be taken to be *de re* – in the usual understanding of the term. However, such an asymmetry is pretty awkward when it comes to explaining behaviour in terms of belief: the Russian astronomers will behave exactly alike, whether they happen to live on our planet or on Twin Earth; so they would have to be in the same (internal) belief states which are related to external events in completely analogous ways.[38]

---

[37]For more on this approach to intentional identity, see van Rooy and Zimmermann [25].

[38]This paper is essentially a revised and more self-contained version of Zimmermann [30], which was a comment on Dekker [3]; see Dekker [2] for a continuation of the discussion. For rewarding discussions of the problems addressed here, I am indebted to Nicholas Asher, Robin Cooper, Jeroen Groenendijk, Irene Heim, David Israel, Hans Kamp, Ruth Kempson, Robert van Rooy, Wolfgang Spohn, and especially Paul Dekker; and for detailed comments on the pre-final version of this paper thanks are due to Michelle Weir, an anonymous referee and, again, Paul Dekker.

References

[1] Cresswell, Maxwell J.; Stechow, Arnim von (1982): 'De Re Belief Generalized'. *Linguistics and Philosophy* 5, 503–35.

[2] Dekker, Paul (1996): Reference and Representation. ILLC Research Report LP-96-12, University of Amsterdam.

[3] Dekker, Paul (1997): 'On Context and Identity'. In: H. Kamp and B. Partee (eds.), vol. I, 87–115.

[4] van der Does, Jaap (1993): 'Formalizing E-type Anaphora'. In: P. Dekker and M. Stokhof (eds.), *Proceedings of the Ninth Amsterdam Colloquium.* Part I. ILLC, University of Amsterdam. 229–48.

[5] Dressler, Alan (1987): 'The Large-Scale Streaming of Galaxies'. *Scientific American* 257, 46–54.

[6] Edelberg, Walter (1992): 'Intentional Identitiy and the Attitudes.' *Linguistics and Philosophy* 15, 561–96.

[7] Frege, Gottlob (1892): 'Über Sinn und Bedeutung'. *Zeitschrift für Philosophie und philosophische Kritik* 100, 25–50.

[8] Geach, Peter (1967): 'Intentional Identity'. *Journal of Philosophy* 64, 627–32.

[9] Groenendijk, Jeroen; Stokhof, Martin; Veltman, Frank (1997a): 'Coreference and Modality in the Context of Multi-Speaker Discourse'. In: H. Kamp and B. Partee (eds.), vol. I, 195–215.

[10] Groenendijk, Jeroen; Stokhof, Martin; Veltman, Frank (1997b): 'Coreference and Quantification. Is There Another Choice?'. In: H. Kamp and B. Partee (eds.), vol. I, 217-240.

[11] Haas-Spohn, Ulrike (1991): 'Kontextveränderung'. In: A. von Stechow & D. Wunderlich (eds.): *Semantik.* Semantics. Berlin & New York. 229–50.

[12] Haas-Spohn, Ulrike (1995): *Versteckte Indexikalität und subjektive Bedeutung.* Berlin.

[13] Heim, Irene (1982): *The Semantics of Definite and Indefinite Noun Phrases.* University of Massachusetts dissertation.

[14] Heim, Irene (1990): 'E-Type Pronouns and Donkey Anaphora'. *Linguistics and Philosophy* 13, 137–77.

[15] Kamp, Hans (1981): 'A theory of truth and semantic representation'. In: J. Groenendijk, et al. (eds.): *Formal Methods in the Study of Language. Part 1.* Amsterdam. 277–322.

[16] Kamp, Hans (1990): 'Prolegomena to a Structural Account of Belief and Other Attitudes'. In: C. A. Anderson, J. Owens (eds.): *Propositional Attitudes. Stanford.* 27–90.

[17] Kamp, Hans; Partee, Barbara (eds.): *Context Dependence in the Analysis of Linguistic Meaning.* Institut für maschinelle Sprachverarbeitung, University of Stuttgart 1996.

[18] Kaplan, David (1969): 'Quantifying in'. In: D. Davidson and J. Hintikka (eds.): *Words and Objections: Essays on the Work of W. V. Quine.* Dordrecht. 178-214.

[19] Kaplan, David (1989): 'Demonstratives'. In: J. Almog et al. (eds.): *Themes from Kaplan.* Oxford. 481–566.

[20] Lewis, David (1979): 'Attitudes de dicto and de se'. *Philosophical Review* 8, 513–43.

[21] Lewis, David (1986): *On the Plurality of Worlds.* Oxford.

[22] Montague, Richard (1970): 'Universal grammar'. *Theoria* 36, 373–98.

[23] Perry, John (1980): 'A Problem About Continued Belief'. *Pacific Philosophical Quarterly* 61, 317–32.

[24] Putnam, Hilary (1975): 'The meaning of "meaning"'. In H. Putnam: Mind, *Language, and Reality. Philosophical Papers.* Vol. 2. Cambridge. 215–71.

[25] van Rooy, Robert; Zimmermann, Thomas E. (1996): 'An Externalist Account of Intentional Identity'. In: U. Egli, K. von Heusinger (eds.): *Reference and Anaphorical Relations.* Fachgruppe Sprachwissenschaft, Universität Konstanz, Arbeitspapier Nr. 79. 123–36.

[26] Spohn, Wolfgang (to appear): 'The Intentional versus the Propositional Conception of the Objects of Belief'. In: L. Villegas-Forero et al. ((eds.): *Truth in Perspective.*

[27] Stalnaker, Robert (1978): 'Assertion'. In: P. Cole (ed.): *Syntax and Semantics 9: Pragmatics.* New York. 315–32.

[28] Stalnaker, Robert (to appear): 'On the Representation of Context'. T. Galloway and J. Spence (eds.): *Proceeedings of SALT VI.* Cornell University.

[29] Zimmermann, Thomas E. (1991): 'Kontextabhängigkeit'. In: A. von Stechow & D. Wunderlich (eds.): *Semantik. Semantics.* Berlin & New York. 156–229.

[30] Zimmermann, Thomas E. (1997): 'Subjects in Perspective. Comments on Paul Dekker's "On Context and Identity"'. In: H. Kamp and B. Partee (eds.), vol. II, 213–31.

Constrained Functions and Semantic Information

R. Zuber

CNRS, Paris

1 Introduction

As formal semantics of natural language, and in particular the study of noun phrases as generalized quantifiers shows, not all functions with a given domain and range are necessary to represent semantic information. Thus concerning for instance quantifiers of the type $\langle 1, 1 \rangle$, i.e. binary relations between sets, there are various non-logical constraints (conservativity, monotony, etc.) which can severely limit the number of functions which are used to interpret or to represent semantic information (Barwise and Cooper 1981, Keenan and Stavi 1986). The purpose of this paper is to show how some specific and well-known kinds of semantic information such as presupposed and asserted information can be associated with particular constraints on functions denoted by natural language function-expressions (in the sense of categorial grammar). As an introduction to the problem I will first discuss propositional presuppositions and assertions attached to functions interpreting in most cases simple noun phrases, i.e. to quantifiers of the type $\langle 1 \rangle$. These presuppositions and assertions will be called *quantifier-induced*. Then, using the notion of restricting and monotone functions, I will define more general notions of *function-induced* assertion and presupposition and show how they apply to any grammatical category (or for that matter to any logical type). The basic idea expressed informally is that presupposed information of an expression interpreted by a function is independent of the argument of that function and asserted information in some sense depends only on the argument and not, or at least less so, on the function itself. To take a classical example such as *the king of France*, one notices that the existence of the object denoted by this noun phrase is (almost) independent of the property it has. In other words a sentence of the form *The king of France VP* presupposes the existence of the king of France independently of the VP (which must be, however, an extensional

Logic, Language, and Computation, Vol. II, edited by Lawrence S. Moss, Jonathan Ginzburg, and Maarten de Rijke.

VP). In this sense one can say that it is the NP *the king of France* which presupposes the existence of the king and not the sentence in which this NP occurs. Similar considerations are valid for assertions.

As will be specified in the third section, all the functions we will be interested in are functions from Boolean algebras into Boolean algebras, since we consider, following the work of Keenan and his associates (cf. Keenan and Faltz 1985) that all logical types (denotations of natural language expressions) form Boolean algebras (in fact atomic and complete Boolean algebras). In other words, for any functional category C the set D_C of its logically possible denotations forms a Boolean algebra with functions as elements and with Boolean operations defined *pointwise.* This in particular means that the relation of *generalized entailment* is defined for any (functional or non-functional) category C: it is just the Boolean order in the Boolean algebra D_C. For simplicity denotational Boolean algebras D_A, D_B, etc will be usually noted as algebras A, B, etc.

Among various functions, elements of denotational algebras, we distinguish two classes of so-called *monotone functions*: the class $MONI(A, B)$ of monotone increasing functions from the algebra A into algebra B and the class $MOND(A, B)$ of monotone decreasing functions (from the algebra A into the algebra B). By definition $f \in MONI(A, B)$ iff for all X and Y in the domain of f (i.e. for all $X, Y \in A$), if $X \leq Y$ then $f(X) \leq f(Y)$. Similarly, $f \in MOND(A, B)$ iff for all X and Y in the domain of f, if $X \leq Y$ then $f(Y) \leq f(X)$. These definitions are as general as possible and they may easily be completed or adapted for more specific functions, with possible additional abbreviations. In our considerations we will begin by operating on the expressions of some simple logical language in which symbols for generalized quantifiers occur. In particular, sentences of the form $Q(P)$ or $D(S)(P)$, where Q is a quantifier of the type $\langle 1 \rangle$, S and P are properties, and D is a determiner or a quantifier of the type $\langle 1, 1 \rangle$, will often be used. Determiners can be considered as binary relations (on power sets) and consequently they can be monotone with respect to their first or their second "argument". The above definitions of monotonicity extends easily to this case. Furthermore, determiners will be assumed to satisfy the property of *conservativity*: a determiner D is conservative iff for any property X and Y, $D(X)(Y)$ is equivalent to $D(X)(X \cap Y)$. Although in later sections I will speak about assertion and presupposition as relations between linguistic expressions of a given category, I will not specify the syntax of the corresponding object language. In fact it will be either a simple logical language of Boolean algebras or just English. In other words although the relations of assertion and presupposition hold between expressions of a given category, I will often consider that they hold between the denotations of these expressions and thus between different elements of a given denotational algebra corresponding to the category considered. This ambiguous simplification seems to me to be a useful way of presenting the basic ideas of the proposal and will not lead to confusion. Finally I want to stress that various examples of natural language constructions will be given mainly for illustrative purposes.

2 Quantifier-induced semantic information

2.1 Quantifier-induced presupposition

In this sub-section I will introduce the notion of presupposition induced by a quantifier interpreting the noun phrase on the subject position (or, for that matter, induced by the subject NP). Consequently I will consider consider some specific presuppositions of the sentence of the form $Q(P)$. The observation made above, according to which the existential-like presuppositions of simple sentences with intransitive verbs depend on the subject noun phrase and are independent of (extensional) verb phrases in the corresponding sentence, leads to the following "negationless" definition of subject-induced presupposition (cf. Zuber, 1997d):

(D1) A sentence $Q(P)$ *QI-presupposes* (or has as a *QI-quantifier induced-presupposition*) sentence T iff $Q(X)$ entails T for all properties X.

Notice that since in the above definition the entailment should hold for all properties, and thus also for the negation of the property P, the QI-presupposition is also a *classical* presupposition.

Before various known classes of quantifiers are analysed from the point of view of the presuppositions they induce according to the above definition, notice the following proposition:

Proposition 1 *If a sentence S QI-presupposes a sentence T, and T entails a sentence U, then S QI-presupposes U.*

Let me now illustrate definition (D1) by considering various classes of quantifiers and determiners in their relation to QI-presupposition. Since I am interested mainly in formal properties of quantifier-induced information I will not discuss all empirically possible cases (some other cases are discussed in Zuber, 1996b, 1997b). Notice first that the following proposition guarantees the existence of QI-presuppositions for determiners which are monotone increasing with respect to the second argument or $MONI(2)$ determiners (assumed to be conservative):

Proposition 2 *If $D \in \mathrm{MONI}(2)$, then $D(S)(P)$ QI-presupposes $D(S)(S)$.*

Proof Let X be arbitrary: By conservativity sentence $D(S)(X)$ is equivalent to $D(S)(S \cap X)$, which by monotony entails $D(S)(S)$. So $D(S)(P)$ QI-presupposes $D(S)(S)$. ⊣

Since sentence $D(S)(S)$ is not trivially true for determiners D which are weak, Proposition 2 defines a class of non-trivial QI-presuppositions for monotonic increasing determiners. Among them are well known determiners giving rise to existential presuppositions such as those from which definite descriptions and possessive clauses are formed. It is possible to distinguish two other classes of determiners which, although non monotonic, give also rise to QI-assertions. These are classes *GEXT* of generalized existential determiners and *EXPT* of (positive) exception determiners (which is a subclass of generalized universal quantifiers, cf. Keenan 1993). By definition $F \in GEXT$ iff there exists a property $P \neq \emptyset$ such that $F(X)(Y)$ is true iff $P \subseteq (X \cap Y)$. Concerning *GEXT* we will use the following condition:

Proposition 3 *If $D \in GEXT$, then for all S, P, $D(S)(P)$ is true only if $S \cap P \neq \emptyset$.*

For this class *GEXT* we have the following presuppositional property:

Proposition 4 *If $D \in GEXT$, then $D(S)(P)$ QI-presupposes $SOME(S)(S)$.*

Notice furthermore, given Proposition 4, that generalized existential quantifiers with exception clauses like **No . . . except Bill** or **No . . . except n** have more specific QI-presuppositions which entail *SOME*$(S)(S)$. They are specified in the following propositions:

Proposition 5 *The sentence* **No(S) except n are (P)** *QI-presupposes* **at least n(S)(S)**.

Proposition 6 *Sentence* **No(S) except Bill (is) (P)** *QI-presupposes* **Bill is S**.

Let us consider now QI-presuppositions related to positive exception determiners or rather to generalized universal quantifiers interpreting noun phrases with exception clauses. Although there are many families of such noun phrases, differing by the exception complement in the exception clause (cf. Moltmann 1995, von Fintel 1993, Zuber 1996b, 1997c), I will consider here only the case of exception clauses with a proper name *PrN* as the exception complement; other cases can be handled similarly. In order to do this we need first a semantic description of sentences with such noun phrases; it is indicated by the following property:

Proposition 7 *The sentence* **All (S) except PrN (is)(P)** *is true iff* **S-P={PrN}**, *where* **{PrN}** *is the unit set whose element is* **PrN**.

From this property the following proposition follows:

Proposition 8 *The sentence* **All(S) except PrN (is)(P)** *QI-presupposes* **PrN (is)(S)**.

Thus *All logicians except Bill are happy* QI-presupposes *Bill is a logician.*

2.2 Quantifier-induced assertion

Observations similar to those concerning QI-presuppositions lead to the following definition of QI-assertion (quantifier induced assertion):

(D2) Sentence $D(S)(P)$ *QI-asserts* sentence $D'(S')(P)$ iff $D(S)(X)$ entails $D'(S')(X)$ for all X

An assertion of a given sentence S has been sometimes "classically" defined as being a consequence of S and of the (internal) negation of S. Thus definition 2, as seen in the case of QI-presupposition, is a generalization of the "classical" notion of assertion, since from the fact that entailment should hold for all properties X, it follows that it holds also for the negation of the property P. Clearly any sentence QI-asserts itself. Furthermore, we have also the following properties of QI-assertions:

Proposition 9 *If a sentence S QI-asserts a sentence T and T QI-asserts a sentence U then S QI-asserts U.*

Proposition 10 *If S QI-asserts T, and T QI-presupposes U, then S QI-presupposes U.*

Thus QI-assertion is transitive and any QI-presupposition of a QI-assertion is a QI-presupposition of the asserting sentence.

The case of QI-assertion needs a development similar to the one of QI-presuppositions since the definition (D2) does not guarantee the existence of QI-assertions. It can be shown, however, that in this case it is also possible to distinguish various properties of nominal determiners, and thus various classes of such determiners which give rise to particular QI-assertions.
As a first case we will consider monotone increasing with respect to the first argument determiners, $MONI(1)$, sometimes called *persistent*, for which one has the following proposition:

Proposition 11 *If $D \in \mathrm{MONI}(1)$, then $D(S)(P)$ QI-asserts $D(S')(P)$ for all S' such that $S \subseteq S'$.*

Thus *A young dentist is sleeping* QI-asserts *A dentist is sleeping.*

Proposition 11 concerns a large sub-class of generalized existential quantifiers, roughly speaking, those which do not contain "exception clauses". One notices, however, that the whole class of *GEXT* quantifiers, even those which are not persistent (which contain an exception clause) give rise not only to QI-presuppositions but also to QI-assertions. Limiting ourselves to those which have the exception complement in the form of a proper name we get the following proposition:

Proposition 12 *Sentence* **No(S) except PrN (is)(P)** *QI-asserts* **PrN (is)(P)**.

The sub-class of universal generalized quantifiers which have been denoted by *EXPT* above give also rise to particular QI-assertions. They can easily be deduced from the definition of QI-assertion and the statements indicating the semantics of sentences with such constructions. The following proposition specifies such assertions for two families of universal generalized quantifiers:

Proposition 13 *A sentence of the form* **All(S) except Bill (are)(P)** *or of the form* **All(S) except n (are)(P)** *QI-asserts* **not-All(S)(are)(P)**.

To conclude this section let me mention that there is another class of complex determiners which also give rise to specific QI-presuppositions and QI-assertions (cf. Zuber 1996b, 1997c): these are determiners formed by *inclusion clauses* of the type *including NP* which interpret complex NP's like *Most students, including Bill* or *Some logicians including five from Albania.* Their asserted and presupposed information can be analysed in a quite similar way. For instance *Most students, including Bill, drink* QI-presupposes *Bill is a student* and QI-asserts *Most students drink.* I will come back to these determiners in what follows.

3 Functions inducing assertions and presuppositions

Recall that one of our goals is to generalize the notions of presupposition and assertion in such a way that we could speak about presupposition and assertion of (theoretically) *any* functional category. More precisely, we want to say that if an expression of a category C has a specific form, then it can have a (generalized) presupposition and a (generalized) assertion expressed by an expression of the same category C. To do this we need to add the following preliminary observations.

As we have seen in the preceding sections, QI-assertions and QI-presuppositions are just particular entailments; they are entailments which should hold between a family of sentences (having, roughly speaking, the same subject and in which the VP varies) and a particular sentence. So first we have to generalize the notion of entailment in such a way that we could say that it can hold between two expressions of the same category C, for any functional category C. As indicated, this generalisation, well-known to algebraists, has also been introduced, with empirical justification, into linguistic semantics (cf. Keenan 1983a, Keenan and Faltz 1985): it is just the partial order in the Boolean algebras formed by the denotations of expressions of the given category C.

The second observation concerns the form which asserting and presupposing expressions should have. A look at various examples and properties from the preceding sections may help: one notices that expressions which induce assertions and presuppositions are functional expressions interpreted by functions. We have seen this in particular with quantifiers of type $\langle 1 \rangle$ which interpret (or are denoted by) NPs on subject position. I will argue that the main presupposing and asserting expressions are modified expressions, i.e. expressions formed by a functional expression called a *modifier* which is applied to an argument. So first I will present some functions which interpret modifiers in modified expressions.

Modifiers are expressions which combine with expressions in category C to form expressions in category C. So they have the category C/C for various choices of C. Thus, semantically, modifiers are interpreted by functions from an algebra (interpreting the category C) into itself. Algebras of such functions will be denoted by CfC. Again, as in the case of determiners, not all elements of CfC algebras are necessary for the semantic interpretation of natural language expressions. For instance, concerning (extensional) adjectival and adverbial modifiers one can suppose (cf. Keenan 1983a, Keenan and Faltz 1985) that the set F of all functions which interpret such modifiers are (positively) restricting in the following sense: $F(a) \leq a$, where $F \in CfC$ and $a \in C$. All restricting functions rf from a Boolean algebra B to itself form a Boolean algebra called a (positive) restricting algebra which will be denoted by $BrfB$.

Restricting algebras are a special case of factor algebras which are defined as follows. Let B be a Boolean algebra and a a non-zero element of B. Then the set B/a defined as

$$B/a = \{x : x \leq a\}$$

is a Boolean algebra, where the zero element, the meet and the join are as in B, and the unit is a itself and the complement $cplx = a \cap nonx$ (where $nonx$ is the complement in B). B/a is called the factor algebra (of B) relative to (or generated by) a. The element a to which an algebra is relativised determines a relative negation and as such will determine a non-trivial presupposition.

Consider now a Boolean algebra BfB, for an arbitrary B. Then the factor algebra $(BfB)/id$ is just the restricting algebra $BrfB$, where id is the identity function. An important sub-class of restricting functions for our purpose is constituted by so-called *intersecting functions.* They are defined as follows, writing 1_B for the unit element of B:

(D3) $f \in BfB$ is *intersecting* iff $f(a) = a \cap f(1_B)$, for all $a \in B$.

The set of all intersecting functions from a Booelan algebra into itself forms a Boolean algebra. Intersecting functions were studied by Keenan (1983a, 1983b). He also showed that they interpret some adjectives (so-called *absolute adjectives*) and relative clauses. It can also be argued that they can be used for the semantic description of *NPs with inclusion clauses.* I will come back to this problem later on.

Restricting and intersecting functions have their *negative* counterparts: these are *negatively restricting* and their important sub-class *negatively intersecting* functions (cf. Zuber 1997b):

(D4) A function $f \in BfB$ is *negatively restricting* iff $f(a) \leq a'$, for all $a \in B$ (where a' is the complement of a in B).

(D5) A function $f \in BfB$ is *negatively intersecting* iff the following holds: $f(a) = a' \cap f(0)$.

Negatively restricting functions are related to factor algebras as well: they correspond to factor algebras relative to negative identity function $n - id$ (i.e. $f(x) = x'$).

Negatively intersecting functions also form a Boolean algebra, a sub-algebra of the corresponding negatively restricting algebra. They can be used to interpret NP's with exception clauses (cf. Zuber 1996b, 1997c).

Finally let me mention a sub-class called *mixing functions* which are related in an interesting way to intersecting functions. For any algebra B, a function $f \in BfB$ is *mixing* iff for all x, $f(x) = (x \cap f(1_B)) \cup (x' \cap f(0_B))$. By a simple calculation one shows that mixing functions form a sub-algebra of the algebra BfB. Now notice the following. Restricting functions need not be monotone: in general positively restricting functions are not monotone increasing and negatively restricting are not in general monotone decreasing. However, (positively) intersecting functions are monotone increasing and negatively intersecting are monotone decreasing. Furthermore, monotone increasing mixing functions are just positively restricting ones and monotone decreasing mixing functions are just negatively restricting ones.

4 Generalized function-induced information

Now we are in a position to formulate and illustrate with various examples a general definition of presupposition and assertion which applies to any category. This definition is a generalisation of the quantifier-induced semantic information discussed in Section 2. The generalisation is based on the observation that a quantifier is a specific function which has properties as its arguments. Then, according to definition (D1) a sentence has as quantifier induced presupposition another sentence iff this second sentence is implied for any value of the quantifier interpreting the subject of the presupposing sentence. But this move can be done with any function F: if there is a value A which is entailed for any argument of the function F then we will say that F induces the presupposition A. More precisely we have the following definition:

(D6) Expression $F(E)$, with $F \in AfB$ *FI-presupposes* $\alpha \in B$ (or α is *an FI-presupposition of* $F(E)$) iff for all $X \in A$, $F(X) \leq \alpha$. We will say in this case that the function F induces the presupposition α.

Less formally, but with a close analogy to definition (D1), and to the classical definition of presupposition, $F(E)$ FI-presupposes α iff for all X of the same type (or strictly speaking, category) as E, one has $F(X) \leq \alpha$.

Definition (D6) has various interesting consequences, in addition to being a generalisation of the classical definition. For instance, according to this definition, any expression has a trivial presupposition equal to the unit of the algebra to which the expression belongs; this presupposition is induced by the identity function. Furthermore we have also:

Proposition 14 *If β is a FI-induced presupposition of α and $\beta \leq \gamma$ then γ is a FI-presupposition of α.*

A similar property was established for sentential presuppositions, even if they are defined in a quite different framework like the one using a three-valued logic (cf. Keenan 1973). So Proposition 14 can be considered as a generalisation of this property.

Of course definition (D6) is very general and neither it nor its consequences presented above guarantee the existence of non-trivial presuppositions. So, as in the case of quantifier induced presuppositions we have to examine particular cases of functions and have to show that they may induce non-trivial presuppositions. I will give below some propositions which show which functions induce non-trivial presuppositions:

Proposition 15 *If $F \in \text{MONI}(B, C)$, then F induces $F(1_B)$ as a presupposition, where 1_B is the unit of the algebra B.*

Proposition 16 *If $F \in \text{MOND}(B, C)$, then F induces $F(0_B)$ as a presupposition, where 0_B is the zero element of the algebra B.*

Proposition 17 *If $F \in BfB$ is a mixing function, then F induces $F(O_B) \cup F(1_B)$ as a presupposition.*

Proposition 15 is analogous to Proposition 2 for quantifier induced presuppositions. To prove it, it is enough to notice that if $F \in BfB$ is monotone increasing ($MONI(B, B)$), then for all $X \in B$ one has $F(X) \leq F(1_B)$. Similarly for Proposition 16: if F is monotone decreasing ($MOND(B, B)$), then for all X, $F(X) \leq F(0_B)$.

The definition of generalized assertion can be given in a similar way and also by analogy with the definition (D2). Instead of taking quantifiers as function inducing assertion we take just any function (of the appropriate type). We have:

(D7) The expression $F(E)$, where $F \in AfB$ and $E \in A$, *FI-asserts* $G(E)$ iff for all X, one has $F(X) \leq G(X)$.

Notice that according to this definition FI-assertion is a generalized entailment. Moreover, the function interpreting the function expression of the asserting expression (the function F above) must entail the function interpreting the function expression in the asserted expression (the function G above). It follows from this that any expression has itself as assertion induced by the identity function.

In order to see whether the definition guarantees the existence of nontrivial assertions it is enough to consider some particular functions discussed above. For them we have the following propositions, which follow directly from the definitions of restricting and negatively restricting functions:

Proposition 18 *If F is restricting, then $F(E)$ FI-asserts E.*

Proposition 19 *If F is negatively restricting then $F(E)$ FI-asserts E', where E' is the complement of E.*

To illustrate these propositions we can use determiners with inclusion and determiners with exclusion clauses. Since the function interpreting the inclusion clause *including NP* is interpreted by a (positively) restricting function we can now say that for instance the determiner *Most ... including Leo* FI-asserts *Most.* Similarly, given that the exclusion clause *except NP* is interpreted by a negatively restricting function, the determiner *All ... except Leo* FI-asserts *not-All.*

Since positively intersecting functions are monotone increasing and positively restricting we have the following proposition for them:

Proposition 20 *If* $F \in BfB$ *is positively intersecting, and* $A \in B$, *then* $F(A)$ *FI-presupposes* $F(1_B)$ *and FI-asserts* A.

We can illustrate this proposition by noting that the commun noun *French car* FI-presupposes *French object* and FI-asserts *car* given that the adjective *French* is an absolute adjective and as such it is interpreted by an intersecting function (cf. Keenan 1983b).

Similarly for negatively intersecting functions: since they are negatively restricting and monotone decreasing the following proposition holds for them:

Proposition 21 *If* $F \in BfB$ *is negatively intersecting, and* $A \in B$, *then* $F(A)$ *FI-presupposes* $F(0_B)$ *and FI-asserts* A'.

Notice that intersecting functions not only induce presuppositions and assertions, but also, given their definition, that presupposed and asserted information induced in this way is maximal in the sense that the (Boolean) intersection of presupposition and assertion is equivalent to the expression which asserts and presupposes.

The last series of propositions I would like to present concerns, to speak somewhat metaphorically, the cross-categorial transfer of presupposed information. In other words I want to show how presuppositions of one category can be in some cases "transferred" or transformed into presuppositions of another category. This phenomenon was illustrated in the section 2 where we saw that some nominal presuppositions of subject NPs are transformed into sentential presuppositions. Now we can illustrate this phenomenon is a general case, independently of the type of functions involved, by considering functions which are composed of a restricting or intersecting function applied to an argument. For them we have the following proposition:

Proposition 22 *Let* G *be an element of* BfC. *If* G *FI-presupposes* H *and* $H \in \mathrm{MONI}(B,C)$, *then* $G(E)$ *FI-presupposes* $H(1_B)$, *for any* $E \in B$.

Proposition 23 *Let* G *be an element of* BfC. *If* G *FI-presupposes* H *and* $H \in \mathrm{MOND}(B,C)$, *then* $G(A)$ *FI-presupposes* $H(0_B)$, *for any* $A \in B$.

Proof of Proposition 22: If G FI-presupposes H, then there exists a Boolean algebra B_1, such that for some $A \in B_1$, and some $F \in B_1fB$, G is of the form $F(A)$ and F induces H as a presupposition. This means that $F(X) \leq H$, for all $X \in B_1$. Since H is monotone increasing we have $F(A)(Y) \leq H(1_B)$, for all $Y \in B$. But this means that $F(A)(E)$, which is equal to $G(E)$, FI-presupposes $H(1_B)$. $\dashv$

Proposition 22 can again be illustrated by noun phrases with inclusion phrases. As we have already seen, the noun phrase (or the corresponding quantifier) like *Most students, including Leo* FI-presupposes the noun phrase *a student who is Leo.* This quantifier, considered as a function from properties to truth-values is a monotone increasing function. Since the unit element of the algebra of properties is interpreted as the universal property *exists*, the application of Proposition 22 to this case gives the result that *Most students, including Leo, are sleeping* FI-presupposes *A student who is Leo, exists*, which entails that Leo is a student. This last presupposition is the same as the one we obtained "directly" in section 2, as a QI-induced presupposition.

5 Concluding remarks

Constraints on interpreting functions give rise to specific semantic information. This function induced information has a cross-categorial character in the sense that it is independent of the category or type considered but only on the specific property of the function considered; it is entirely determined by the specific property the considered function has. I was mainly concerned here with such properties as monotony and restrictiveness. Interestingly enough it appeared that monotony can be correlated with information traditionally known as presupposed information while restrictiveness corresponds to asserted information. Both types of information can be attached to complex expressions of any category provided that their function expressions are interpreted by a function satisfying specific constraints. It is possible that other properties of functions may give rise to other types of information. It is interesting, however, that precisely the properties of monotony and restrictiveness are related to presupposed and asserted information since they are often considered as being related to semantic universals (Barwise and Cooper 1981, Keenan 1983b, Keenan and Stavi 1986). In other words my proposal suggests possible origins for presupposition and assertion: they have something to do with non-logical, albeit universal, constraints such as conservativity, monotony or restrictiveness.

Notice furthermore, that according to the proposal made here, presuppositions and assertions are just generalized entailments satisfying some additional conditions. Consequently there is no need for non-standard logical systems such as default logic, multi-valued logic or logic with truth-value gaps in order to account for various phenomena giving rise to assertions and presuppositions. As has been shown, it is possible to give non-contextual, purely extensional definitions of presupposition and assertion. These definitions, which do not use explicitly the notion of negation, are formulated in a simple and natural framework of Boolean algebras and account for a huge class of relevant linguistic structures giving rise to such functional information.

In order to apply my proposal to the semantic analysis of various syntactically complex expressions many additional steps should be carried out. Expressions for which function-induced information was defined are supposed to be of the simple form: function-argument. Obviously an expression can contain more functions and in particular the argument can be itself a functional expression. This means that various rules of composition of the function-induced information are necessary. Some such rules are indicated in Propositions 22 and 23. Many other, also concerning assertions, remain to be explicitly formulated. The use of the Boolean framework can but facilitate such an enterprise.

It is by now well-established by that functions play an essential role in natural language semantics. Given the fact that the number and the complexity of functions from algebras into algebras grow with the complexity of algebras to which they apply, it should not be surprising that functions denoted by natural language expressions are restricted in various ways. I have made use of positively and negatively restricting functions and of mixing functions which all form Boolean sub-algebras of the algebra of unary functions from a given algebra into itself. Various properties of such sub-algebras are related, and possibly are consequences of some more general "classical" algebraic results (cf. Nowak 1997). However, a full algebraic characterisation of functions needed in natural language semantics cannot to be done without taking into account precise and well-established linguistic data. I have frequently used samples of exclusive phrases as illustrations. It is interesting to note that these phrases denote atoms of the corresponding denotational algebras (cf. Keenan 1993) and that they are special cases of various other typical "natural languages expressions" also denoting atoms or co-atoms as well (cf. Zuber 1997c). The fact that some specific natural language expressions may denote atoms of "higher order" algebras is again not surprising, since the number of logically possible functions in such "higher order" algebras is huge. This can be considered as just another restriction on the functions interpreting natural language expressions. Accordingly, there are many restrictions on functions which remain to be studied*.

*This paper is a natural sequel to Zuber (1996a), which was prepared for the conference at which it was presented. However, the few additional months I had at my disposal after the conference and before writing the present paper allowed me to extend and generalize the basic ideas of Zuber (1996a) in the present paper and to add some additional technical results. My thanks to Ross Charnock, Ed Keenan, Brenda Laca, Daniel Lacombe, Alex Orenstein and Richard Zuber. They have been on the whole helpful and most of them are not responsible for the majority of the remaining errors. None the less, the main responsibility lies with one of them.

References

[1] Barwise, J. and Cooper, R. (1981) Generalized Quantifiers and Natural Language, *Linguistics and Philosophy* 4:2, 159-220

[2] von Fintel, K. (1993) Exceptive constructions, *Natural Language Semantics* 1, 123-148

[3] Keenan, E. L. (1973) Presupposition in Natural Logic, *The Monist* 53:3, 344-377

[4] Keenan, E. L. (1983a) Boolean Algebra for linguists, in Mordechay, S. (ed.) *UCLA Working Papers in Semantics*, UCLA, 1-75

[5] Keenan, E. L. (1983b) Facing the truth, *Linguistics and Philosophy* 6, 335-371

[6] Keenan, E. L. (1993) Natural Language, Sortal Reducibility and Generalized Quantifiers, *J. of Symbolic logic* 58:1, 314-325

[7] Keenan, E. L. and Faltz, L. M. (1985) *Boolean Semantics for Natural Language*, D. Reidel Publishing Company, Dordrecht

[8] Moltmann, F. (1995) Exception sentences and polyadic quantification, *Linguistics and Philosophy* 18:3, 223-280

[9] Nowak, M. (1997) A general approach to the algebras of unary functions in a Boolean semantics of natural language, *Bulletin of the Section of Logic* 26:2, 192-108

[10] Zuber, R. (1996a) Function-induced information, *Proceedings of the Second Conference on Information-Theoretic Approaches to Logic, Language and Computation*, Regent's College, London, 18-24 July 1996, 233-238

[11] Zuber, R. (1996b) Two Semantic Components of Noun Phrases, in Dubach Green, A. and Motapanyane, V. (eds.) *Proceedings of ESCOL '96*, Cornell University Press, 347-354

[12] Zuber, R. (1997a) Uogólnione kwantyfikatory, presupozycje i asercje, *Nowa krytyka* 8

[13] Zuber, R. (1997b) On negatively restricting Boolean algebras, *Bulletin of the Section of Logic*, 26:1, 50-54

[14] Zuber, R. (1997c) Some algebraic properties of higher order modifiers, in Becker, T. and Krieger, H-U. *Proceedings of the Fifth Meeting on Mathematics of Language*, Deitsches Forschungszentrum für Künstliche Intelligenz GmbH, 161-168

[15] Zuber, R. (1997d) Defining presupposition without negation, in Murawski, R. and Pogonowski, J. (eds.) *Euphony and Logos*, Editions Rodopi, Amsterdam/Atlanta, 457-468